AF618437

Industrial and Applied Mathematics

The Industrial and Applied Mathematics series publishes high-quality research-level monographs, lecture notes and contributed volumes focusing on areas where mathematics is used in a fundamental way, such as industrial mathematics, bio-mathematics, financial mathematics, applied statistics, operations research and computer science.

More information about this series at http://www.springer.com/series/13577

Aleksandr Krivoshein · Vladimir Protasov
Maria Skopina

Multivariate Wavelet Frames

Aleksandr Krivoshein
Department of Applied Mathematics and Control Processes
Saint Petersburg State University
Saint Petersburg
Russia

Maria Skopina
Department of Applied Mathematics and Control Processes
Saint Petersburg State University
Saint Petersburg
Russia

Vladimir Protasov
Department of Mechanics and Mathematics
Lomonosov Moscow State University
Moscow
Russia

ISSN 2364-6837 ISSN 2364-6845 (electronic)
Industrial and Applied Mathematics
ISBN 978-981-10-3204-2 ISBN 978-981-10-3205-9 (eBook)
DOI 10.1007/978-981-10-3205-9

Library of Congress Control Number: 2016957146

Printed on acid-free paper

This Springer imprint is published by Springer Nature
The registered company is Springer Nature Singapore Pte Ltd.
The registered company address is: 152 Beach Road, #21-01/04 Gateway East, Singapore 189721, Singapore

Preface

Wavelet theory lies at the intersection of pure and computational mathematics, as well as of audio and graphics signal processing, including compression and transmission of information. Wavelet bases have several advantages compared with other bases used as approximation tools. One of them is the so-called time-frequency localization property: Wavelet basis functions as well as their Fourier transformations rapidly decay at infinity. Through this property, in the decomposition into the basis of signals, frequency characteristics of which vary according to time or space, many expansion coefficients with unnecessary at this spatial or temporal area harmonics are small and can be discarded, thereby providing data compression.

Wavelet frames (framelets) are actively used for the same purposes. Moreover, they are very efficient in the image recovery from incomplete observed data, including the tasks of inpainting and image/video enhancement. In the recovery of missing data from incomplete and/or damaged and noisy samples, application of wavelet methods based on frames is more advanced due to the redundancy of frame systems.

Multivariate wavelet systems with matrix dilations (so-called nonseparable wavelets) have been increasingly used for digital processing of multidimensional signals such as images, videos, tomography, and seismic and other signals. Nonseparable wavelets turn out to be more natural in signal processing because multidimensional signals are usually nonseparable. Nonseparable filter banks have better characteristics than their separable counterparts (which consist of products of 1-D filter banks along each dimension). The number of degrees of freedom is also much bigger for nonseparable filter banks. In tomography, the 2-D separable wavelets impose a rectangular tiling of the frequency plane, which is not well suited to the radial band-limited assumption of the image. The application of nonseparable multiresolution tomography to 2-D wavelets allows us to respect the geometry of the system by tiling the frequency plane in a diamond-shaped fashion that is more suitable to the radial band-limited assumptions. Local tomography using these nonseparable bases shows an improvement in terms of PSNR. Another successful application of nonseparable wavelets was in 3-D rotational angiography.

This book presents a systematic study of the theory and methods for the construction of multivariate wavelet frames with matrix dilation, in particular, orthogonal and biorthogonal bases which are a special case of frames. Construction of multivariate nonseparable wavelet frames, especially bases with desirable properties, is a challenging problem. Though a general scheme of construction is well known, its practical implementation in the multidimensional setting is difficult.

We describe methods for the construction of wavelet frames with a matrix dilation providing an arbitrary approximation order and other important features. We also discuss possible conditions under which a frame constitutes a basis. Applied mathematicians and engineers are especially interested in the construction of compactly supported wavelet systems. We give algorithmic methods for the construction of dual and tight compactly supported wavelet frames. Another important feature is symmetry. Different kinds of symmetry of wavelets are very much desirable in various applications, since they preserve linear-phase properties and also allow symmetric boundary conditions in wavelet algorithms which normally perform better. We discuss how to provide H-symmetry, where H is an arbitrary symmetry group, for wavelet bases and frames. The so-called frame-like wavelet systems and their approximation properties are also studied. The frame-like systems inherit many advantages of frames and can be used in applications instead of frames, although their construction is much simpler. The smoothness of wavelets is also important in engineers' problems. To provide smoothness of a wavelet system, one has to begin with a smooth generating refinable function. We extend the matrix method of computing the regularity of refinable function from the univariate case to multivariate refinement equations with arbitrary dilation matrices. This makes it possible to find the exact values of the Hölder exponent of refinable functions and to make a very refine analysis of their moduli of continuity.

The book will be useful for engineers working in signal processing, who can use algorithmic methods for wavelet construction or extract ready-made examples of masks. It can also be interesting for specialists working in functional analysis, approximations, numerical PDE, and related areas. Most of the material is available for graduate students familiar with the basics of functional and real analysis on the level of a standard university course.

The list of references contains a lot of papers which are related to the topics we discuss. Not all of them are cited, and not all results and ideas of these papers are presented in the book.

The authors acknowledge Saint Petersburg State University for a research Grant #9.38.198.2015, RFBR, research projects No. 15-01-05796a, No. 14-01-00332, No. 16-04-00832, the grants of Dynasty foundation and Volkswagen foundation.

Saint Petersburg, Russia A. Krivoshein
Moscow, Russia V. Protasov
Saint Petersburg, Russia M. Skopina

Contents

About the Authors

A. Krivoshein received his Ph.D. from St. Petersburg State University in 2013 and is currently an associate professor at the Department of Applied Mathematics and Control Processes there. His main research interests include wavelets and their applications to signal processing. He is the author of 10 research papers.

V. Protasov is a professor at the University of L'Aquila and of the Department of Mechanics and Mathematics, Moscow State University, where he received his Ph.D. in 1999. He was awarded his D.Sc. at St. Petersburg Department of V.A. Steklov Institute of Mathematics (2006). He is a Professor (2015) and a Corresponding Member (2016) of the Russian Academy of Science. His research areas include functional analysis, Fourier analysis, wavelets, matrix theory, optimization, convex geometry, and combinatorics. He is the author of more than 80 research papers, 2 monographs, and more than 20 popular and educational publications, including 4 books. He is a member of the editorial boards of the journals: Applied Mathematics and Computation, Sbornik: Mathematics, Analysis Mathematica, and Quantum.

M. Skopina is a professor at the Department of Applied Mathematics and Control Processes, St. Petersburg State University. She received her Ph.D. there in 1980 and her D.Sc. from St. Petersburg Department of V.A. Steklov Institute of Mathematics, the Russian Academy of Sciences in 2000. Her main research interests are wavelets, Fourier series, approximation theory, and abstract harmonic analysis. She is the author of more than 70 research papers and the monograph "Wavelet Theory," which was published in Russian in 2005 and translated into English by the American Mathematical Society (AMS) in 2011. She is a member of the St. Petersburg Mathematical Society and the AMS. She is also an associate editor for the International Journal of Wavelets, Multiresolution and Information Processing (IJWMIP) and has organized four international conferences on "Wavelets and Applications."

Basic Notation

$\mathbb{N}$ is the set of positive integers, $\mathbb{R}^d$ is the d-dimensional Euclidean space, $x = (x_1, \ldots, x_d)$, $y = (y_1, \ldots, y_d)$ are its elements (vectors), $\mathbf{0} = (0, \ldots, 0)$, $(x, y) = x_1 y_1 + \cdots + x_d y_d$, $|x| = \sqrt{(x, x)}$, $\mathbb{R} = \mathbb{R}^1$, $\mathbb{Z}^d$ is the integer lattice in $\mathbb{R}^d$, $\mathbb{Z} = \mathbb{Z}^1$, $\mathbb{T}^d = [0, 1)^d$ is the d-dimensional torus. For $x, y \in \mathbb{R}^d$, we write $x \geq y$, if $x_j \geq y_j$ for all $j = 1, \ldots, d$ and we write $x > y$, if $x \geq y$ and $x \neq y$; $\mathbb{Z}_+^d := \{x \in \mathbb{Z}^d : x \geq 0\}$, $\mathbb{Z}_+ = \mathbb{Z}_+^1$.

If $\alpha, \beta \in \mathbb{Z}_+^d$, $a, b \in \mathbb{R}^d$, we set $[\alpha] = \sum\limits_{j=1}^{d} \alpha_j$, $\alpha! = \prod\limits_{j=1}^{d} \alpha_j!$, $\begin{pmatrix} \alpha \\ \beta \end{pmatrix} = \frac{\alpha!}{\beta!(\alpha!-\beta)}$, $a^b = \prod\limits_{j=1}^{d} a_j^{b_j}$, $D^\alpha f = \frac{\partial^{[\alpha]} f}{\partial^{\alpha_1} x_1 \ldots \partial^{\alpha_d} x_d}$; $\delta_{\alpha\beta}$ is the Kronecker delta, which is equal to 1 for $\alpha = \beta$; otherwise, it is equal to 0.

If $A, B \subset \mathbb{R}^d$, then $A \pm B = \{x = a \pm b : \ a \in A, b \in B\}$. By ConvA we denote the convex hull of the set A.

$\mathbb{C}^d$ is the d-dimensional complex Euclidean space with the inner product given by $\langle x, y \rangle = \sum_{j=1}^{d} x_j \overline{y_j}$ for $x = (x_1, \ldots, x_d) \in \mathbb{C}^d$ and $y = (y_1, \ldots, y_d) \in \mathbb{C}^d$.

The Lebesgue measure in $\mathbb{R}^d$ is denoted by μ. By a function on $\mathbb{R}^d$ we mean a complex-valued Lebesgue measurable function on $\mathbb{R}^d$. For $1 \leq p \leq \infty$, $L_p(\mathbb{R}^d)$ is the usual Banach space of functions f on $\mathbb{R}^d$ (of the equivalence classes) such that $\|f\|_p := \left(\int_{\mathbb{R}^d} |f(x)|^p dx\right)^{1/p} < \infty$ for $1 \leq p < \infty$ and $\|f\|_\infty := \operatorname{vraisup}_{x \in \mathbb{R}^d} |f(x)|$ for $p = \infty$. For $p = 2$, $L_2(\mathbb{R}^d)$ is a Hilbert space with the inner product given by $\langle f, g \rangle := \int_{\mathbb{R}^d} f(x) \overline{g(x)} dx$.

The support of a function f, i.e., the minimal (with respect to inclusion) closed set such that f is equal to zero almost everywhere on the complement of this set and is denoted by $\operatorname{supp} f$.

χ_e is the characteristic function of a set $e \subset \mathbb{R}^d$; it takes the value 1 at the points $t \in E$ and 0 at all other points.

span $\{f_n, n \in \mathbb{N}\}$ is the set of finite linear combinations of a system $\{f_n\}_{n=1}^{\infty}$ with complex coefficients, span$A$ is the linear span of a set $A \subset \mathbb{R}^d$.

$\mathcal{S}$ is the Schwartz space on $\mathbb{R}^d$, i.e., the space of infinitely differentiable and rapidly decreasing functions on $\mathbb{R}$:

$$\mathcal{S} = \left\{ f \in C^{\infty}(\mathbb{R}^d) \mid \|D^{\alpha} f(x)(1+|x|)^k\|_{\infty} < \infty \quad \forall \alpha \in \mathbb{Z}_+^d, \ \forall k \geq 0 \right\}.$$

The topology of the space $\mathcal{S}$ is defined as follows:

$$f_j \to 0 \Leftrightarrow \|D^{\alpha} f_j(x)(1+|x|)^k\|_{\infty} \to 0 \quad \forall \alpha \in \mathbb{Z}_+^d, \ \forall k \geq 0.$$

$\mathcal{S}'$ is the space of linear continuous functionals on the space $\mathcal{S}$ (the space of tempered distributions).

$\widehat{f}(x) = \int\limits_{\mathbb{R}^d} f(t) e^{-2\pi i (x,t)}\, dt$ is the Fourier transform of a function f from $L_1(\mathbb{R}^d)$; the same notation $\widehat{f}$ is used for the Fourier transform of f which is in $L_2(\mathbb{R}^d)$, or in the space $\mathcal{S}'$ of tempered distributions; $\mathcal{F}$ and $\mathcal{F}^{-1}$ are the operators taking a function to its direct and inverse Fourier transforms, respectively.

For $s \in \mathbb{N}$, $p \geq 1$, the Sobolev space $W_p^s = W_p^s(\mathbb{R}^d)$ consists of all the functions $f \in L_p(\mathbb{R}^d)$ such that

$$\int\limits_{\mathbb{R}^d} |D^{\alpha} f(x)|^p \, dx < \infty \quad \forall \alpha \in \mathbb{Z}^d, [\alpha] \leq s.$$

For $s > 0$, $p = 2$, the Sobolev space $W_2^s = W_2^s(\mathbb{R}^d)$ consists of all the functions $f \in L_2(\mathbb{R}^d)$ such that

$$\int\limits_{\mathbb{R}^d} |\widehat{f}(\xi)|^2 (1+|\xi|^2)^s d\xi < \infty.$$

For a continuous function $f : \mathbb{R}^d \to \mathbb{R}$, we denote

$$\omega(f,t) = \sup_{\|h\| \leq t} \|f(\cdot + h) - f(\cdot)\|_{C(\mathbb{R}^d)}, \qquad t > 0,$$

the modulus of continuity of f. The Hölder exponent is the supremum of numbers $\alpha \geq 0$ such that $\omega(\varphi, t) \leq C t^{\alpha}$.

$\widehat{f}(k) = \int\limits_{\mathbb{T}^d} f(t) e^{-2\pi i (k,t)}\, dt$ is the k-th Fourier coefficient, $k \in \mathbb{Z}^d$, of a function $f \in L_1(\mathbb{T}^d)$ with respect to the trigonometric system.

If f is a trigonometric polynomial, then $spec(f)$ denotes the spectrum of f, i.e., the set of $k \in \mathbb{Z}^d$ such that $\widehat{f}(k) \neq 0$.

For a countable index set $\mathbb{K}$, $\ell_p(\mathbb{K})$, $1 \le p \le \infty$, is the Banach space of sequences of complex numbers $c = \{c_n\}_{n\in\mathbb{K}}$ with norm $\|c\|_{\ell_p} = \left(\sum_{n\in\mathbb{K}} |c_n|^p\right)^{1/p}$ for $1 \le p < \infty$ or $\|c\|_{\ell_\infty} = \sup_{n\in\mathbb{K}} |c_n|$ for $p = \infty$; $\mathbb{K}$, $\ell_0(\mathbb{K})$ denotes the linear subspace of $\ell_1(\mathbb{K})$ consisting of finite sequences; $\ell_p := \ell_p(\mathbb{N})$.

The inner product of elements f, g of a Hilbert space is denoted by $\langle f, g\rangle$, the norm of element f of a Hilbert space is denoted by $\|f\|$.

For an operator T in a Hilbert space, T^* is the operator adjoint to T; T^{-1} is a norm of the operator.

If A is a $d \times d$ matrix, then $\|A\|$ is its Euclidean operator norm from $\mathbb{R}^d$ to $\mathbb{R}^d$, A^T is its transpose, A^* is its Hermitian conjugate matrix, $\det A$ is the determinant of A; I_d is the identity $d \times d$ matrix.

M denotes a dilation matrix in $\mathbb{R}^d$, i.e., an integer $d \times d$ matrix whose eigenvalues are strictly greater than 1; $m = |\det M|$.

For a function ψ defined on $\mathbb{R}^d$ and a dilation matrix M in $\mathbb{R}^d$, $j \in \mathbb{Z}$, $k \in \mathbb{Z}^d$, $\psi_{jk} = m^{j/2}\psi(M^j \cdot + k)$.

If $\psi^{(\nu)}$, $\nu = 1, \ldots, r$, are functions defined on $\mathbb{R}^d$ and M is a dilation matrix in $\mathbb{R}^d$, $j \in \mathbb{Z}$, $k \in \mathbb{Z}^d$, then $\{\psi_{jk}^{(\nu)}\}_{j,k,\nu} := \{\psi_{jk}^{(\nu)}, j \in \mathbb{Z}, k \in \mathbb{Z}^d, \nu = 1, \ldots, r\}$

Chapter 1
Bases and Frames in Hilbert Spaces

Abstract Concepts of the Riesz basis and frame in a Hilbert space are introduced and studied.

1.1 Riesz Bases

In this section, we introduce and discuss an important notion for the wavelet theory, namely the notion of the Riesz basis for a Hilbert space, which is a special case of the notion of frame and generalizes the notion of orthogonal basis preserving its most essential properties.

Let us denote by ℓ_2 the Hilbert space of sequences of complex numbers $c = \{c_n\}_{n=1}^{\infty}$ such that $\sum\limits_{n=1}^{\infty} |c_n|^2 < \infty$ with the scalar product $\langle c^1, c^2\rangle := \sum\limits_{n=1}^{\infty} c_n^1\overline{c_n^2}$, where $c^1 = \{c_n^1\}_{n=1}^{\infty}$, $c^2 = \{c_n^2\}_{n=1}^{\infty}$ and $c^1, c^2 \in \ell_2$.

Definition 1.1.1 Let H be a Hilbert space. A system $\{f_n\}_{n=1}^{\infty} \subset H$ is called *a Riesz system* with constants $A, B > 0$ if for any $c = \{c_n\}_{n=1}^{\infty} \in \ell_2$ the series $\sum\limits_{n=1}^{\infty} c_n f_n$ converges in H and

$$A\|c\|_{\ell_2}^2 \leq \left\|\sum_{n=1}^{\infty} c_n f_n\right\|_H^2 \leq B\|c\|_{\ell_2}^2. \tag{1.1}$$

If a Riesz system is a basis, then it is called a Riesz basis.

Theorem 1.1.2 *Let H be a Hilbert space, and let $\{f_n\}_{n=1}^{\infty}$ be a Riesz system in H with constants A, B. Then,*

(i) $\{f_n\}_{n=1}^{\infty}$ is a Riesz basis for the space

$$V := \left\{f = \sum_{n=1}^{\infty} c_n f_n,\ \sum_{n=1}^{\infty} |c_n|^2 < \infty\right\};$$

A. Krivoshein et al., *Multivariate Wavelet Frames*,
Industrial and Applied Mathematics, DOI 10.1007/978-981-10-3205-9_1

(ii) $V = \overline{\operatorname{span}\{f_n, n \in \mathbb{N}\}}$;

(iii) For any element $f \in V$, *the following inequality holds:*

$$A\|f\|_H^2 \le \sum_{n=1}^{\infty} |\langle f, f_n\rangle|^2 \le B\|f\|_H^2. \tag{1.2}$$

Proof By definition, any element of the space V is the sum of a series $\sum_{n=1}^{\infty} c_n f_n$. Therefore, to prove (i), we only have to verify the uniqueness of such a representation. Suppose that for some $f = \sum_{n=1}^{\infty} c_n f_n \in V$, there exists another representation $\sum_{n=1}^{\infty} c_n' f_n$. Let us fix $k \in \mathbb{N}$. Given $\varepsilon > 0$ choose $N \ge k$ such that

$$\left\| \sum_{n=N+1}^{\infty} c_n f_n \right\|_H < \varepsilon, \quad \left\| \sum_{n=N+1}^{\infty} c_n' f_n \right\|_H < \varepsilon.$$

Using (1.1), we have

$$|c_k - c_k'| \le \sqrt{\sum_{n=1}^{N} |c_n - c_n'|^2} \le \frac{1}{\sqrt{A}} \left\| \sum_{n=1}^{N} (c_n - c_n') f_n \right\|_H \le$$
$$\frac{1}{\sqrt{A}} \left(\left\| \sum_{n=1}^{\infty} (c_n - c_n') f_n \right\|_H + \left\| \sum_{n=N+1}^{\infty} c_n f_n \right\|_H + \left\| \sum_{n=N+1}^{\infty} c_n' f_n \right\|_H \right) < \frac{2\varepsilon}{\sqrt{A}}.$$

The arbitrariness of ε implies that $c_k = c_k'$, $\forall k \in \mathbb{N}$.

Let us introduce the operator $\mathcal{T} : \ell_2 \longrightarrow H$ that takes each element $c = \{c_n\}_{n=1}^{\infty} \in \ell_2$ to the element $f = \mathcal{T}c = \sum_{n=1}^{\infty} c_n f_n$. It is clear that $\mathcal{T}$ is a one-to-one map from ℓ_2 to V. Furthermore, by (1.1), we have

$$\|\mathcal{T}\|^2 \le B, \quad \|\mathcal{T}^{-1}\|^2 \le A^{-1}, \tag{1.3}$$

i.e., $\mathcal{T}$ is an isomorphism from ℓ_2 to V. In particular, this implies that V is closed. Taking into account the inclusions

$$\operatorname{span}\{f_n, n \in \mathbb{N}\} \subset V, \quad V \subset \overline{\operatorname{span}\{f_n, n \in \mathbb{N}\}},$$

we obtain (ii). Let $e_n = (0, \dots, 0, 1, 0, \dots)$ be the nth basis vector in ℓ_2, then $f_n = \mathcal{T}e_n$. Since $\{e_n\}_{n=1}^{\infty}$ is an orthonormal basis for ℓ_2, using (1.3) we obtain

$$\sum_{n=1}^{\infty} |\langle f, f_n \rangle|^2 = \sum_{n=1}^{\infty} |\langle f, \mathcal{T} e_n \rangle|^2 = \sum_{n=1}^{\infty} |\langle \mathcal{T}^* f, e_n \rangle|^2 = \|\mathcal{T}^* f\|_{\ell_2}^2 \leq B\|f\|_H^2,$$

$$\|f\|_H^2 = \|\mathcal{T}^{*-1}\mathcal{T}^* f\|_H^2 \leq \frac{1}{A}\|\mathcal{T}^* f\|_{\ell_2} = \frac{1}{A}\sum_{n=1}^{\infty} |\langle \mathcal{T}^* f, e_n \rangle|^2 = \frac{1}{A}\sum_{n=1}^{\infty} |\langle f, f_n \rangle|^2,$$

which proves (iii). ◊

Remark 1.1.3 If in the latter theorem, instead of the assumption that $\{f_n\}_{n=1}^{\infty}$ is a Riesz system, we only assume that the right-hand inequality in (1.1) holds (in this case, $\{f_n\}_{n=1}^{\infty}$ is said to be a *Bessel system*), then we can claim that for any $f \in H$, the following inequality holds:

$$\sum_{n=1}^{\infty} |\langle f, f_n \rangle|^2 \leq B\|f\|^2.$$

To prove this assertion, it suffices to repeat the reasoning of Theorem 1.1.2 for any finite set of functions $\{f_n\}_{n=1}^{N}$ (here, we do not have to prove the closedness of the space $V_N = \left\{f = \sum_{n=1}^{N} c_n f_n\right\}$) and note that for any $f \in H$, we have

$$\sum_{n=1}^{N} |\langle f, f_n \rangle|^2 = \sum_{n=1}^{N} |\langle P_N f, f_n \rangle|^2,$$

where P_N is the orthogonal projector on V_N.

We have proved that as in the case of orthogonal bases, expansions with respect to a Riesz basis are linear combinations with coefficients from ℓ_2, while the Fourier coefficients of any element satisfy relation (1.2), which generalizes the Parseval's equality. Let us show that Riesz bases have another remarkable feature of orthogonal bases: The way of enumerating their elements is of no significance. Such bases are called unconditional.

Theorem 1.1.4 *Any Riesz basis for a Hilbert space H is an unconditional basis.*

Proof By Theorem 1.1.2, any element $f \in H$ can be represented in the form $f = \sum_{n=1}^{\infty} c_n f_n$, $\sum_{n=1}^{\infty} |c_n|^2 < \infty$. Denote by s_n the partial sums of this representation and by s_n' the partial sums of the series obtained from $\sum_{n=1}^{\infty} c_n f_n$ by some rearrangement of its terms. Let $\mathcal{T}$ be the operator defined in Theorem 1.1.2. We set $e = \mathcal{T}^{-1} f$, $\sigma_n = \mathcal{T}^{-1} s_n$, and $\sigma_n' = \mathcal{T}^{-1} s_n'$. As it was shown above, $\mathcal{T}$ is an isomorphism from ℓ_2 to H. Hence, it follows from the relation $f = \lim_{n\to\infty} s_n$ that $e = \lim_{n\to\infty} \sigma_n = \lim_{n\to\infty} \sigma_n'$, which implies that $\lim_{n\to\infty} s_n' = f$. ◊

In view of the above theorem, the way of taking partial sums in expansions with respect to Riesz systems does not matter. Hence, the terms in these sums can be enumerated by any countable set. In particular, in what follows, we shall assume that the summation index ranges over $\mathbb{Z}^d$.

Next, we discuss for which functions their integer shifts form a Riesz basis. Recall that $\widehat{f}$ denotes the Fourier transform of f.

Lemma 1.1.5 *Let $\varphi, \widetilde{\varphi} \in L_2(\mathbb{R}^d)$, then the series $\sum_{k\in\mathbb{Z}^d} \widehat{\varphi}(\xi+k)\overline{\widehat{\widetilde{\varphi}}(\xi+k)}$ absolutely converges almost everywhere and its sum is summable on $\mathbb{T}^d$. Moreover, the sum $\sum_{k\in\mathbb{Z}^d} |\widehat{\varphi}(\xi+k)\widehat{\widetilde{\varphi}}(\xi+k)|$ is summable on $\mathbb{T}^d$.*

Proof By the Plancherel theorem, $\widehat{\varphi}$ and $\widehat{\widetilde{\varphi}}$ are in $L_2(\mathbb{R}^d)$, which yields that the function $\widehat{\varphi}\overline{\widehat{\widetilde{\varphi}}}$ is summable on $\mathbb{R}^d$. Consider the series

$$\sum_{k\in\mathbb{Z}^d} \int_{\mathbb{T}^d} |\widehat{\varphi}(\xi+k)\widehat{\widetilde{\varphi}}(\xi+k)|\,d\xi = \sum_{k\in\mathbb{Z}^d} \int_{\mathbb{T}^d+k} |\widehat{\varphi}(\xi)\widehat{\widetilde{\varphi}}(\xi)|\,d\xi = \int_{\mathbb{R}^d} |\widehat{\varphi}(\xi)\widehat{\widetilde{\varphi}}(\xi)|\,d\xi.$$

Since the right-hand side of this equality is finite, by the Levi theorem, the function $\sum_{k\in\mathbb{Z}^d} |\widehat{\varphi}(\xi+k)\widehat{\widetilde{\varphi}}(\xi+k)|$ is finite almost everywhere and

$$\int_{\mathbb{T}^d} \sum_{k\in\mathbb{Z}^d} |\widehat{\varphi}(\xi+k)\widehat{\widetilde{\varphi}}(\xi+k)|\,d\xi = \sum_{k\in\mathbb{Z}^d} \int_{\mathbb{T}^d} |\widehat{\varphi}(\xi+k)\widehat{\widetilde{\varphi}}(\xi+k)|\,d\xi,$$

which yields all statements of the lemma. ◊

Theorem 1.1.6 *Let $\varphi \in L_2(\mathbb{R}^d)$. For a system of functions $\{\varphi(\cdot+n)\}_{n\in\mathbb{Z}^d}$ to be a Riesz system with constants A, B, it is necessary and sufficient that*

$$A \le \sum_{k\in\mathbb{Z}^d} |\widehat{\varphi}(\xi+k)|^2 \le B \tag{1.4}$$

for almost all $\xi \in \mathbb{R}^d$.

Proof Sufficiency. Let $c = \{c_n\}_{n\in\mathbb{Z}^d} \in \ell_2(\mathbb{Z}^d)$. It follows from (1.4) and the Parseval's equality that

$$\int_{\mathbb{T}^d} \left|\sum_{n\in\mathbb{Z}^d} c_n e^{2\pi i(n,\xi)}\right|^2 \sum_{k\in\mathbb{Z}^d} |\widehat{\varphi}(\xi+k)|^2\,d\xi \ge A \sum_{n\in\mathbb{Z}^d} |c_n|^2, \tag{1.5}$$

$$\int_{\mathbb{T}^d} \left|\sum_{n\in\mathbb{Z}^d} c_n e^{2\pi i(n,\xi)}\right|^2 \sum_{k\in\mathbb{Z}^d} |\widehat{\varphi}(\xi+k)|^2\,d\xi \le B \sum_{n\in\mathbb{Z}^d} |c_n|^2, \tag{1.6}$$

Using Lebesgue's dominated convergence theorem to interchange integration and summation over k, we have

$$\int_{\mathbb{T}^d} \left|\sum_{n\in\mathbb{Z}^d} c_n e^{2\pi i(n,\xi)}\right|^2 \sum_{k\in\mathbb{Z}^d} |\widehat{\varphi}(\xi+k)|^2 \, d\xi = \int_{\mathbb{R}^d} \left|\sum_{n\in\mathbb{Z}^d} c_n e^{2\pi i(n,\xi)} \widehat{\varphi}(\xi)\right|^2 d\xi. \quad (1.7)$$

Thus, $\sum\limits_{n\in\mathbb{Z}^d} c_n e^{2\pi i(n,\cdot)} \widehat{\varphi}(\cdot) \in L_2(\mathbb{R}^d)$. By the Plancherel theorem, it follows that $\sum\limits_{n\in\mathbb{Z}^d} c_n \varphi(\cdot+n) \in L_2(\mathbb{R}^d)$ and

$$\left\|\sum_{n\in\mathbb{Z}^d} c_n \varphi(\cdot+n)\right\|^2 = \int_{\mathbb{R}^d} \left|\sum_{n\in\mathbb{Z}^d} c_n e^{2\pi i(n,\xi)} \widehat{\varphi}(\xi)\right|^2 d\xi. \quad (1.8)$$

Combining this equality with (1.7) and using (1.5), (1.6), we get

$$A\sum_{n\in\mathbb{Z}^d} |c_n|^2 \le \left\|\sum_{n\in\mathbb{Z}^d} c_n \varphi(\cdot+n)\right\|^2 \le B\sum_{n\in\mathbb{Z}^d} |c_n|^2,$$

which means that condition (1.1) is satisfied for the system $\{\varphi(\cdot+n)\}_{n\in\mathbb{Z}^d}$.

Necessity. Let the functions $\varphi(\cdot+n)$, $n\in\mathbb{Z}^d$, form a Riesz system. Then, $\sum\limits_{n\in\mathbb{Z}^d} c_n \varphi(\cdot+n) \in L_2(\mathbb{R}^d)$ for any $c=\{c_n\}\in \ell_2(\mathbb{Z}^d)$; hence, by the Plancherel theorem, $\sum\limits_{n\in\mathbb{Z}} c_n e^{2\pi i(n,\cdot)} \widehat{\varphi}(\cdot) \in L_2(\mathbb{R}^d)$, and equality (1.8) holds. Since the right-hand side of (1.8) is finite, by the Levi theorem we have equality (1.7). Using (1.7), (1.8), we rewrite condition (1.1) in the following form: For any $g=\sum\limits_{n\in\mathbb{Z}^d} c_n e^{2\pi i(n,\cdot)} \in L_2(\mathbb{T}^d)$, the inequality

$$A\|g\|^2 \le \int_{\mathbb{T}^d} |g(\xi)|^2 \sigma(\xi)\, d\xi \le B\|g\|^2, \quad (1.9)$$

holds, where

$$\sigma(\xi) := \sum_{k\in\mathbb{Z}^d} |\widehat{\varphi}(\xi+k)|^2 .$$

The function σ is 1-periodic, and by Lemma 1.1.5, it is summable on $\mathbb{T}^d$. Since

$$\sup_{\|g\|_2\ne 0} \int_{\mathbb{T}^d} \frac{|g|^2}{\|g\|_2^2}\sigma = \sup_{\substack{\|h\|_1=1\\ h\ge 0}} \int_{\mathbb{T}^d} h\sigma = \sup_{\|h\|_1=1} \int_{\mathbb{T}^d} h\sigma,$$

by the Riesz representation theorem, we have

$$\operatorname*{vraisup}_{\mathbb{T}^d} \sigma = \|\sigma\|_\infty = \sup_{\|g\|_2 \neq 0} \int_{\mathbb{T}^d} \frac{|g|^2}{\|g\|_2^2} \sigma \leq B,$$

i.e., the right-hand inequality in (1.4) holds. Let us assume that the left-hand one does not hold. Then, there exists a set $E \subset \mathbb{T}^d$ of positive measure such that $\sigma \leq A - \varepsilon$, $\varepsilon > 0$ on E. Putting $g = \chi_E$, we obtain

$$\int_{\mathbb{T}^d} |g|^2 \sigma = \int_E \sigma \leq (A - \varepsilon)\mu E = (A - \varepsilon)\|g\|_2^2 < A\|g\|_2^2,$$

which contradicts with (1.9).◊

Remark 1.1.7 It is readily seen from the proof of this theorem that a system of functions $\{\varphi(\cdot + n)\}_{n \in \mathbb{Z}^d}$ is Bessel if and only if for almost all $\xi \in \mathbb{R}^d$

$$\sum_{k \in \mathbb{Z}^d} |\widehat{\varphi}(\xi + k)|^2 \leq B. \tag{1.10}$$

Proposition 1.1.8 *Let a function $\varphi \in L_2(\mathbb{R}^d)$ be compactly supported. Then, the system $\{\varphi(\cdot + n)\}_{n \in \mathbb{Z}^d}$ is a Bessel system.*

Proof Suppose $\operatorname{supp} \varphi \subset [-N, N]^d$, and let $c = \{c_k\}_{k \in \mathbb{Z}^d} \in \ell_2(\mathbb{Z}^d)$. Then,

$$\int_{\mathbb{T}^d + l} \left| \sum_{k \in \mathbb{Z}^d} c_k \varphi(x + k) \right|^2 dx = \int_{\mathbb{T}^d + l} \left| \sum_{k \in [-N,N]^d - l} c_k \varphi(x + k) \right|^2 dx \leq$$
$$\sum_{k \in [-N,N]^d - l} |c_k|^2 \sum_{k \in [-N,N]^d - l} \int_{\mathbb{T}^d + l} |\varphi(x + k)|^2 \, dx \leq \|\varphi\|^2 \sum_{k \in [-N,N]^d - l} |c_k|^2.$$

It follows that

$$\int_{\mathbb{R}^d} \left| \sum_{k \in \mathbb{Z}^d} c_k \varphi(x + k) \right|^2 dx \leq (2N + 1)^d \|\varphi\|^2 \sum_{k \in \mathbb{Z}^d} |c_k|^2. \quad \Diamond$$

Let us now study some properties of spaces spanned by the integer shifts of a function.

Proposition 1.1.9 *Let $\varphi \in L_2(\mathbb{R}^d)$, and let the system $\{\varphi(\cdot + n)\}_{n \in \mathbb{Z}^d}$ be a Riesz basis for a closed subspace V of the space $L_2(\mathbb{R}^d)$. A function $f \in L_2(\mathbb{R}^d)$ belongs to V if and only if there exists a function $m_f \in L_2(\mathbb{T}^d)$ such that*

$$\widehat{f}(\xi) = m_f(\xi)\widehat{\varphi}(\xi) \tag{1.11}$$

for almost all $\xi \in \mathbb{R}^d$*. In addition, the coefficients of the expansion of* f *with respect to the system* $\{\varphi(\cdot + n)\}_{n\in\mathbb{Z}^d}$ *coincide with the corresponding Fourier coefficients of the function* m_f.

Proof By Theorem 1.1.2, we have

$$V = \left\{ f = \sum_{n\in\mathbb{Z}^d} c_n \varphi(\cdot + n),\ \sum_{n\in\mathbb{Z}^d} |c_n|^2 < \infty \right\}.$$

Let $\sum_{n\in\mathbb{Z}^d} |c_n|^2 < \infty$,

$$f(x) = \sum_{n\in\mathbb{Z}^d} c_n \varphi(x+n). \tag{1.12}$$

Since taking the Fourier transform and passing to the limit commute in $L_2(\mathbb{R}^d)$, and the series in the right-hand side of (1.12) converges in $L_2(\mathbb{R}^d)$, it follows that applying the Fourier transform to both sides of this equality, we get

$$\widehat{f}(\xi) = \sum_{n\in\mathbb{Z}^d} c_n \widehat{\varphi(\cdot + n)}(\xi) = \sum_{n\in\mathbb{Z}^d} c_n e^{2\pi i(n,\xi)} \widehat{\varphi}(\xi). \tag{1.13}$$

To prove the necessity, it remains to set

$$m_f(\xi) = \sum_{n\in\mathbb{Z}^d} c_n e^{2\pi i(n,\xi)} \tag{1.14}$$

and note that $m_f \in L_2(\mathbb{T}^d)$ by the Parseval's equality. Conversely, if (1.11) holds and $m_f \in L_2(\mathbb{T}^d)$, then multiplying (1.14) by $\widehat{\varphi}(\xi)$, we obtain equality (1.13), which is equivalent to (1.12), as it was shown above. ◊

Proposition 1.1.10 *Let* φ *and* V *be as in Proposition 1.1.9. A function* $f \in L_2(\mathbb{R}^d)$ *is orthogonal to* V *if and only if*

$$\sum_{l\in\mathbb{Z}^d} \widehat{f}(\xi + l)\overline{\widehat{\varphi}(\xi + l)} = 0$$

for almost all $\xi \in \mathbb{R}^d$.

Proof The relation $f \perp V$ is equivalent to

$$\int_{\mathbb{R}^d} f(x)\overline{\varphi(x+k)}\,dx = 0$$

for all $k \in \mathbb{Z}^d$. On the other hand, by the Plancherel theorem,

$$\int_{\mathbb{R}^d} f(x)\overline{\varphi(x+k)}\,dx = \int_{\mathbb{R}^d} \widehat{f}(\xi)\overline{\widehat{\varphi}(\xi)}e^{-2\pi i(k,\xi)}\,d\xi =$$

$$\sum_{n\in\mathbb{Z}^d} \int_{\mathbb{T}^d+n} \widehat{f}(\xi)\overline{\widehat{\varphi}(\xi)}e^{-2\pi i(k,\xi)}\,d\xi = \int_{\mathbb{R}^d} \sum_{n\in\mathbb{Z}^d} \widehat{f}(\xi+n)\overline{\widehat{\varphi}(\xi+n)}e^{-2\pi i(k,\xi)}\,d\xi \tag{1.15}$$

for any $k \in \mathbb{Z}^d$. It is possible to change the order of summation and integration due to Lebesgue's dominated convergence theorem and Lemma 1.1.5. By the same lemma, the series on the right-hand side is in $L_1(\mathbb{T}^d)$, and due to the uniqueness theorem, all the Fourier coefficients are equal to zero if and only if this function is equal to zero almost everywhere. ◊

1.2 Frames

In this section, we introduce the notion of frame which is a generalization of the notion of a Riesz basis and preserves some of its advantages.

Definition 1.2.1 Let H be a Hilbert space. A system $\{f_n\}_{n=1}^{\infty} \subset H$ is called *a frame*, if there exist constants $A, B > 0$ such that the inequality

$$A\|f\|^2 \leq \sum_{n=1}^{\infty} |\langle f, f_n\rangle|^2 \leq B\|f\|^2 \tag{1.16}$$

holds for any $f \in H$.

The constants A and B are called the upper and lower frame bounds, respectively. If the frame bounds coincide, then the frame is said to be *tight*. Tight frames with unit bounds ($A = B = 1$) are called Parseval's frames, and sometimes, they are also called orthogonal-like systems, which is justified by the following statement.

Proposition 1.2.2 *For a system $\{f_n\}_{n=1}^{\infty} \subset H$ to be a Parseval's frame, it is necessary and sufficient that*

$$f = \sum_{n=1}^{\infty} \langle f, f_n\rangle f_n, \tag{1.17}$$

for any $f \in H$.

Proof The sufficiency is obvious. Let us prove the necessity. Let $f \in H$, then by (1.16) the equality

$$\sum_{n=1}^{\infty} |\langle f, f_n\rangle|^2 = \|f\|^2 \tag{1.18}$$

holds. First of all, we note that the series on the right-hand side of (1.17) converges. Indeed, we have

$$\left\|\sum_{n=N}^{M}\langle f, f_n\rangle f_n\right\|^2 = \left|\sum_{n=N}^{M}\langle f, f_n\rangle\langle f_n, g_{NM}\rangle\right|^2 \le \sum_{n=N}^{M}|\langle f, f_n\rangle|^2\sum_{n=1}^{\infty}|\langle f_n, g_{NM}\rangle|^2,$$

where $\|g_{NM}\| \le 1$. It follows from the definition of frame that the right-hand side tends to zero as $N, M \to \infty$. Let us check equality (1.17). We can assume that $\|f\| = 1$. Set

$$s_N = \sum_{n=1}^{N}\langle f, f_n\rangle f_n, \quad \sigma_N = \sum_{n=1}^{N}|\langle f, f_n\rangle|^2,$$

then it is obvious that

$$\|f - s_N\|^2 = 1 - 2\sigma_N + \|s_N\|^2.$$

Since (1.18) implies that $\lim\limits_{N\to\infty}\sigma_N = 1$, it remains to check that $\|s_N\| \to 1$. Given $\varepsilon > 0$, we choose N such that $\sigma_N \ge 1 - \varepsilon$. Since $\|f\| = 1$, by the Cauchy–Schwarz inequality we have

$$\|s_N\| \ge |\langle s_N, f\rangle| = \left|\sum_{n=1}^{N}\langle f, f_n\rangle\langle f_n, f\rangle\right| = \sigma_N \ge 1 - \varepsilon.$$

On the other hand,

$$\|s_N\|^2 = |\langle s_N, g_N\rangle|^2 = \left|\sum_{n=1}^{N}\langle f, f_n\rangle\langle f_n, g_N\rangle\right|^2$$

where $\|g_N\| \le 1$. Applying the Cauchy–Schwarz inequality and (1.18) to the right-hand side, we obtain $\|s_N\| \le 1$.◊

Corollary 1.2.3 *If a Parseval's frame is a basis, then this basis is orthonormal.*

To prove this fact, it suffices to set $f = f_k$ in (1.17) and to take into account the uniqueness of this expansion.

Thus, Parseval's frames can be considered as a generalization of orthonormal bases for redundant systems (i.e., systems that are not minimal). Now, we show that the notion of frame generalizes the notion of Riesz basis for redundant systems.

Lemma 1.2.4 *Let $\{f_n\}_{n=1}^{\infty}$ be a frame in a Hilbert space H with an upper bound B, and $c = \{c_n\}_{n=1}^{\infty}$ be in ℓ_2. Then, the series $\sum_{n=1}^{\infty} c_n f_n$ converges in H unconditionally, and*

$$\left\|\sum_{n=1}^{\infty} c_n f_n\right\| \le B\|c\|_{\ell_2}. \tag{1.19}$$

Proof Given $\varepsilon > 0$, choose N such that $\sum_{n=N}^{\infty} |c_n|^2 < \varepsilon$. If Ω is a finite set of positive integers which are greater than N, then

$$\left\| \sum_{n\in\Omega} c_n f_n \right\| = \left| \left\langle \sum_{n\in\Omega} c_n f_n, g_\Omega \right\rangle \right| = \left| \sum_{n\in\Omega} c_n \langle f_n, g_\Omega \rangle \right|,$$

where $\|g_\Omega\| \leq 1$. It follows from the Cauchy–Schwarz inequality and (1.16) that

$$\left| \sum_{n\in\Omega} c_n \langle f_n, g_\Omega \rangle \right|^2 \leq \sum_{n\in\Omega} |c_n|^2 \sum_{n=1}^{\infty} |\langle f_n, g_\Omega \rangle|^2 \leq B \|g_\Omega\|^2 \sum_{n\in\Omega} |c_n|^2 < B\varepsilon.$$

This yields unconditional convergence of the series $\sum_{n=1}^{\infty} c_n f_n$. Similarly, replacing Ω by $\mathbb{N}$, we obtain (1.19). $\Diamond$

Corollary 1.2.5 *Let $\{f_n\}_{n=1}^{\infty}$ and $\{g_n\}_{n=1}^{\infty}$ be frames in a Hilbert space H. If*

$$\sum_{n=1}^{\infty} \langle f, g_n \rangle \langle f_n, g \rangle = \langle f, g \rangle \quad \forall f, g \in H, \tag{1.20}$$

then every $f \in H$ can be decomposed as

$$f = \sum_{n=1}^{\infty} \langle f, g_n \rangle f_n, \tag{1.21}$$

where the series converges unconditionally in H.

The proof follows immediately from Lemma 1.2.4.

Frames $\{f_n\}_{n=1}^{\infty}$, $\{g_n\}_{n=1}^{\infty}$ as in Corollary 1.2.5 are called *dual*.

Theorem 1.2.6 *A system $\{f_n\}_{n=1}^{\infty}$ is a frame with bounds A, B in a Hilbert space H if and only if it is an orthogonal projection of some Riesz basis with constants A, B in a Hilbert space H^* that contains H. In particular, if elements f_n, $n = 1, 2, \ldots,$ form a basis for the space H, then it is a Riesz basis.*

Proof Sufficiency. Suppose H^* is a Hilbert space, H is its subspace, and $\{F_n\}_{n=1}^{\infty}$ is a Riesz basis for H^* with constants A, B. Denote by P the operator of orthogonal projection of H^* on H and set $f_n = PF_n$, $n = 1, 2, \ldots$. By Theorem 1.1.2, the system $\{F_n\}_{n=1}^{\infty}$ forms a frame with bounds A, B. It remains to note that if $f \in H$, then $\|f\|_{H^*} = \|f\|_H$ and $\langle f, F_n \rangle = \langle f, f_n \rangle$.

Necessity. Let the system $\{f_n\}_{n=1}^{\infty}$ satisfy (1.16) and $c = \{c_n\}_{n=1}^{\infty}$ be in ℓ_2. By Lemma 1.2.4, the linear operator

$$T: \ c \to \sum_{n=1}^{\infty} c_n f_n$$

is well defined and bounded. Let h be the kernel of the operator $\mathcal{T}$, i.e., the set of points $c \in \ell_2$ such that $\sum_{n=1}^{\infty} c_n f_n = 0$. Since the operator $\mathcal{T}$ is continuous, h is a closed subspace of the space ℓ_2. Denote by $h^{\perp}$ its orthogonal complement. Let us define the operator $\mathcal{Q}$ that maps H to ℓ_2 and takes every $f \in H$ to the sequence $\{\langle f, f_n\rangle\}_{n=1}^{\infty}$. It follows from the left inequality (1.16) that $\mathcal{Q}$ is a one-to-one map of H to $\mathcal{Q}(H) =: h'$; therefore, there exists the inverse operator $\mathcal{Q}^{-1}$ of h' to H. Moreover, (1.16) implies that the operators $\mathcal{Q}$, $\mathcal{Q}^{-1}$ are bounded, where

$$\|\mathcal{Q}\|^2 \leq B, \quad \|\mathcal{Q}^{-1}\|^2 \leq 1/A, \tag{1.22}$$

i.e., $\mathcal{Q}$ is an isomorphism of H to h'. Let us show that $h' = h^{\perp}$. If $c \in h$, $c' \in h'$, then there exists $f \in H$ such that $c'_n = \langle f, f_n\rangle$ for all $n = 1, 2, \ldots$, and consequently,

$$\langle c, c'\rangle = \sum_{n=1}^{\infty} c_n \overline{\langle f, f_n\rangle} = \left\langle \sum_{n=1}^{\infty} c_n f_n, f \right\rangle = 0.$$

Therefore, $h' \subset h^{\perp}$. If $h' \neq h^{\perp}$, and then, there exists a nonzero element $c \in h^{\perp} \ominus h'$, and $\sum_{n=1}^{\infty} c_n \langle f, f_n\rangle = 0$ for any f from H, in particular, for $f = f_0 := \sum_{n=1}^{\infty} c_n f_n$, and thus, $f_0 = 0$, which contradicts the assumption $c \in h^{\perp}$, $c \neq 0$.

Denote by $\mathcal{T}_1$ the restriction of the operator $\mathcal{T}$ to $h^{\perp}$. For any $c \in h^{\perp}$ and $g \in H$, we have the equality

$$\langle \mathcal{T}_1 c, g\rangle = \left\langle \sum_{n=1}^{\infty} c_n f_n, g \right\rangle = \sum_{n=1}^{\infty} c_n \langle f_n, g\rangle = \langle c, \mathcal{Q} g\rangle,$$

which implies that $\mathcal{T}_1 = \mathcal{Q}^*$. Thus, for any $c \in h^{\perp}$: $\mathcal{Q}^* c = \mathcal{T}_1 c = \sum_{n=1}^{\infty} c_n f_n$; using (1.22), we get

$$A \sum_{n=1}^{\infty} |c_n|^2 \leq \left\| \sum_{n=1}^{\infty} c_n f_n \right\|^2 \leq B \sum_{n=1}^{\infty} |c_n|^2, \quad \text{for } c \in h^{\perp}. \tag{1.23}$$

Let $\{\alpha_n\}_{n \in \Lambda}$ be an orthogonal basis for h such that $\|\alpha_n\| = \sqrt{A}$ for all $n \in \Lambda$. The set Λ may be empty, finite, or countable. Assume first that Λ is countable. Then, for any $c \in \ell_2$, the following equality holds:

$$\left\| \sum_{n=1}^{\infty} c_n \alpha_n \right\|^2 = A \sum_{n=1}^{\infty} |c_n|^2. \tag{1.24}$$

Set $\tilde{h} := \{\sum_{n=1}^{\infty} c_n \alpha_n,\ c \in h\}$. It is clear that the orthogonal complement to the space $\tilde{h}$ in h consists of elements of the form $\sum_{n=1}^{\infty} c_n \alpha_n$, where $c \in h^{\perp}$. Denote by β_n the orthogonal projection of α_n onto $\tilde{h}$. For any $c \in h$, we have

$$\sum_{n=1}^{\infty} c_n(\alpha_n - \beta_n) = \sum_{n=1}^{\infty} c_n \alpha_n - \sum_{n=1}^{\infty} c_n \beta_n \in \tilde{h},$$

which implies that

$$\sum_{n=1}^{\infty} c_n(\alpha_n - \beta_n) = 0 \ \text{ for } \ c \in h. \tag{1.25}$$

Similarly, we get

$$\sum_{n=1}^{\infty} c_n \beta_n = 0 \ \text{ for } \ c \in h^{\perp}. \tag{1.26}$$

Combining (1.24) and (1.25), we ascertain that the following equality

$$\left\| \sum_{n=1}^{\infty} c_n \beta_n \right\|^2 = A \sum_{n=1}^{\infty} |c_n|^2 \tag{1.27}$$

holds for all $c \in h$.

Define the space H^* as the direct sum of the spaces H and h. The space H^* consists of pairs $(f, c) =: f \oplus c$, where $f \in H$, $c \in h$, and the scalar product of elements $F = f \oplus c$, $F' = f' \oplus c'$ is given by the formula

$$\langle F, F' \rangle_{H^*} = \langle f, f' \rangle_H + \langle c, c' \rangle_h.$$

Let us show that the elements $F_n := f_n \oplus \beta_n$ form a Riesz basis in H^*. Any $c \in \ell_2$ can be represented in the form $c = c^1 + c^2$, where $c^1 \in h^{\perp}$, $c^2 \in h$. Using (1.26) and the definition of the space h, we get

$$\left\| \sum_{n=1}^{\infty} c_n F_n \right\|^2 = \left\| \sum_{n=1}^{\infty} c_n f_n \right\|^2 + \left\| \sum_{n=1}^{\infty} c_n \beta_n \right\|^2 = \left\| \sum_{n=1}^{\infty} c_n^1 f_n \right\|^2 + \left\| \sum_{n=1}^{\infty} c_n^2 \beta_n \right\|^2.$$

Using (1.23) and (1.27), we obtain

$$A \sum_{n=1}^{\infty} |c_n|^2 \leq \left\| \sum_{n=1}^{\infty} c_n F_n \right\|^2 \leq B \sum_{n=1}^{\infty} |c_n|^2, \ \text{ for } \ c \in \ell_2.$$

If Λ is finite, we set

$$\beta_n = \frac{1}{\sqrt{A}} \sum_{k \in \Lambda} \overline{(\alpha_k)_n} \alpha_k.$$

It is not hard to check that relations (1.26), (1.27) hold.

It remains to note that the elements f_n, $n = 1, 2, \ldots$, form a basis in H if the space h is empty. Hence, H^* coincides with H and $F_n = f_n$, $n = 1, 2, \ldots$, in this case. ◊

Note that the operators $\mathcal{Q}$ and $\mathcal{T}$ from the above proof are usually referred as the *analysis operator* and the *synthesis operator*, respectively.

Corollary 1.2.7 *If a system $\{f_n\}_{n=1}^{\infty}$ forms a frame with bounds A, B in a Hilbert space H, then in H there exists a dual frame $\{g_n\}_{n=1}^{\infty}$ with bounds $1/B$, $1/A$.*

Proof Let $\{F_n\}_{n=1}^{\infty}$ be the Riesz basis for H^* constructed in Theorem 1.2.6, and let P be the orthogonal projection of H^* on H. Thus, $f_n = PF_n$, $n = 1, 2, \ldots$. For any $F \in H^*$, we have the following expansion:

$$F = \sum_{n=1}^{\infty} \langle F, G_n \rangle F_n,$$

where $\{G_n\}_{n=1}^{\infty}$ is the dual system to $\{F_n\}_{n=1}^{\infty}$. In particular, if $f \in H$, we have

$$f = \sum_{n=1}^{\infty} \langle f, G_n \rangle F_n = \sum_{n=1}^{\infty} \langle f, PG_n \rangle F_n = \sum_{n=1}^{\infty} \langle f, PG_n \rangle f_n. \tag{1.28}$$

Set $g_n = PG_n$, $n = 1, 2, \ldots$, and verify that $\{g_n\}_{n=1}^{\infty}$ is the required frame. In view of Theorem 1.2.6, it suffices to show that the system $\{G_n\}_{n=1}^{\infty}$ is a Riesz basis for H^*. Let $c \in \ell_2$. Using Theorem 1.1.2 and the fact that the systems $\{F_n\}_{n=1}^{\infty}$, $\{G_n\}_{n=1}^{\infty}$ are biorthonormal, we obtain

$$\left\| \sum_{n=N}^{M} c_n G_n \right\| = \sup_{\|F\| \le 1} \left| \left\langle \sum_{n=N}^{M} c_n G_n, F \right\rangle \right| \le$$
$$\sup_{\|c'\|_{\ell_2}^2 \le 1/A} \left| \left\langle \sum_{n=N}^{M} c_n G_n, \sum_{k=1}^{\infty} c'_k F_k \right\rangle \right| = \sup_{\|c'\|_{\ell_2}^2 \le 1/A} \left| \sum_{n=N}^{M} c_n c'_n \right| \le \frac{1}{\sqrt{A}} \sqrt{\sum_{n=N}^{M} |c_n|^2}. \tag{1.29}$$

It follows that the series $\sum\limits_{n=1}^{\infty} c_n G_n =: G$ converges in norm. By Theorem 1.1.2, the system $\{F_n\}_{n=1}^{\infty}$ is a frame with bounds A, B. Therefore,

$$\frac{1}{B} \sum_{n=1}^{\infty} |\langle G, F_n \rangle|^2 \le \|G\|^2 \le \frac{1}{A} \sum_{n=1}^{\infty} |\langle G, F_n \rangle|^2.$$

To prove that $\{G_n\}_{n=1}^{\infty}$ is a Riesz basis with bounds $1/B$, $1/A$, it remains to take into account that $c_n = \langle G, F_n\rangle$. On the other hand, it follows from (1.28) that (1.20) holds for all $f, g \in H$, i.e., the frames $\{f_n\}_{n=1}^{\infty}$, $\{g_n\}_{n=1}^{\infty}$ are dual. ◊

Corollary 1.2.8 *Let $\{f_n\}_{n=1}^{\infty}$, $\{\tilde{f}_n\}_{n=1}^{\infty}$ be biorthonormal systems in a Hilbert space H. If one of these systems forms a frame in H, then both systems are Riesz bases in H.*

Proof Let the system $\{f_n\}_{n=1}^{\infty}$ satisfy (1.16). By Corollaries 1.2.7, 1.2.5, there exists a frame $\{g_n\}_{n=1}^{\infty}$ such that (1.21) holds for any $f \in H$. The coefficients of this expansion can be found by multiplying scalarly (1.21) by $\tilde{f}_n$, $n = 1, 2, \ldots$. It follows that the expansion with respect to the system $\{f_n\}_{n=1}^{\infty}$ is unique, and the inequality

$$\langle f, g_n\rangle = \langle f, \tilde{f}_n\rangle, \quad n = 1, 2, \ldots \tag{1.30}$$

holds for all $f \in H$. Thus, $\{f_n\}_{n=1}^{\infty}$ is a basis for H. By Theorem (1.2.6), it is a Riesz basis. Since $\{g_n\}_{n=1}^{\infty}$ is a frame, it follows from (1.30) that the system $\{\tilde{f}_n\}_{n=1}^{\infty}$ is also a frame, and as we have already proved, it is a Riesz basis. ◊

Chapter 2
MRA-Based Wavelet Bases and Frames

Abstract Construction of multivariate wavelet systems based on a multiresolution analysis (MRA) is presented and discussed.

2.1 Dilation Matrix

The goal of the book is to study multivariate wavelet frames which consist of the shifts and dilations of several functions. Multiplication of each component of the argument by the same factor means the multiplication of the vector argument by a diagonal matrix with equal diagonal elements, i.e., the same dilation along each coordinate axis. It is natural to consider other matrices as coefficients of dilation (so-called nonseparable wavelets). Moreover, nonseparable wavelets are preferable in signal processing because multidimensional signals by their nature are nonseparable. Nonseparable filter banks (masks in our terminology) have better characteristics than their separable counterparts (where the matrix is similar to the identity matrix and generating wavelet functions are the tensor products of one-dimensional wavelet functions). The number of degrees of freedom is also much bigger for nonseparable filter banks. In tomography, the 2-D separable wavelets impose a rectangular tiling of the frequency plane, which is not well suited to the radial band-limited assumption of the image. The application of nonseparable multiresolution tomography to 2-D wavelets allows to respect the geometry of the system by tiling the frequency plane in a diamond-shaped fashion that is more respectful to the radial band-limited assumptions. Another successful application of nonseparable wavelets is in 3-D rotational angiography.

In this section, a class of matrices used as dilations will be introduced. First, we discuss some properties of square integer matrices.

If M is a nonsingular integer $d \times d$ matrix, we say that vectors $k, n \in \mathbb{Z}^d$ are *congruent modulo* M and write $k \equiv n \pmod{M}$, if $k - n = Ml$, $l \in \mathbb{Z}^d$. The integer lattice $\mathbb{Z}^d$ is partitioned into cosets with respect to the congruence introduced above. Any set containing only one representative of each coset is called *a set of digits of the matrix* M. When it does not matter which set of digits is chosen, we shall assume that it is chosen arbitrarily and denote it by $D(M)$.

A. Krivoshein et al., *Multivariate Wavelet Frames*,
Industrial and Applied Mathematics, DOI 10.1007/978-981-10-3205-9_2

Proposition 2.1.1 *Let M be a nonsingular integer $d \times d$ matrix. Then, the number of cosets with respect to the congruence modulo M is equal to $|\det M|$, the set $H(M) := \mathbb{Z}^d \cap M[0,1)^d$ is a set of digits.*

Proof Let us show that the number of integer points in the set $M[0,1)^d$ is equal to its volume. Denote by a and b the volume of the set $M[0,1)^d$ and the number of its integer points, respectively. Consider a d-dimensional cube with side N and find the number of parallelepipeds of the form $M([0,1)^d + k)$, $k \in \mathbb{Z}^d$, inside the cube $[0,N]^d$ in two ways. Let S denote the sum of all such parallelepipeds. Obviously, for large N, the volume of S is $N^d + o(N^d)$. The number of integer points in S is $N^d + o(N^d)$. The volume of a parallelepiped is a, and the number of integer points in it is b. Expressing the number of parallelepipeds in S in two ways and equating these expressions, we get

$$\frac{N^d + o(N^d)}{a} = \frac{N^d + o(N^d)}{b}.$$

Dividing both sides by N^d and passing to the limit as N tends to infinity, we get $a = b$. Let us show that the set $H(M)$ does not contain any congruent elements. Suppose integer points $s_1, s_2 \in M[0,1)^d$ are congruent modulo M, then $s_1 - s_2 = Ml$, where $l \in \mathbb{Z}^d$. Multiplying both sides of this equality by M^{-1}, we deduce that the difference between two elements from $[0,1)^d$ is equal to an integer vector, which is impossible unless $l = \mathbf{0}$; thus, $s_1 = s_2$. Show that the set $H(M)$ contains representatives of all cosets of the integer lattice. Let an integer vector $s \notin M[0,1)^d$ represent a coset. The vector $M^{-1}s$ belongs to a cube of the form $[0,1)^d + k$, $k \in \mathbb{Z}^d$ (since such cubes cover the whole space $\mathbb{R}^d$). Set $s' = M(M^{-1}s - k)$. It is clear that s' and s are congruent modulo M and $s' \in M[0,1)^d$; i.e., in $M[0,1)^d$, we have found a representative of the coset that contains s. Thus, $H(M)$ contains exactly one representative of each coset; therefore, the number of cosets is a. It remains to note that $a = |\det M|$.◊

Lemma 2.1.2 *Let M be a nonsingular integer $d \times d$ matrix, and let $D(M)$ be its set of digits. Then, the equality $p = Mk + s$ gives a one-to-one mapping of the set of all $p \in \mathbb{Z}^d$ to the set of pairs (k, s), $k \in \mathbb{Z}^d$, $s \in D(M)$.*

Proof First, let $D(M) = H(M)$. Any $p \in \mathbb{Z}^d$ can be represented in the form $p = Mk + Mr$, where the vector k is the integer part of the vector $M^{-1}p$ and r is its fractional part. Setting $s = Mr$, we find the required representation, since $s \in M[0,1)^d$ and $s \in \mathbb{Z}^d$ (the difference $p - Mk$ is integer). Expressing the digit $s \in H(M)$ by the digit $s' \in D(M)$ from the same coset, we obtain the required representation for an arbitrary $D(M)$. The uniqueness of the representation for any $D(M)$ is obvious.◊

Lemma 2.1.3 *Let M be a nonsingular integer $d \times d$ matrix, $|\det M| > 1$. Then, the set $\{r + M^j p\}$ for all possible $r \in D(M^j)$ and $p \in D(M)$ is a set of digits of the matrix M^{j+1}.*

Proof The number of all possible pairs (r, p) for $r \in D(M^j)$ and $p \in D(M)$ is $|\det M|^{j+1}$, i.e., it suffices to prove that for different pairs, we cannot obtain any vectors congruent modulo M^{j+1}. Let $r, r_1 \in D(M^j)$ and $p, p_1 \in D(M)$. Suppose $r + M^j p$ and $r_1 + M^j p_1$ are congruent modulo M^{j+1}, i.e., $(r - r_1) + M^j(p - p_1) = M^{j+1}n$ for some $n \in \mathbb{Z}^d$. Multiplying both sides of this equality by M^{-j} from left, we get $M^{-j}(r - r_1) = -(p - p_1) + Mn \in \mathbb{Z}^d$, i.e., $r \equiv r_1 \pmod{M^j}$. However, both r and r_1 belong to $D(M^j)$, which contains only one representative of each coset; hence, $r = r_1$. The equality $(p - p_1) = Mn$ implies that $p = p_1$. Thus, $r + M^j p$ and $r_1 + M^j p_1$ can be congruent modulo M^{j+1} if and only if $r = r_1$ and $p = p_1$.◊

We shall say that a set $K \subset \mathbb{R}^d$ is *congruent to a set* $L \subset \mathbb{R}^d$ *(modulo* $\mathbb{Z}^d$*)* if the set K can be partitioned into a finite number of mutually disjoint measurable subsets K_n, $n = 1, \dots, N$, in such a way that there exist integer vectors $l_1, \dots, l_N$ such that $L = \bigcup_{n=1}^{N} (K_n + l_n)$, where $(K_n + l_n) \cap (K_{n_1} + l_{n_1}) = \emptyset$ for $n \neq n_1$.

It is obvious that if a set K is congruent to a set L, then L is congruent to K. Congruence modulo $\mathbb{Z}^d$ means that by shifting parts of the set K by integer vectors, we can "compose" the set L.

Lemma 2.1.4 *Let M be a nonsingular integer $d \times d$ matrix. Then, the set*

$$K := \bigcup_{r \in D(M)} (M^{-1}[0, 1)^d + M^{-1}r)$$

is congruent to $[0, 1)^d$ *modulo* $\mathbb{Z}^d$*, and for any function* $f \in L_1(\mathbb{T}^d)$*, the following equality holds:*

$$\int_{[0,1)^d} f = \sum_{r \in D(M)} \int_{M^{-1}[0,1)^d + M^{-1}r} f.$$

Proof For $n \in \mathbb{Z}^d$, set

$$K_n := \big([0, 1)^d + n\big) \bigcap \bigcup_{r \in D(M)} (M^{-1}[0, 1)^d + M^{-1}r).$$

It is obvious that $K = \bigcup_{n \in \mathbb{Z}^d} K_n$, where $K_n \cap K_{n_1} = \emptyset$ for $n \neq n_1$. As K is bounded, the number of nonempty sets K_n is finite. Let us show that $K_n \cap (K_{n_1} - l) = \emptyset$ for any $l \in \mathbb{Z}^d$. Let $u \in K_n \subset K$, then $u = M^{-1}v + M^{-1}r$, where $v \in [0, 1)^d$, $r \in D(M)$. Suppose there exists a vector $u_1 \in K_{n_1}$, $n \neq n_1$ such that $u = u_1 - l$, $l \in \mathbb{Z}^d$, or (what is the same) $u - u_1 \in \mathbb{Z}^d$. Let $u_1 = M^{-1}v_1 + M^{-1}r_1$, then the difference $u - u_1$ has the form

$$M^{-1}(v - v_1) + M^{-1}(r - r_1) = l, \quad l \in \mathbb{Z}^d.$$

Multiplying this equality by M from left, we obtain $v - v_1 + r - r_1 = Ml$, where $l \in \mathbb{Z}^d$. Since the vectors r, r_1, and Ml are integer and the vectors v and v_1 belong to $[0, 1)^d$, this equality is satisfied only for $v = v_1$. However, since both vectors r, r_1 belong to $D(M)$, they cannot be congruent modulo M. Hence, $r = r_1$, i.e., $u = u_1$, which contradicts the assumption $n \neq n_1$.

It remains to verify that $[0, 1)^d = \bigcup_n (K_n - n)$. Note that by definition, the set $K_n - n$ belongs to $[0, 1)^d$ for any n. Moreover, $(K_n - n) \bigcap (K_{n_1} - n_1) = \emptyset$. Let us verify that for any u from $[0, 1)^d$, there exist n and $w \in K_n$ such that $u = w - n$. Multiplying the vector u by the matrix M from left, we represent Mu in the form $Mu = p + v$, where $p \in \mathbb{Z}^d$, $v \in [0, 1)^d$. The vector p is congruent to one of the digits, i.e., there exist $r \in D(M)$ and $l \in \mathbb{Z}^d$ such that $p = r + Ml$. Expressing u from this relation, we get $u = M^{-1}r + l + M^{-1}v$; therefore, $u - l \in K$. Consequently, $u - l \in K_n$, where $n = -l$. Setting $w := u - l = u + n$, $w \in K_n$, we obtain $u = w - n$.

The second statement of the lemma immediately follows from 1-periodicity of the function f with respect to each variable.◊

Lemma 2.1.5 *Let M be a nonsingular integer $d \times d$ matrix, $m = |\det M|$. Then, the $m \times m$ matrix*

$$\mathcal{P} = \left\{ \frac{1}{\sqrt{m}} e^{2\pi i (s, M^{*-1} q)} \right\}_{s \in D(M), q \in D(M^*)}$$

is unitary. In particular,

$$\sum_{q \in D(M^*)} e^{2\pi i (s, M^{*-1} q)} = \begin{cases} m, & \text{if } s \equiv \mathbf{0} \pmod{M}, \\ 0, & \text{if } s \not\equiv \mathbf{0} \pmod{M}, \end{cases}$$

$$\sum_{s \in D(M)} e^{2\pi i (s, M^{*-1} q)} = \begin{cases} m, & \text{if } q \equiv \mathbf{0} \pmod{M^*}, \\ 0, & \text{if } q \not\equiv \mathbf{0} \pmod{M^*}. \end{cases}$$

Proof Denote the columns of the matrix $\mathcal{P}$ by $\mathcal{P}_q$, $q \in D(M^*)$, i.e.,

$$\mathcal{P}_q := \left\{ \frac{1}{\sqrt{m}} e^{2\pi i (s, M^{*-1} q)} \right\}_{s \in D(M)},$$

and note that $\mathcal{P}_q$ does not depend on the choice of $D(M)$. Let us show that the columns of $\mathcal{P}$ are orthonormal, i.e., $\langle \mathcal{P}_q, \mathcal{P}_{\widetilde{q}} \rangle = \delta_{q,\widetilde{q}}$, for all $q, \widetilde{q} \in D(M^*)$. Let $q, \widetilde{q} \in D(M^*)$, $r \in \mathbb{Z}^d$. Taking into account that the set

$$D'(M) := \{s' = s + r,\ s \in D(M)\}$$

is a set of digits of M, we have

$$e^{2\pi i(r,M^{*-1}q)}\langle \mathcal{P}_q, \mathcal{P}_{\widetilde{q}}\rangle = \frac{1}{m}\sum_{s\in D(M)} e^{2\pi i((s+r),M^{*-1}q)}e^{-2\pi i(s,M^{*-1}\widetilde{q})}$$

$$= \frac{1}{m}\sum_{s\in D'(M)} e^{2\pi i(s,M^{*-1}q)}e^{-2\pi i((s-r),M^{*-1}\widetilde{q})} = e^{2\pi i(r,M^{*-1}\widetilde{q})}\langle \mathcal{P}_q, \mathcal{P}_{\widetilde{q}}\rangle.$$

Therefore,

$$(e^{2\pi i(r,M^{*-1}q)} - e^{2\pi i(r,M^{*-1}\widetilde{q})})\langle \mathcal{P}_q, \mathcal{P}_{\widetilde{q}}\rangle = 0.$$

If $q \neq \widetilde{q}$, then $M^{*-1}(q-\widetilde{q}) \notin \mathbb{Z}^d$, and hence, there exists $r \in \mathbb{Z}^d$ such that $(r, M^{*-1}(q-\widetilde{q})) \notin \mathbb{Z}^d$. It follows that $e^{2\pi i(r,M^{*-1}q)} \neq e^{2\pi i(r,M^{*-1}\widetilde{q})}$, which yields $\langle \mathcal{P}_q, \mathcal{P}_{\widetilde{q}}\rangle = 0$. If $q = \widetilde{q}$, then $(\mathcal{P}_q, \mathcal{P}_{\widetilde{q}}) = m \cdot \frac{1}{m} = 1.\Diamond$

Now, we introduce a class of matrices which will be used as dilations. To provide the completeness property of the systems consisting from the shifts and dilations of several functions, repeated multiplication by the matrix must result in dilation along all directions. That is why, we shall consider $d \times d$ integer matrices M with eigenvalues greater than one in modulus and call them *dilation matrices*. For such matrices, the relation

$$\lim_{n\to+\infty} \|M^{-n}\| = 0, \tag{2.1}$$

holds, which provides the above-mentioned condition. Moreover, we have

$$\sum_{n=1}^{\infty} \|M^{-n}\|^{\delta} < \infty \tag{2.2}$$

for any $\delta > 0$. To explain this, we remind that the spectral radius of an operator A is the number

$$\rho(A) := \lim_{n\to\infty} \|A^n\|^{1/n}.$$

The whole spectrum of any linear bounded operator A is contained inside the circle of radius $\rho(A)$ centered at the origin, and there is at least one point of the spectrum on the boundary of this circle. The proof of this fact is presented, for example, in this book [1] (see p. 281). In a finite-dimensional space, a spectrum of a matrix coincides with the set of eigenvalues of the matrix. So, if M is a $d \times d$ matrix, whose eigenvalues are greater than 1 in modulus, then the matrix M^{-1} has finitely many eigenvalues which are less than 1 in modulus; hence, $\rho(M^{-1}) < 1$. It follows that for any $\delta > 0$, the sequence $\|M^{-n}\|^{\delta}$ is majorized by a geometric progression, which implies relation (2.2).

Within the whole book, we consider that M is a fixed dilation matrix and $m = |\det M|$. Obviously, m is integer and $m > 1$. Note also that (2.1) yields

$$\lim_{n\to\infty} |M^n x| = \infty \tag{2.3}$$

for any $x \in \mathbb{R}^d, x \neq \mathbf{0}$, i.e., the operator associated with this matrix provides dilation in all directions when applied repeatedly.

In the sequel, we shall use the following notations:

$$f_{jn} := m^{j/2} f\left(M^j \cdot +n\right), \quad n \in \mathbb{Z}^d, j \in \mathbb{Z} \tag{2.4}$$

and

$$\left\{f_{jn}\right\}_{j,n} = \left\{f_{jn} : j \in \mathbb{Z}, n \in \mathbb{Z}^d\right\}.$$

2.2 Shift-Invariant Systems and Bracket Product

In this section, we study basic properties of shift-invariant systems, i.e., systems consisting of the functions φ_{jk}, $k \in \mathbb{Z}^d$, for some $\varphi \in L_2(\mathbb{R}^d)$ and some $j \in \mathbb{Z}$.

Let us consider $f, g \in L_2(\mathbb{R}^d)$. The *bracket product* of functions f and g is defined by

$$[f, g](x) := \sum_{k \in \mathbb{Z}^d} f(x+k)\overline{g(x+k)}.$$

Due to Lemma 1.1.5, the series is absolutely convergent almost everywhere and $[f, g]$ is a 1-periodic function in $L_1(\mathbb{T}^d)$. In the case $f = g$, the bracket product of function f with itself is

$$[f, f](x) = \sum_{k \in \mathbb{Z}^d} |f(x+k)|^2.$$

The bracket product is very convenient to describe the properties of the shift-invariant systems. First of all, we note that if $\varphi \in L_2(\mathbb{R}^d)$ and $[\widehat{\varphi}, \widehat{\varphi}] \in L_\infty(\mathbb{R}^d)$, then the system $\{\varphi_{jn}\}_{j,n}$ is a Bessel system with constant $B = \|[\widehat{\varphi}, \widehat{\varphi}]\|_\infty$, which yields from Remark 1.1.7.

Let us summarize some simple properties of the bracket product in the following lemma.

Lemma 2.2.1 *Let f and g be functions in $L_2(\mathbb{R}^d)$. Then, the following statements are valid:*

(i) $|[f, g]|^2 \leq [f, f][g, g]$ *a.e. on* $\mathbb{R}^d$;
(ii) *For any 1-periodic function τ, we have $[\tau f, g] = \tau[f, g]$, $[f, \tau g] = \overline{\tau}[f, g]$ a.e. on $\mathbb{R}^d$;*
(iii) *For any 1-periodic function τ such that $\tau f \in L_2(\mathbb{R}^d)$, the following equality holds:*

$$\int_{\mathbb{R}^d} \tau(x) f(x)\overline{g(x)} dx = \int_{\mathbb{T}^d} \tau(x)[f, g](x) dx;$$

if $g = \tau f$, then

$$\int_{\mathbb{R}^d} |\tau(x)f(x)|^2 dx = \int_{\mathbb{T}^d} |\tau(x)|^2 [f, f](x) dx;$$

(iv) For a 1-periodic function $\tau \in L_2(\mathbb{T}^d)$ *and function* f *such that* $[f, f] \in L_\infty(\mathbb{R}^d)$, *their product* τf *is in* $L_2(\mathbb{R}^d)$ *and*

$$\|\tau f\|_2^2 \le \|\tau\|^2_{L_2(\mathbb{T}^d)} \|[f, f]\|_\infty;$$

(v) For $k \in \mathbb{Z}^d$, $j \in \mathbb{Z}$, $\langle f, g_{jk}\rangle$ *is the kth Fourier coefficient of* $[\widehat{f}(M^{*j}\cdot), \widehat{g}]$, *i.e.,*

$$\langle f, g_{jk}\rangle = m^{j/2} \int_{\mathbb{T}^d} [\widehat{f}(M^{*j}\cdot), \widehat{g}](\xi) e^{-2\pi i(k,\xi)} d\xi. \tag{2.5}$$

(vi) If function f *is compactly supported, then* $[\widehat{f}, \widehat{f}]$ *is equivalent to the trigonometric polynomial*

$$\sum_{k\in\mathbb{Z}^d} \langle f, f_{0k}\rangle e^{2\pi i(k,\xi)},$$

in particular, $[\widehat{f}, \widehat{f}] \in L_\infty(\mathbb{T}^d)$.

Proof Item (i) follows from the Cauchy–Schwarz inequality. Item (ii) is obvious. In item (iii), $\tau[f, g]$ is integrable by Lemma 1.1.5. Using Lebesgue's dominated convergence theorem together with Lemma 1.1.5, we get item (iii).

Since $\||\tau|^2[f, f]\|_{L_1(\mathbb{T}^d)} \le \|[f, f]\|_\infty \|\tau\|_{L_2(\mathbb{T}^d)}$, then $|\tau|^2[f, f] \in L_1(\mathbb{T}^d)$. Therefore, Lebesgue's dominated convergence theorem yields that

$$\||\tau|^2[f, f]\|^2_{L_1(\mathbb{T}^d)} = \int_{\mathbb{T}^d} [\tau f, \tau f](x) dx = \int_{\mathbb{R}^d} |\tau(x)f(x)|^2 dx = \|\tau f\|_2^2.$$

This proves item (iv). Item (v) can be established by the following equalities: $\langle f, g_{jk}\rangle = \langle f_{-j,\mathbf{0}}, g_{0k}\rangle = \langle \widehat{f_{-j,\mathbf{0}}}, \widehat{g_{0k}}\rangle$, $\widehat{g_{0k}} = e^{2\pi i(k,\cdot)}\widehat{g}$, $\widehat{f_{-j,\mathbf{0}}} = m^{j/2}\widehat{f}(M^{*j}\cdot)$ together with the previous items. Item (vi) follows immediately from (v).◊

It is very easy now to give a criterium of biorthonormality of two systems consisting of integer shifts in terms of the bracket product.

Proposition 2.2.2 *Let* $\varphi, \widetilde{\varphi} \in L_2(\mathbb{R}^d)$. *The systems* $\{\varphi_{0n}\}_{n\in\mathbb{Z}^d}$ *and* $\{\widetilde{\varphi}_{0n}\}_{n\in\mathbb{Z}^d}$ *are biorthonormal if and only if*

$$[\widehat{\varphi}, \widehat{\widetilde{\varphi}}] = 1 \tag{2.6}$$

for almost all $\xi \in \mathbb{R}^d$. *In particular, the system* $\{\varphi_{0n}\}_{n\in\mathbb{Z}^d}$ *is orthonormal if and only if*

$$[\widehat{\varphi}, \widehat{\varphi}] = 1 \tag{2.7}$$

for almost all $\xi \in \mathbb{R}^d$.

Proof Biorthonormality of the systems $\{\varphi_{0n}\}_{n\in\mathbb{Z}^d}$, $\{\widetilde{\varphi}_{0n}\}_{n\in\mathbb{Z}^d}$ means that

$$\langle \varphi, \widetilde{\varphi}_{0k} \rangle = \int_{\mathbb{R}^d} \varphi(x)\overline{\widetilde{\varphi}(x+k)}\,dx = \delta_{k\mathbf{0}}, \quad \forall k \in \mathbb{Z}^d, \tag{2.8}$$

On the other hand, $\langle \varphi, \widetilde{\varphi}_{0n} \rangle$ is the kth Fourier coefficient of $[\widehat{\varphi}, \widehat{\widetilde{\varphi}}]$, due to Lemma 2.2.1, item (v). Hence, by the uniqueness theorem, (2.8) is equivalent to the fact that $[\widehat{\varphi}, \widehat{\widetilde{\varphi}}] = 1$ for almost all $\xi \in \mathbb{R}^d$.◊

Lemma 2.2.3 *Suppose* $f, \varphi, \widetilde{\varphi} \in L_2(\mathbb{R}^d)$, *such that* $[\widehat{\varphi}, \widehat{\varphi}] \in L_\infty(\mathbb{R}^d)$ *and* $[\widehat{\widetilde{\varphi}}, \widehat{\widetilde{\varphi}}] \in L_\infty(\mathbb{R}^d)$, $j \in \mathbb{Z}$. *Then,*

$$\left\| \sum_{k\in\mathbb{Z}^d} \langle f, \widetilde{\varphi}_{jk} \rangle \varphi_{jk} \right\|_2^2 \le B\widetilde{B}\|f\|_2^2, \tag{2.9}$$

where $B = \|[\widehat{\varphi}, \widehat{\varphi}]\|_\infty$ *and* $\widetilde{B} = \left\|[\widehat{\widetilde{\varphi}}, \widehat{\widetilde{\varphi}}]\right\|_\infty$, *and*

$$\mathcal{F}\left(\sum_{k\in\mathbb{Z}^d} \langle f, \widetilde{\varphi}_{jk} \rangle \varphi_{jk} \right) = [\widehat{f}(M^{*j}\cdot), \widehat{\widetilde{\varphi}}](M^{*-j}\cdot)\widehat{\varphi}(M^{*-j}\cdot). \tag{2.10}$$

Proof Note that $\|\varphi_{j\mathbf{0}}\|_2 = \|\varphi\|_2$. Thus, using Remarks 1.1.3 and 1.1.7 together with $\langle f, \widetilde{\varphi}_{jk} \rangle = \langle f_{-j,\mathbf{0}}, \widetilde{\varphi}_{0k} \rangle$, we get

$$\left\| \sum_{k\in\mathbb{Z}^d} \langle f, \widetilde{\varphi}_{jk} \rangle \varphi_{0k} \right\|_2^2 = B \sum_{k\in\mathbb{Z}^d} |\langle f_{-j,\mathbf{0}}, \widetilde{\varphi}_{0k} \rangle|^2 \le B\widetilde{B}\|f_{-j,\mathbf{0}}\|_2^2 = B\widetilde{B}\|f\|_2^2.$$

Using the Carleson theorem (with the convergence of Fourier series over the cubic partial sums), equality (2.10) follows from (2.5) and

$$\mathcal{F}\left(\sum_{k\in\mathbb{Z}^d} \langle f, \widetilde{\varphi}_{jk} \rangle \varphi_{jk} \right)(\xi) = \sum_{k\in\mathbb{Z}^d} \langle f, \widetilde{\varphi}_{jk} \rangle e^{2\pi i (k, M^{*-j}\xi)} \widehat{\varphi_{j\mathbf{0}}}(\xi) =$$

$$[\widehat{f}(M^{*j}\cdot), \widehat{\widetilde{\varphi}}](M^{*-j}\xi)\,\widehat{\varphi}(M^{*-j}\xi). \quad \Diamond$$

Lemma 2.2.4 *Let $\varphi, \widetilde{\varphi}, f, g \in L_2(\mathbb{R}^d)$, such that $[\widehat{\varphi}, \widehat{\varphi}] \in L_\infty(\mathbb{R}^d)$ and $[\widehat{\widetilde{\varphi}}, \widehat{\widetilde{\varphi}}] \in L_\infty(\mathbb{R}^d)$, $j \in \mathbb{Z}$. Then,*

$$\sum_{k\in\mathbb{Z}^d} \langle f, \widetilde{\varphi}_{jk}\rangle\langle \varphi_{jk}, g\rangle = m^j \int_{\mathbb{T}^d} [\widehat{f}(M^{*j}\cdot), \widehat{\widetilde{\varphi}}](\xi)\overline{[\widehat{g}(M^{*j}\cdot), \widehat{\varphi}](\xi)}\, d\xi. \tag{2.11}$$

Proof Note that $\{\langle f, \widetilde{\varphi}_{jk}\rangle\}_{k\in\mathbb{Z}^d} \in \ell_2(\mathbb{Z}^d)$ and $\{\langle g, \varphi_{jk}\rangle\}_{k\in\mathbb{Z}^d} \in \ell_2(\mathbb{Z}^d)$ by Remarks 1.1.3 and 1.1.7. Therefore, by item (v) in Lemma 2.2.1, the functions $[\widehat{f}(M^{*j}\cdot), \widehat{\widetilde{\varphi}}]$ and $[\widehat{g}(M^{*j}\cdot), \widehat{\varphi}]$ are in $L_2(\mathbb{T}^d)$. Using Parseval's identity, we obtain (2.11).◊

Lemma 2.2.5 *Let $\varphi, \widetilde{\varphi}, f, g \in L_2(\mathbb{R}^d)$, such that $[\widehat{\varphi}, \widehat{\varphi}] \in L_\infty(\mathbb{R}^d)$ and $[\widehat{\widetilde{\varphi}}, \widehat{\widetilde{\varphi}}] \in L_\infty(\mathbb{R}^d)$, $j \in \mathbb{Z}$. Then,*

$$\sum_{k\in\mathbb{Z}^d} \langle f, \widetilde{\varphi}_{jk}\rangle\langle \varphi_{jk}, g\rangle = \int_{\mathbb{R}^d} \widehat{\varphi}\left(M^{*-j}\xi\right)\overline{\widehat{\widetilde{\varphi}}\left(M^{*-j}\xi\right)}\widehat{f}(\xi)\overline{\widehat{g}(\xi)}\, d\xi + R_j, \tag{2.12}$$

where

$$|R_j| \le \int_{\mathbb{R}^d} \sum_{l\in\mathbb{Z}^d, l\neq\mathbf{0}} |\widehat{f}(\xi)| \left|\widehat{g}\left(\xi + M^{*j}l\right)\right| \left|\widehat{\widetilde{\varphi}}\left(M^{*-j}\xi\right)\right| \left|\widehat{\varphi}\left(M^{*-j}\xi + l\right)\right| d\xi. \tag{2.13}$$

Proof First of all, we note that the integrand in the right-hand side of (2.12) is summable because $\widehat{\varphi}, \widehat{\widetilde{\varphi}}$ are essentially bounded and $\widehat{f}, \widehat{g} \in L_2(\mathbb{R}^d)$. Using (2.11), we obtain

$$\sum_{k\in\mathbb{Z}^d} \langle f, \widetilde{\varphi}_{jk}\rangle\langle \varphi_{jk}, g\rangle = m^j \int_{\mathbb{T}^d} [\widehat{f}(M^{*j}\cdot), \widehat{\widetilde{\varphi}}](\xi)\overline{[\widehat{g}(M^{*j}\cdot), \widehat{\varphi}](\xi)}\, d\xi =$$

$$m^j \int_{\mathbb{T}^d} \sum_{k\in\mathbb{Z}^d} \widehat{f}(M^{*j}(\xi + k))\overline{\widehat{\widetilde{\varphi}}(\xi + k)}\overline{[\widehat{g}(M^{*j}\cdot), \widehat{\varphi}](\xi)}\, d\xi =$$

$$m^j \int_{\mathbb{R}^d} \widehat{f}(M^{*j}\xi)\overline{\widehat{\widetilde{\varphi}}(\xi)}\overline{[\widehat{g}(M^{*j}\cdot), \widehat{\varphi}](\xi)}\, d\xi =$$

$$\int_{\mathbb{R}^d} \sum_{l\in\mathbb{Z}^d} \widehat{f}(\xi)\overline{\widehat{g}\left(\xi + M^{*j}l\right)\widehat{\widetilde{\varphi}}\left(M^{*-j}\xi\right)}\widehat{\varphi}\left(M^{*-j}\xi + l\right) d\xi.$$

The order of summation and integration can be interchanged here due to the Lebesgue's dominated convergence theorem because $[\widehat{g}(M^{*j}\cdot), \widehat{\varphi}]$ is an essentially bounded function and $\sum_{k\in\mathbb{Z}^d} |\widehat{f}(M^{*j}(\xi + k))\widehat{\widetilde{\varphi}}(\xi + k)|$ is summable on $\mathbb{T}^d$ by Lemma 1.1.5. To complete the proof, it remains to extract the summand with $l = \mathbf{0}$ from the latter sum.◊

Theorem 2.2.6 *Let $\varphi, \widetilde{\varphi}, f, g \in L_2(\mathbb{R}^d)$, such that $[\widehat{\varphi}, \widehat{\varphi}] \in L_\infty(\mathbb{R}^d)$ and $[\widehat{\widetilde{\varphi}}, \widehat{\widetilde{\varphi}}] \in L_\infty(\mathbb{R}^d)$, $j \in \mathbb{Z}$. Suppose that $\widehat{\varphi}$ and $\widehat{\widetilde{\varphi}}$ are continuous at the origin. Then,*

$$\lim_{j\to+\infty} \sum_{k\in\mathbb{Z}^d} \langle f, \widetilde{\varphi}_{jk}\rangle\langle\varphi_{jk}, g\rangle = \widehat{\varphi}(\mathbf{0})\overline{\widehat{\widetilde{\varphi}}(\mathbf{0})}\langle f, g\rangle. \tag{2.14}$$

Proof First, we assume that the functions f and g are such that $\widehat{f}$ and $\widehat{g}$ are continuous and compactly supported. Note that Lemma 2.2.5 is valid. The first summand in (2.12) has the following limit:

$$\lim_{j\to+\infty} \int_{\mathbb{R}^d} \widehat{\varphi}\left(M^{*-j}\xi\right) \overline{\widehat{\widetilde{\varphi}}\left(M^{*-j}\xi\right)} \widehat{f}(\xi)\overline{\widehat{g}(\xi)}\, d\xi = \widehat{\varphi}(\mathbf{0})\overline{\widehat{\widetilde{\varphi}}(\mathbf{0})}\langle f, g\rangle,$$

where the limit and the integral can be interchanged by the Lebesgue's dominated convergence theorem. The second summand can be estimated as follows:

$$|R_j| \le \|\widehat{\varphi}\|_\infty \|\widehat{\widetilde{\varphi}}\|_\infty \int_{\mathbb{R}^d} \sum_{l\in\mathbb{Z}^d, l\neq\mathbf{0}} |\widehat{f}(\xi)| \left|\widehat{g}\left(\xi + M^{*j}l\right)\right| d\xi.$$

Since $\widehat{f}$ and $\widehat{g}$ are compactly supported, the latter sum is finite for any $j \in \mathbb{Z}_+$ and $R_j = 0$ whenever j is big enough, i.e., $\lim_{j\to+\infty} R_j = 0$. Thus, (2.14) follows from Lemma 2.2.5 in this case.

Now, we assume that $f \in L_2(\mathbb{R}^d)$ and g is such that $\widehat{g}$ is continuous and compactly supported. Then, f can be approximated in L_2-norm by $\widetilde{f}$ such that $\widehat{\widetilde{f}}$ is compactly supported and continuous. Therefore,

$$\lim_{j\to+\infty} \sum_{k\in\mathbb{Z}^d} \langle \widetilde{f}, \widetilde{\varphi}_{jk}\rangle\langle\varphi_{jk}, g\rangle = \widehat{\varphi}(\mathbf{0})\overline{\widehat{\widetilde{\varphi}}(\mathbf{0})}\langle \widetilde{f}, g\rangle$$

and

$$\sum_{k\in\mathbb{Z}^d} \langle f, \widetilde{\varphi}_{jk}\rangle\langle\varphi_{jk}, g\rangle = \sum_{k\in\mathbb{Z}^d} \langle \widetilde{f}, \widetilde{\varphi}_{jk}\rangle\langle\varphi_{jk}, g\rangle + \sum_{k\in\mathbb{Z}^d} \langle f - \widetilde{f}, \widetilde{\varphi}_{jk}\rangle\langle\varphi_{jk}, g\rangle \to \widehat{\varphi}(\mathbf{0})\overline{\widehat{\widetilde{\varphi}}(\mathbf{0})}\langle f, g\rangle,$$

because, by Lemma 2.2.3,

$$\left\| \sum_{k\in\mathbb{Z}^d} \langle f - \widetilde{f}, \widetilde{\varphi}_{jk}\rangle\langle\varphi_{jk}, g\rangle \right\|_2 \le \sqrt{B\widetilde{B}}\|g\|_2\|f - \widetilde{f}\|_2,$$

where $B = \|[\widehat{\varphi}, \widehat{\varphi}]\|_\infty$ and $\widetilde{B} = \|[\widehat{\widetilde{\varphi}}, \widehat{\widetilde{\varphi}}]\|_\infty$.

Similarly, approximating $g \in L_2(\mathbb{R}^d)$ by $\widetilde{g}$ such that $\widehat{\widetilde{g}}$ is compactly supported and continuous, we obtain (2.14) for the general case.$\Diamond$

Lemma 2.2.7 *Let $\varphi \in L_2(\mathbb{R}^d)$, $f \in L_2(\mathbb{R}^d)$ and $[\widehat{\varphi}, \widehat{\varphi}] \in L_\infty(\mathbb{R}^d)$. Then,*

$$\lim_{j\to-\infty} \sum_{k\in\mathbb{Z}^d} |\langle f, \varphi_{jk}\rangle|^2 = 0.$$

Proof First, we assume that f is continuous on $\mathbb{R}^d$ and $\operatorname{supp} f \subset B_R := \{x \in \mathbb{R}^d : |x| < R\}$, $R > 0$. Using the Cauchy–Schwarz inequality, we obtain

$$\sum_{k\in\mathbb{Z}^d} |\langle f, \varphi_{jk}\rangle|^2 \le m^j \sum_{k\in\mathbb{Z}^d} \left(\int_{B_R} |f(x)||\varphi(M^j x + k)|\,dx \right)^2 \le$$

$$m^j \|f\|_2^2 \sum_{k\in\mathbb{Z}^d} \int_{B_R} |\varphi(M^j x + k)|^2\,dx \le \|f\|_2 \sum_{k\in\mathbb{Z}^d} \int_{M^j B_R + k} |\varphi(y)|^2\,dy. \qquad (2.15)$$

By (2.1), the diameter of the set $M^j B_R$ tends to zero as $j \to -\infty$. Hence, for negative and big enough by modulus j, the sets $M^j B_R + k$, $k \in \mathbb{Z}^d$ are pairwise disjoint. Then, setting

$$S(R, j) = \bigcup_{k\in\mathbb{Z}^d} (M^j B_R + k),$$

we can rewrite the right-hand side of (2.15) as

$$\|f\|_2 \int_{S(R,j)} |\varphi(y)|^2\,dy = \|f\|_2 \int_{\mathbb{R}^d} \chi_{R,j}(y)|\varphi(y)|^2\,dy,$$

where $\chi_{R,j}$ is a characteristic function of the set $S(R, j)$. If $y \notin \mathbb{Z}^d$, then

$$\lim_{j\to-\infty} \chi_{R,j}(y) = 0.$$

Therefore, by Lebesgue's dominated convergence theorem, we get

$$\lim_{j\to-\infty} \int_{\mathbb{R}^d} \chi_{R,j}(y)|\varphi(y)|^2\,dy = 0.$$

Now, let $f \in L_2(\mathbb{R}^d)$. Given $\varepsilon > 0$, we find a compactly supported continuous $\widetilde{f}$, such that $\|f - \widetilde{f}\|_2 < \varepsilon$. Using Minkowski's inequality and Remark 1.1.3, we obtain

$$\left(\sum_{k\in\mathbb{Z}^d}|\langle f,\varphi_{jk}\rangle|^2\right)^{\frac{1}{2}} \le \left(\sum_{k\in\mathbb{Z}^d}|\langle \widetilde{f},\varphi_{jk}\rangle|^2\right)^{\frac{1}{2}} + \sqrt{B}\,\varepsilon,$$

where $B = \|[\widehat{\varphi},\widehat{\varphi}]\|_\infty$.$\Diamond$

2.3 Multiresolution Analysis

A notion of multiresolution analysis (MRA in the sequel) was introduced by Y. Meyer and S. Mallat as a tool for the construction of orthogonal and biorthogonal wavelet bases. Later, it will be clear that there are many difficulties in the implementation of this method in the multidimensional case. However, it appeared that the method allows to construct wavelet frames instead, in some situations. Anyway, we start with a detailed study of MRAs and a method for the construction MRA-based wavelet bases, which makes clear why some modifications of this method are natural for frame constructions and under which conditions the frame is a basis.

Remind that a dilation matrix M is supposed to be fixed, and notation (2.4) will be used here and in the sequel.

Definition 2.3.1 A collection of closed spaces $V_j \subset L_2(\mathbb{R}^d)$, $j \in \mathbb{Z}$, is called a *multiresolution analysis (MRA) in $L_2(\mathbb{R}^d)$ with a dilation matrix M*, if the following conditions (axioms) hold:

MR1. $V_j \subset V_{j+1}$ for all $j \in \mathbb{Z}$;

MR2. $\bigcup\limits_{j\in\mathbb{Z}} V_j$ is dense in $L_2(\mathbb{R}^d)$;

MR3. $\bigcap\limits_{j\in\mathbb{Z}} V_j = \{0\}$;

MR4. $f \in V_0 \iff f\left(M^j\cdot\right) \in V_j$ for all $j \in \mathbb{Z}$;

MR5. There exists a function $\varphi \in V_0$ such that the sequence $\{\varphi_{0n}\}_{n\in\mathbb{Z}^d}$ forms a Riesz basis for V_0.

The function φ from axiom **MR5** is called *scaling* (for the MRA). We shall also say that φ generates the MRA.

The axioms **MR4**, **MR5** imply two obvious properties:

MR6. For any $j \in \mathbb{Z}$, the functions φ_{jn}, $n \in \mathbb{Z}^d$, form a Riesz basis for V_j with the same constants A, B as the functions φ_{0n}, $n \in \mathbb{Z}^d$;

MR7. If $f \in V_j$, then $f\left(\cdot + M^{-j}n\right) \in V_j$, for all $n \in \mathbb{Z}^d$.

Let φ be a scaling function for an MRA. Using properties **MR1**, **MR6**, and Theorem 1.1.2, we can decompose φ with respect to the system $\{\varphi_{1n}\}_{n\in\mathbb{Z}^d}$ (which is a Riesz basis for V_1), i.e.,

$$\varphi(x) = \sqrt{m}\sum_{n\in\mathbb{Z}^d} h_n\varphi(Mx+n), \quad \sum_{n\in\mathbb{Z}^d}|h_n|^2 < \infty. \tag{2.16}$$

Thus, any scaling function φ satisfies a functional equation (2.16), which is called *refinement*. If a function is a solution of a refinement equation, it is called *refinable*.

Applying the Fourier transform to both sides of (2.16), we get

$$\widehat{\varphi}(\xi) = m^{1/2} \sum_{n\in\mathbb{Z}^d} h_n \int_{\mathbb{R}^d} \varphi(Mx+n)e^{-2\pi i(x,\xi)}\,dx.$$

After the change of variables in the integral, the equality takes the form

$$\widehat{\varphi}(\xi) = m_0(M^{*-1}\xi)\widehat{\varphi}(M^{*-1}\xi), \tag{2.17}$$

where

$$m_0(\xi) = m^{-1/2} \sum_{n\in\mathbb{Z}^d} h_n e^{2\pi i(n,\xi)}.$$

The function m_0 is called a *mask* or a *refinable mask*. Obviously, a mask is 1-periodic with respect to each variable. It follows from (2.16) and the Parseval's equality that $m_0 \in L_2(\mathbb{T}^d)$.

Let us discuss the problem of constructing an MRA. We start with finding an appropriate generating function $\varphi \in L_2(\mathbb{R}^d)$ among the functions whose Fourier transform satisfies (1.4). Set

$$V_j := \overline{\text{span}\,\{\varphi_{jn},\ \ n \in \mathbb{Z}^d\}}. \tag{2.18}$$

Axiom **MR5** follows from Theorems 1.1.2 and 1.1.6. By the definition of the spaces V_j, **MR4** also holds, and **MR1** is equivalent to the fact that φ is a solution of the refinement equation (2.16). Hence, it is necessary that a required function φ satisfies (2.16) or, what is the same, its Fourier transform satisfies (2.17).

Theorem 2.3.2 *Suppose $\varphi \in L_2(\mathbb{R}^d)$ and (1.4) is satisfied. Then,* **MR3** *holds true for the collection of the spaces V_j, $j \in \mathbb{Z}$, defined by equality (2.18).*

Proof Suppose that there exists a function $f \in L_2(\mathbb{R}^d)$ such that $f \in V_j$ for all $j \in \mathbb{Z}$. It follows from (1.4), Proposition 1.1.6, and Theorem1.1.2 that

$$A\|f\|_2 \le \left(\sum_{k\in\mathbb{Z}^d} |\langle f, \varphi_{jk}\rangle|^2\right)^{1/2}.$$

Due to Lemma 2.2.7, the right-hand side tends to zero as $j \to -\infty$. Hence, $\|f\|_2 = 0$, which means that **MR3** holds true.◊

It remains to check whether **MR2** holds true. We shall get a positive answer, provided that φ satisfies an additional insignificant condition; e.g., it suffices to assume that $\widehat{\varphi}$ is continuous and nonzero at the origin. But first, we prove two auxiliary statements.

Lemma 2.3.3 *Suppose $\varphi, f \in L_2(\mathbb{R}^d)$ and (1.4) holds. Let $\{V_j\}_{j\in\mathbb{Z}}$ be the collection of spaces generated by the function φ. The relation $f \in V_j$ holds if and only if there exists a function $m_f \in L_2(\mathbb{T}^d)$ such that*

$$\widehat{f}(\xi) = m_f(M^{*-j}\xi)\widehat{\varphi}(M^{*-j}\xi)$$

for almost all $\xi \in \mathbb{R}^d$.

The statement of this lemma follows immediately from Propositions 1.1.9 and **MR4**.

To formulate another statement (which is called Wiener's theorem for $L_2(\mathbb{R}^d)$), we need to introduce a notion and a notation. A subspace $X \subset L_2(\mathbb{R}^d)$ is called invariant under shifts if $f(\cdot + t) \in X$ for all $f \in X$ and all $t \in \mathbb{R}^d$. If $X \subset L_2(\mathbb{R}^d)$, then by $\widehat{X}$ we denote the set of the Fourier transforms of all functions from X.

Theorem 2.3.4 (Wiener) *A closed subspace X of the space $L_2(\mathbb{R}^d)$ is invariant under shifts if and only if $\widehat{X} = L_2(\Omega)$ for some measurable set $\Omega \subset \mathbb{R}^d$.*

Proof Sufficiency. Let $\widehat{X} = L_2(\Omega)$, $\Omega \subset \mathbb{R}^d$, $f \in X$, $f_t := f(\cdot + t)$, then $\widehat{f_t} = \widehat{f} e_t$, where $e_t(x) := e^{2\pi i (x,t)}$, which implies that $\widehat{f_t} \in L_2(\Omega)$. By the definition of $\widehat{X}$, this means that $f_t \in X$, i.e., the space X is invariant under shifts.

Necessity. Let now X be invariant under shifts. Then, $\widehat{X}$ is invariant under multiplication by the functions e_t, $t \in \mathbb{R}^d$. Denote by P the orthogonal projection on $\widehat{X}$. For any $f, g \in L_2(\mathbb{R}^d)$ and any $t \in \mathbb{R}^d$, since $(Pg)\, e_t \in \widehat{X}$, we have $f - Pf \perp (Pg)\, e_t$; i.e., the following equality holds:

$$\int\limits_{\mathbb{R}^d} (f(x) - Pf(x))\overline{Pg(x)} e^{-2\pi i (x,t)}\, dx = 0.$$

Taking into account that the function $F := (f - Pf)\overline{Pg}$ is summable on $\mathbb{R}^d$, we get $\widehat{F} \equiv 0$, which implies that $\|F\|_2 = 0$. Thus,

$$(f - Pf)\overline{Pg} = 0 \quad \text{a.e. on } \mathbb{R}^d.$$

It follows that

$$f\,Pg = Pf\,Pg \quad \text{a.e. on } \mathbb{R}^d.$$

The right-hand side of this equality is symmetric with respect to the functions f, g, hence interchanging the roles of f and g, we get

$$f\,Pg = g\,Pf \quad \text{a.e. on } \mathbb{R}^d \tag{2.19}$$

for any $f, g \in L_2(\mathbb{R}^d)$. For g, take any positive function from $L_2(\mathbb{R}^d)$ and set $\sigma = Pg/g$. From (2.19), we get

$$Pf = \sigma f \quad \text{a.e. on } \mathbb{R}^d \tag{2.20}$$

for all $f \in L_2(\mathbb{R}^d)$, which implies

$$\sigma^2 f = \sigma P f = P^2 f = P f = \sigma f.$$

Therefore, $\sigma^2 = \sigma$ almost everywhere; i.e., almost all values of the function σ are equal to 0 or 1. Denote by Ω the set, on which $\sigma = 1$. If $f \in \widehat{X}$, then $Pf = f$, and from (2.20), we get

$$f = \sigma f \quad \text{a.e. on } \mathbb{R}^d \tag{2.21}$$

It follows that $f = 0$ almost everywhere on $\mathbb{R}^d \setminus \Omega$, i.e., $f \in L_2(\Omega)$. We have proved the inclusion $\widehat{X} \subset L_2(\Omega)$. If $f \in L_2(\Omega)$, then (2.21) holds, which together with (2.20) imply $Pf = f$, i.e., $f \in \widehat{X}$; hence, $\widehat{X} \subset L_2(\Omega) \subset \widehat{X}$.◊

Theorem 2.3.5 *Let (1.4) be true for $\varphi \in L_2(\mathbb{R}^d)$. Then,* **MR2** *holds for the collection of spaces $\{V_j\}_{j\in\mathbb{Z}}$, which are generated by the function φ and satisfy* **MR1** *if and only if*

$$\bigcup_{j\in\mathbb{Z}} \operatorname{supp} \widehat{\varphi}(M^{*j}\cdot) = \mathbb{R}^d. \tag{2.22}$$

Proof First of all, we note that the spaces V_j generated by the function φ satisfy axioms **MR4** and **MR5**; hence, they satisfy axiom **MR7** as well. Therefore, each space V_j is invariant under the shifts $t = M^{-j}n$, $n \in \mathbb{Z}^d$. Let us show that the space $\overline{\cup_{j\in\mathbb{Z}} V_j}$ is invariant under shifts. Any $t \in \mathbb{R}^d$ can be approximated by vectors of the form $M^{-j}n$, $n \in \mathbb{Z}^d$, $j \in \mathbb{N}$, for arbitrarily large j. Indeed, it follows from (2.1) that for any $\varepsilon > 0$, there exists $j_0 \in \mathbb{N}$ such that $|M^{-j}x| < \varepsilon$ for all $j \geq j_0$ and all $x \in [0, 1)^d$. If $j \geq j_0$, then choosing $n \in \mathbb{Z}^d$ in such a way that $M^j t - n \in [0, 1)^d$, we obtain $|t - M^{-j}n| \leq \varepsilon$. If $f \in \bigcup_{j\in\mathbb{Z}} V_j$, then due to axiom **MR1**, $f \in V_j$ for all $j \geq j_1$. Since $\|f(\cdot + r) - f\|_2 \to 0$ as $|r| \to 0$, $r \in \mathbb{R}^d$, we get that $f(\cdot + t) \in \overline{\bigcup_{j\in\mathbb{Z}} V_j}$, $t \in \mathbb{R}^d$. If $g \in \overline{\bigcup_{j\in\mathbb{Z}} V_j}$, then approximating g by the functions $f \in \bigcup_{j\in\mathbb{Z}} V_j$, using the above arguments and the invariance of L_2-norm under shifts, we deduce that $g(\cdot + t) \in \overline{\bigcup_{j\in\mathbb{Z}} V_j}$.

Let $X := \overline{\bigcup_{j\in\mathbb{Z}} V_j}$. By Theorem 2.3.4, $\widehat{X} = L_2(\Omega)$ for some set $\Omega \subset \mathbb{R}^d$. Therefore, $X = L_2(\mathbb{R}^d)$ if and only if $\Omega = \mathbb{R}^d$. Set

$$\varphi_j := \varphi(M^j\cdot), \quad \Omega_0 := \bigcup_{j\in\mathbb{Z}} \operatorname{supp} \widehat{\varphi}_j.$$

Let us prove that $\Omega = \Omega_0$. Since $\varphi_j \in V_j$, $j \in \mathbb{Z}$, we have $\operatorname{supp} \widehat{\varphi}_j \subset \Omega$; hence, $\Omega_0 \subset \Omega$. Suppose now that $\Omega \setminus \Omega_0$ contains a set of positive measure Ω_1. By Lemma 2.3.3, the Fourier transform of any element from V_j is equal to zero almost everywhere on Ω_1. Consequently, the same is true for any element from $\bigcup_{j\in\mathbb{Z}} V_j$. Passing to the limit, we deduce that the Fourier transform of any element from X is

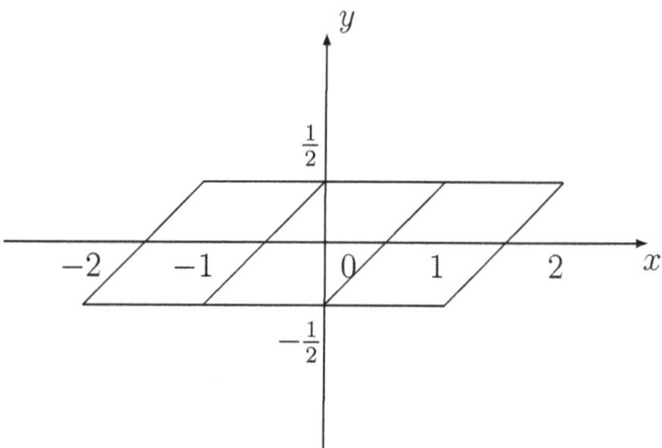

Fig. 2.1 The set P is the small central parallelogram, and the set MP is the large central parallelogram (Example 2.3.7)

equal to zero almost everywhere on Ω_1, which contradicts the fact that $\widehat{X} \supset L_2(\Omega_1)$. It remains to note that $\operatorname{supp} \widehat{\varphi}_j = \operatorname{supp} \widehat{\varphi}(M^{*-j}).\Diamond$

Remark 2.3.6 Analyzing the proof of Theorem 2.3.5 and the auxiliary facts, it is easy to see that the theorem remains true without the assumption that φ satisfies (1.4) if we consider the spaces

$$V_j := \overline{\operatorname{span}\{\varphi_{jn},\ n \in \mathbb{Z}^d\}}.$$

as the spaces generated by this function.

Finally, we give some examples of refinable functions generating MRA.

Example 2.3.7 Let $d = 2$, $M = \begin{pmatrix} -1 & 4 \\ -1 & 1 \end{pmatrix}$, the set $D(M)$ consists of the digits $s_0 = (0, 0)$, $s_1 = (1, 0)$, $s_2 = (-1, 0)$, and let φ be the characteristic function of the parallelogram P with vertices $(0, 1/2)$, $(1, 1/2)$, $(0, -1/2)$, $(-1, -1/2)$ (see Fig. 2.1). Since

$$MP = P \cup (P + s_1) \cup (P + s_2),$$

and the interiors of the sets P, $P + s_1$, $P + s_2$ are mutually disjoint, the function φ is refinable, and evidently, its integer shifts form an orthogonal system. Thus, φ generates an MRA. This MRA is naturally called Haar MRA by analogy with the classic Haar MRA whose scaling function is the characteristic function of $[0, 1].\Diamond$

Example 2.3.8 Let $M = 2I_d$. Define the function B_N, $N \ge d$, by its Fourier transform

$$\widehat{B_N}(\xi) = 2^d \prod_{k=1}^{N} \frac{1 - e^{2\pi i (a_k, \xi)}}{2\pi i (a_k, \xi)},$$

where $a_k \in \mathbb{R}^d$, $0 < |a_k| < 2$, $k = 1, \dots, N$, and the rank of the matrix $A := (a_1, \dots, a_N)$ is d. The equality

$$\widehat{B_N}(\xi) = 2^d \prod_{k=1}^{N} \frac{(1 - e^{\pi i(a_k,\xi)})(1 + e^{\pi i(a_k,\xi)})}{2\pi i(a_k,\xi)} =$$

$$\widehat{B_N}(\xi/2) \prod_{k=1}^{N} \frac{1 + e^{\pi i(a_k,\xi)}}{2}$$

implies that the function B_N satisfies the refinement equation with mask

$$m_0(\xi) = \prod_{k=1}^{N} \frac{1 + e^{2\pi i(a_k,\xi)}}{2}.$$

By Theorem 1.1.6, in order to show that integer shifts of the function B_N form a Riesz basis, it suffices to verify that the sum $\sum_{n\in\mathbb{Z}^d} \left|\widehat{B_N}(\xi+n)\right|^2$ is bounded almost everywhere and bounded away from zero. It is clear that for $|a_k| < 2$, there exists a constant $C = C(A) > 0$ such that

$$\left|\widehat{B_N}(\xi)\right|^2 = \left|2^d \prod_{k=1}^{N} \frac{\sin \pi(a_k,\xi)}{\pi(a_k,\xi)}\right|^2 \geq C$$

for all $\xi \in [-1/2, 1/2]^d$; hence,

$$\sum_{n\in\mathbb{Z}^d} \left|\widehat{B_N}(\xi+n)\right|^2 \geq C > 0.$$

There are at least d linearly independent vectors in the set $a_1, \ldots, a_N$ (we assume that these are $a_1, \ldots, a_d$); therefore,

$$\sum_{n\in\mathbb{Z}^d} \left|\widehat{B_N}(\xi+n)\right|^2 \leq \left(\frac{2}{\pi}\right)^{2d} \sum_{n\in\mathbb{Z}^d} \prod_{k=1}^{d} \left|\frac{\sin \pi((a_k,\xi)+(a_k,n))}{(a_k,\xi)+(a_k,n)}\right|^2.$$

It is not hard to verify that the series on the right-hand side converges uniformly on $[-1/2, 1/2]^d$, so we get

$$\sum_{n\in\mathbb{Z}^d} \left|\widehat{B_N}(\xi+n)\right|^2 \leq C'.$$

Thus, B_N generates an MRA.

The functions B_N are the multivariate generalization of B-splines, and they are called *box splines*. Clearly, B_d is the characteristic function of the set $A[-1/2, 1/2]^d$; i.e., B_d is a spline of order zero. It can be shown that for an arbitrary N, the function B_N is a spline of order $N - d$.$\Diamond$

2.4 MRA-Based Wavelets and Matrix Extension Principle

As already mentioned, MRAs are intended for the construction of orthogonal and biorthogonal wavelet bases. First, we describe such a construction for the orthogonal case. Let $\{V_j\}_{j\in\mathbb{Z}}$ be an MRA generated by a scaling function φ whose integer shifts form an orthonormal system. Denote by W_j the orthogonal complement to V_j in the space V_{j+1}. By the properties of the Hilbert space, V_j can be decomposed into the direct sum $V_{j+1} = V_j \oplus W_j$. Thus, we obtain the collection $\{W_j\}_{j\in\mathbb{Z}}$ of subspaces of the space $L_2(\mathbb{R}^d)$ such that $W_j \perp V_j$, $W_j \perp W_k$ for all $j, k \in \mathbb{Z}$, $k \neq j$ and

$$V_j = V_k \oplus W_k \oplus \cdots \oplus W_{j-1}$$

for all $j, k \in \mathbb{Z}$, $k < j$. It follows from **MR2** that

$$L_2(\mathbb{R}^d) = V_0 \oplus W_0 \oplus W_1 \oplus \ldots,$$

and, by **MR3**, we have

$$V_0 = W_{-1} \oplus W_{-2} \oplus \ldots.$$

Combining these two decompositions, we get

$$L_2(\mathbb{R}^d) = \bigoplus_{j=-\infty}^{\infty} W_j. \tag{2.23}$$

Thus, the space $L_2(\mathbb{R}^d)$ is decomposed into the direct sum of pairwise orthogonal subspaces W_j. It is clear from the construction that an analog of **MR4** holds for the spaces W_j, i.e.,

$$f \in W_j \iff f\left(2^{-j}\cdot\right) \in W_0.$$

The main idea of the construction is that the scaling function φ generates a finite number of functions $\psi^{(\nu)} \in W_0$ called *wavelet functions*, $\nu = 1, \ldots, r$, for some $r \in \mathbb{N}$. These functions inherit many properties of φ, in particular, integer shifts of the functions $\psi^{(\nu)}$ form an orthonormal basis for W_0. Then, similar to property **MR6**, for each $j \in \mathbb{Z}$ the system of functions $\{\psi_{jk}^{(\nu)}\}_{k,\nu}$ is an orthonormal basis for W_j. By (2.23), the union of these systems over all $j \in \mathbb{Z}$ is an orthonormal basis for $L_2(\mathbb{R}^d)$ consisting of shifts and dilations of the functions $\psi^{(\nu)}$. Additionally, any function $f \in L_2(\mathbb{R}^d)$ can be expanded into the series

$$f = \sum_{\nu=1}^{r} \sum_{j\in\mathbb{Z}} \sum_{n\in\mathbb{Z}^d} \langle f, \psi_{jn}^{(\nu)} \rangle \psi_{jn}^{(\nu)}. \tag{2.24}$$

The spaces W_j are called *wavelet spaces*. We say that the orthonormal wavelet basis $\{\psi_{jk}^{(\nu)}\}_{j,k,\nu}$ is *MRA-based*.

So, to construct an orthogonal wavelet basis, one starts with finding a function φ which generates an MRA and such that its integer shifts form an orthonormal system. In Sects. 2.6 and 2.7, we shall discuss a method for the construction of such functions, in particular compactly supported ones, which is important for applications. Unfortunately, it is not easy to implement this method practically. Finding concrete appropriate scaling functions with desirable properties is a complicated problem. The next step is finding an MRA-based orthogonal wavelet system. It will be proved in Sect. 2.5 that such a system always exists, and a constructive method will be given. However, this method does not lead to compactly supported wavelet functions, even if the scaling function is compactly supported. We shall also discuss that the situation with finding compactly supported orthogonal wavelet bases is much worse for most dilation matrices.

Now, we describe the construction of biorthogonal wavelet systems. Let $\{V_j\}_{j\in\mathbb{Z}}$ and $\{\widetilde{V}_j\}_{j\in\mathbb{Z}}$ be multiresolution analyses generated by scaling functions φ and $\widetilde{\varphi}$, respectively. Suppose that the systems $\{\varphi_{0n}\}_n$ and $\{\widetilde{\varphi}_{0n}\}_n$ are biorthonormal. To implement the same idea as in the orthogonal case, we have to define a wavelet space W_j for the first MRA $\{V_j\}_{j\in\mathbb{Z}}$ as the completion to V_j in V_{j+1} orthogonal to the space $\widetilde{V}_j$ of the second MRA $\{\widetilde{V}_j\}_{j\in\mathbb{Z}}$. The wavelet spaces $\widetilde{W}_j$ for the MRA $\{\widetilde{V}_j\}_{j\in\mathbb{Z}}$ are defined similarly. Next, one finds a finite number of *wavelet functions* $\psi^{(\nu)} \in W_0$ and $\widetilde{\psi}^{(\nu)} \in \widetilde{W}_0$ such that the system $\{\psi_{0n}^{(\nu)}\}_{n,\nu}$ forms a Riesz basis for W_0, the system $\{\widetilde{\psi}_{0n}^{(\nu)}\}_{n,\nu}$ forms a Riesz basis for $\widetilde{W}_0$, and these two systems are biorthonormal. Obviously, the wavelet systems $\{\psi_{jn}^{(\nu)}\}_{j,n,\nu}$, $\{\widetilde{\psi}_{jn}^{(\nu)}\}_{j,n,\nu}$ are biorthonormal, and one can hope that dual biorthonormal bases will be constructed in this way. We say that $\{\psi_{jn}^{(\nu)}\}_{j,n,\nu}$, $\{\widetilde{\psi}_{jn}^{(\nu)}\}_{j,n,\nu}$ are *dual MRA-based wavelet systems.*

So, to construct dual biorthogonal wavelet bases, one starts with finding two functions φ, $\widetilde{\varphi}$ which generate MRAs and such that their integer shifts form biorthonormal systems. Again, this is a complicated problem, what will be clear from Sects. 2.6 and 2.7. The next step is finding dual MRA-based wavelet systems. We will see soon that such systems always exist. Moreover, if the functions φ, $\widetilde{\varphi}$ are compactly supported, then compactly supported dual MRA-based wavelet systems can be constructed for arbitrary dilation matrix, at least theoretically. Unfortunately, in contrast to the orthogonal case, one cannot state that any dual MRA-based wavelet systems are dual biorthonormal bases for $L_2(\mathbb{R}^d)$. However, this is true under some additional assumptions on φ, $\widetilde{\varphi}$, what will be discussed in Sect. 3.1.

Now, we assume that φ and $\widetilde{\varphi}$ are scaling functions generating MRAs $\{V_j\}_{j\in\mathbb{Z}}$ and $\{\widetilde{V}_j\}_{j\in\mathbb{Z}}$, respectively. Recall that these functions are refinable, that is,

$$\widehat{\varphi}(\xi) = m_0(M^{*-1}\xi)\widehat{\varphi}(M^{*-1}\xi), \quad \widehat{\widetilde{\varphi}}(\xi) = \widetilde{m}_0(M^{*-1}\xi)\widehat{\widetilde{\varphi}}(M^{*-1}\xi), \tag{2.25}$$

where m_0, $\widetilde{m}_0$ are their masks which are 1-periodic with respect to each variable functions in $L_2(\mathbb{T}^d)$. First of all, we note that biorthogonality of the integer shifts of φ, $\widetilde{\varphi}$ implies some special properties of their masks.

Proposition 2.4.1 *Let $\varphi, \widetilde{\varphi}$ be refinable functions in $L_2(\mathbb{R}^d)$ with masks m_0, $\widetilde{m_0}$, respectively. If the integer shifts of φ, $\widetilde{\varphi}$ form a biorthonormal system, then*

$$\sum_{q\in D(M^*)} m_0(\xi + M^{*^{-1}}q)\overline{\widetilde{m}_0(\xi + M^{*^{-1}}q)} = 1 \tag{2.26}$$

for almost all $\xi \in \mathbb{R}^d$ *and any set of digits* $D(M^*)$ *of the matrix* M^*. *In particular, if the integer shifts of* φ *form an orthonormal system, then*

$$\sum_{q\in D(M^*)} |m_0(\xi + M^{*^{-1}}q)|^2 = 1 \tag{2.27}$$

for almost all $\xi \in \mathbb{R}^d$.

Proof Using Proposition 2.2.2 and (2.25), we have

$$1 = \sum_{n\in\mathbb{Z}^d} \widehat{\varphi}(M^*\xi + n)\overline{\widehat{\widetilde{\varphi}}(M^*\xi + n)} =$$

$$\sum_{n\in\mathbb{Z}^d} m_0(\xi + M^{*^{-1}}n)\widehat{\varphi}(\xi + M^{*^{-1}}n)\overline{\widetilde{m}_0(\xi + M^{*^{-1}}n)\widehat{\widetilde{\varphi}}(\xi + M^{*^{-1}}n)}. \tag{2.28}$$

Using Lemmas 2.1.2, we make the change of the index of summation $n = M^*k + q$, $k \in \mathbb{Z}^d$, $q \in D(M^*)$ on the right-hand side of (2.28), reducing it to the form

$$\sum_{k\in\mathbb{Z}^d}\sum_{q\in D(M^*)} m_0(\xi + \theta_{kq})\overline{\widetilde{m}_0(\xi + \theta_{kq})}\widehat{\varphi}(\xi + \theta_{kq})\overline{\widehat{\widetilde{\varphi}}(\xi + \theta_{kq})},$$

where $\theta_{kq} = M^{*^{-1}}q + k$. Changing the order of summation, using 1-periodicity of the functions $m_0, \widetilde{m}_0$ in each variable and Proposition 2.2.2, we obtain

$$\sum_{q\in D(M^*)} m_0(\xi + \theta_{0q})\overline{\widetilde{m}_0(\xi + \theta_{0q})} \sum_{k\in\mathbb{Z}^d} \widehat{\varphi}(\xi + \theta_{kq})\overline{\widehat{\widetilde{\varphi}}(\xi + \theta_{kq})} =$$

$$\sum_{q\in D(M^*)} m_0(\xi + M^{*^{-1}}q)\overline{\widetilde{m}_0(\xi + M^{*^{-1}}q)}.\Diamond$$

Next, we state an analogous proposition for the case, when the integer shifts of φ form a Riesz system.

Proposition 2.4.2 *Let* φ *be a refinable function in* $L_2(\mathbb{R}^d)$ *with masks* m_0. *If the integer shifts of* φ *form a Riesz system with bounds* A *and* B, *then*

$$\frac{A}{B} \le \sum_{q\in D(M^*)} |m_0(\xi + M^{*^{-1}}q)|^2 \le \frac{B}{A} \tag{2.29}$$

for almost all $\xi \in \mathbb{R}^d$ *and any set of digits* $D(M^*)$ *of the matrix* M^*.

Proof It follows from Proposition 2.4.1 that

$$\sum_{n\in\mathbb{Z}^d} |\widehat{\varphi}(M^*\xi+n)|^2 = \sum_{q\in D(M^*)} |m_0(\xi+{M^*}^{-1}q)|^2 \sum_{k\in\mathbb{Z}^d} |\widehat{\varphi}(\xi+{M^*}^{-1}q+k)|^2.$$

It remains to apply (1.4).◊

Let us next discuss a possibility to find appropriate wavelet functions $\psi^{(\nu)}$, $\widetilde{\psi}^{(\nu)}$ (or only functions $\psi^{(\nu)}$ for the orthogonal case). The following statement gives no answer to the question of the existence of MRA-based dual wavelet systems (an MRA-based wavelet system in the orthogonal case), but it is a key to solve the problem.

Theorem 2.4.3 *Let MRAs $\{V_j\}_{j\in\mathbb{Z}}$, $\{\widetilde{V}_j\}_{j\in\mathbb{Z}}$ be generated by scaling functions φ, $\widetilde{\varphi}$ with masks m_0, $\widetilde{m}_0$, respectively, and the systems $\{\varphi_{0n}\}_{n\in\mathbb{Z}^d}$, $\{\widetilde{\varphi}_{0n}\}_{n\in\mathbb{Z}^d}$ be biorthonormal, and let $D(M^*)=\{q_0,\dots,q_{m-1}\}$. Suppose there exist functions $m_\nu, \widetilde{m}_\nu \in L_2(\mathbb{T}^d)$, $\nu=1,\dots,m-1$, such that the rows of the matrices*

$$\{m_\nu(\xi+{M^*}^{-1}q_k)\}_{\nu,k=0}^{m-1},\quad \{\widetilde{m}_\nu(\xi+{M^*}^{-1}q_k)\}_{\nu,k=0}^{m-1}, \tag{2.30}$$

are biorthonormal for almost all $\xi\in\mathbb{R}^d$. Let functions $\psi^{(\nu)}$, $\widetilde{\psi}^{(\nu)}$, $\nu=1,\dots,m-1$, be defined by

$$\widehat{\psi^{(\nu)}}(\xi) = m_\nu({M^*}^{-1}\xi)\widehat{\varphi}({M^*}^{-1}\xi),\quad \widehat{\widetilde{\psi}^{(\nu)}}(\xi) = \widetilde{m}_\nu({M^*}^{-1}\xi)\widehat{\widetilde{\varphi}}({M^*}^{-1}\xi). \tag{2.31}$$

Then, $\psi^{(\nu)}\in V_1$, $\widetilde{\psi}^{(\nu)}\in\widetilde{V}_1$, $\psi^{(\nu)}\perp\widetilde{V}_0$, $\widetilde{\psi}^{(\nu)}\perp V_0$ for all $\nu=1,\dots,m-1$, and the systems $\{\psi_{jn}^{(\nu)}\}_{j,n,\nu}$, $\{\widetilde{\psi}_{jn}^{(\nu)}\}_{j,n,\nu}$ are biorthonormal.

Proof First of all, we note that Lemma 2.3.3 implies the inclusions $\psi^{(\nu)}\in V_1$, $\widetilde{\psi}^{(\nu)}\in\widetilde{V}_1$, $\nu=1,\dots,m-1$.

For the sake of convenience, let us rename the functions φ, $\widetilde{\varphi}$ by $\psi^{(0)}$, $\widetilde{\psi}^{(0)}$, respectively. We need to verify the equality

$$\langle\psi^{(\nu)}(\cdot+k), \widetilde{\psi}^{(\kappa)}(\cdot+l)\rangle = \delta_{\nu\kappa}\delta_{kl}. \tag{2.32}$$

for all $k,l\in\mathbb{Z}^d$ and all $\nu,\kappa=0,\dots,m-1$ (by assumption, it holds for $\nu=\kappa=0$). Let $\nu\neq 0$. Due to Propositions 2.2.2, 1.1.10, equality (2.32) is equivalent to

$$\sum_{n\in\mathbb{Z}^d} \widehat{\psi^{(\nu)}}(M^*\xi+n)\overline{\widehat{\widetilde{\psi}^{(\kappa)}}(M^*\xi+n)} = \delta_{\nu\kappa} \tag{2.33}$$

for almost all $\xi\in\mathbb{R}^d$.

Using (2.31) and the refinement equations for the functions φ, $\widetilde{\varphi}$, we get

$$\sum_{n\in\mathbb{Z}^d} \widehat{\psi^{(\nu)}}(M^*\xi+n)\overline{\widehat{\widetilde{\psi}^{(\kappa)}}(M^*\xi+n)} =$$

$$\sum_{n\in\mathbb{Z}^d} m_\nu(\xi+M^{*-1}n)\widehat{\varphi}(\xi+M^{*-1}n)\overline{\widetilde{m}_\kappa(\xi+M^{*-1}n)\widehat{\widetilde{\varphi}}(\xi+M^{*-1}n)}. \qquad (2.34)$$

Using Lemma 2.1.2, set the index of summation $n = M^*k+q, k \in \mathbb{Z}^d, q \in D(M^*)$, on the right-hand side of (2.34), which reduces it to the form

$$\sum_{k\in\mathbb{Z}^d}\sum_{q\in D(M^*)} m_\nu(\xi+\theta_{kq})\overline{\widetilde{m}_\kappa(\xi+\theta_{kq})}\widehat{\varphi}(\xi+\theta_{kq})\overline{\widehat{\widetilde{\varphi}}(\xi+\theta_{kq})},$$

where $\theta_{kq} = M^{*-1}q + k$. Changing the order of summation, using 1-periodicity of the functions $m_\nu, \widetilde{m}_\kappa$ in each variable and Proposition 2.2.2, we get

$$\sum_{q\in D(M^*)} m_\nu(\xi+\theta_{\mathbf{0}q})\overline{\widetilde{m}_\kappa(\xi+\theta_{\mathbf{0}q})}\sum_{k\in\mathbb{Z}^d}\widehat{\varphi}(\xi+\theta_{kq})\overline{\widehat{\widetilde{\varphi}}(\xi+\theta_{kq})} =$$

$$\sum_{q\in D(M^*)} m_\nu(\xi+M^{*-1}q)\overline{\widetilde{m}_\kappa(\xi+M^{*-1}q)}.$$

To prove (2.32), it remains to recall that

$$\sum_{q\in D(M^*)} m_\nu(\xi+M^{*-1}q)\overline{\widetilde{m}_\kappa(\xi+M^{*-1}q)} = \delta_{\nu\kappa}$$

for almost all $\xi \in \mathbb{R}^d$.

Since $\psi^{(\nu)} \in V_1$, $\widetilde{\psi}^{(\nu)} \in \widetilde{V}_1$, by the definitions of MRA, for each $j \in \mathbb{Z}$,

$$\psi_{jk}^{(\nu)} \in V_{j+1}, \ \widetilde{\psi}_{jk}^{(\nu)} \in \widetilde{V}_{j+1}$$

for all $k \in \mathbb{Z}^d$, $\nu = 1, \dots, m-1$. It follows from (2.32) that

$$\langle \psi_{jk}^{(\nu)}, \widetilde{\psi}_{jl}^{(\kappa)} \rangle = \delta_{\nu\kappa}\delta_{kl}.$$

Since $\psi_{jk}^{(\nu)} \perp \widetilde{V}_i$, $\widetilde{\psi}_{jk}^{(\nu)} \perp V_i$ for all $i \le j$, for all $k, l \in \mathbb{Z}^d$, $\nu, \kappa = 1, \dots, m-1$,

$$\langle \psi_{jk}^{(\nu)}, \widetilde{\psi}_{il}^{(\kappa)} \rangle = \delta_{\nu\kappa}\delta_{kl}\delta_{ij}.$$

Hence, the systems $\{\psi_{jk}^{(\nu)}\}_{j,k,\nu}$, $\{\widetilde{\psi}_{jk}^{(\nu)}\}_{j,k,\nu}$ are biorthonormal.◊

The periodic functions, $m_\nu, \widetilde{m}_\nu$, $\nu = 1, \dots, m-1$, that appeared in the definition of the wavelet functions $\psi^{(\nu)}$, $\widetilde{\psi}^{(\nu)}$ are called *wavelet masks*.

Theorem 2.4.4 *Let MRAs $\{V_j\}_{j\in\mathbb{Z}}$, $\{\widetilde{V}_j\}_{j\in\mathbb{Z}}$ be generated, respectively, by scaling functions $\varphi, \widetilde{\varphi}$ whose masks $m_0, \widetilde{m}_0$ satisfy (2.26), $\widehat{\varphi}, \widehat{\widetilde{\varphi}}$ be continuous at the origin, and $\widehat{\varphi}(\mathbf{0})\cdot\overline{\widehat{\widetilde{\varphi}}(\mathbf{0})} = 1$. Suppose functions m_ν, $\widetilde{m}_\nu$ and $\psi^{(\nu)}$, $\widetilde{\psi}^{(\nu)}$, $\nu = 1, \dots, m-1$, are as in Theorem 2.4.3.*

(i) If $\varphi = \widetilde{\varphi}$ and $m_\nu = \widetilde{m}_\nu$, $\nu = 1, \dots, m-1$, then $\{\psi^{(\nu)}_{jk}\}_{j,k,\nu}$ is a Parseval's frame.
(ii) If the systems $\{\psi^{(\nu)}_{jk}\}_{j,k,\nu}$, $\{\widetilde{\psi}^{(\nu)}_{jk}\}_{j,k,\nu}$ are Bessel, then they are dual frames.

Proof First, we show that for any functions f, g from $L^2(\mathbb{R}^d)$ and any $j, j' \in \mathbb{Z}$, $j < j'$, the following equality holds:

$$\sum_{k\in\mathbb{Z}^d}\langle f, \varphi_{jk}\rangle\overline{\langle g, \widetilde{\varphi}_{jk}\rangle} + \sum_{i=j}^{j'-1}\sum_{\nu=1}^{m-1}\sum_{k\in\mathbb{Z}^d}\langle f, \psi^{(\nu)}_{ik}\rangle\overline{\langle g, \widetilde{\psi}^{(\nu)}_{ik}\rangle}$$
$$= \sum_{k\in\mathbb{Z}^d}\langle f, \varphi_{j',k}\rangle\overline{\langle g, \widetilde{\varphi}_{j',k}\rangle}. \tag{2.35}$$

Clearly, it suffices to check (2.35) for $j' = j + 1$. Using Lemmas 2.2.4, 2.1.2, Remark 1.1.7, and the refinement equations for $\varphi, \widetilde{\varphi}$, taking into account periodicity of the functions $m_0, m_1, \dots, m_{m-1}$, we have

$$\sum_{k\in\mathbb{Z}^d}\langle f, \varphi_{jk}\rangle\overline{\langle g, \widetilde{\varphi}_{jk}\rangle} + \sum_{\nu=1}^{m-1}\sum_{k\in\mathbb{Z}^d}\langle f, \psi^{(\nu)}_{jk}\rangle\overline{\langle g, \widetilde{\psi}^{(\nu)}_{jk}\rangle} =$$
$$m^j\int_{\mathbb{T}^d}[\widehat{f}(M^{*j}\cdot), \widehat{\varphi}](\xi)\overline{[\widehat{g}(M^{*j}\cdot), \widehat{\widetilde{\varphi}}](\xi)}\,d\xi +$$
$$m^j\sum_{\nu=1}^{m-1}\int_{\mathbb{T}^d}[\widehat{f}(M^{*j}\cdot), \widehat{\psi^{(\nu)}}](\xi)\overline{[\widehat{g}(M^{*j}\cdot), \widehat{\widetilde{\psi}^{(\nu)}}](\xi)}\,d\xi =$$
$$m^j\sum_{\nu=0}^{m-1}\int_{[0,1)^d}\sum_{k\in\mathbb{Z}^d}\widehat{f}\left(M^{*j}(\xi+k)\right)\overline{m_\nu\left(M^{*-1}(\xi+k)\right)\widehat{\varphi}\left(M^{*-1}(\xi+k)\right)}\times$$
$$\overline{\sum_{l\in\mathbb{Z}^d}\widehat{g}\left(M^{*j}(\xi+l)\right)\overline{\widetilde{m}_\nu\left(M^{*-1}(\xi+l)\right)\widehat{\widetilde{\varphi}}\left(M^{*-1}(\xi+l)\right)}}\,d\xi =$$

$$m^j\sum_{\nu=0}^{m-1}\int_{[0,1)^d}\left(\sum_{q\in D(M^*)}[\widehat{f}(M^{*(j+1)}\cdot), \widehat{\varphi}](M^{*-1}(\xi+q))\overline{m_\nu\left(M^{*-1}(\xi+q)\right)}\right)\times$$
$$\overline{\left(\sum_{r\in D(M^*)}[\widehat{g}(M^{*(j+1)}\cdot), \widehat{\widetilde{\varphi}}](M^{*-1}(\xi+r))\overline{\widetilde{m}_\nu\left(M^{*-1}(\xi+r)\right)}\right)}\,d\xi. \tag{2.36}$$

Since matrices (2.30) are square and the rows of these matrices are biorthonormal almost everywhere, their columns are also biorthonormal almost everywhere, i.e.,

$$\sum_{\nu=0}^{m-1} m_\nu\left(M^{*-1}(\xi+q)\right)\overline{\widetilde{m}_\nu\left(M^{*-1}(\xi+r)\right)} = \delta_{qr}, \quad q, r \in D(M^*),$$

for almost all $\xi \in \mathbb{R}^d$. Hence, the right-hand side of (2.36) can be reduced to

$$m^j \int\limits_{[0,1)^d} \sum_{q\in D(M^*)} [\widehat{f}(M^{*(j+1)}\cdot), \widehat{\varphi}](M^{*-1}(\xi+q))\overline{[\widehat{g}(M^{*(j+1)}\cdot), \widehat{\widetilde{\varphi}}](M^{*-1}(\xi+q))}\, d\xi.$$

Using Lemma 2.1.4, we obtain

$$\sum_{k\in\mathbb{Z}^d}\langle f, \varphi_{jk}\rangle\overline{\langle g, \widetilde{\varphi}_{jk}\rangle} + \sum_{\nu=1}^{m-1}\sum_{k\in\mathbb{Z}^d}\langle f, \psi_{jk}^{(\nu)}\rangle\overline{\langle g, \widetilde{\psi}_{jk}^{(\nu)}\rangle} =$$

$$m^{j+1} \sum_{q\in D(M^*)} \int\limits_{M^{*-1}([0,1)^d+q)} [\widehat{f}(M^{*(j+1)}\cdot), \widehat{\varphi}](\xi)\overline{[\widehat{g}(M^{*(j+1)}\cdot), \widehat{\widetilde{\varphi}}](\xi)}\, d\xi =$$

$$m^{j+1}\int\limits_{\mathbb{T}^d} [\widehat{f}(M^{*(j+1)}\cdot), \widehat{\varphi}](\xi)\overline{[\widehat{g}(M^{*(j+1)}\cdot), \widehat{\widetilde{\varphi}}](\xi)}\, d\xi.$$

To prove (2.35) for $j' = j+1$, it remains to use Lemma 2.2.4 and Remark 1.1.7 once more.

It follows from the Cauchy–Schwarz inequality and Lemma 2.2.7 that

$$\lim_{j\to-\infty}\sum_{k\in\mathbb{Z}^d}\langle f, \varphi_{jk}\rangle\overline{\langle g, \widetilde{\varphi}_{jk}\rangle} = 0.$$

Passing to the limit as $j \to -\infty$ in (2.35), for any $j' \in \mathbb{Z}$, we get the equality

$$\sum_{i=-\infty}^{j'}\sum_{\nu=1}^{m-1}\sum_{k\in\mathbb{Z}^d}\langle f, \psi_{ik}^{(\nu)}\rangle\overline{\langle g, \widetilde{\psi}_{ik}^{(\nu)}\rangle} = \sum_{k\in\mathbb{Z}^d}\langle f, \varphi_{j',k}\rangle\overline{\langle g, \widetilde{\varphi}_{j',k}\rangle}.$$

It follows from Theorem 2.2.6 that

$$\lim_{j'\to+\infty}\sum_{i=-\infty}^{j'}\sum_{\nu=1}^{m-1}\sum_{k\in\mathbb{Z}^d}\langle f, \psi_{ik}^{(\nu)}\rangle\overline{\langle g, \widetilde{\psi}_{ik}^{(\nu)}\rangle} = \langle f, g\rangle, \tag{2.37}$$

in particular,

$$\lim_{j'\to+\infty} \sum_{i=-\infty}^{j'} \sum_{\nu=1}^{m-1} \sum_{k\in\mathbb{Z}^d} \langle f, \psi_{ik}^{(\nu)}\rangle \overline{\langle f, \widetilde{\psi}_{ik}^{(\nu)}\rangle} = \|f\|^2,$$

which yields (i).

Suppose now that the systems $\{\psi_{jk}^{(\nu)}\}_{j,k,\nu}$, $\{\widetilde{\psi}_{jk}^{(\nu)}\}_{j,k,\nu}$ are Bessel, i.e., there exist constants B, $\widetilde{B}$ such that

$$\sum_{j,k,\nu} |\langle f, \psi_{jk}^{(\nu)} \rangle|^2 \le B\|f\|^2, \tag{2.38}$$

$$\sum_{j,k,\nu} |\langle f, \widetilde{\psi}_{jk}^{(\nu)} \rangle|^2 \le \widetilde{B}\|f\|^2. \tag{2.39}$$

If $f \in L_2(\mathbb{R}^d)$, $\|f\| \neq 0$, given $\varepsilon > 0$, there exists $j' \in \mathbb{Z}$ such that

$$\sum_{i=-\infty}^{j'} \sum_{\nu=1}^{m-1} \sum_{k\in\mathbb{Z}^d} \langle f, \psi_{ik}^{(\nu)}\rangle \overline{\langle f, \widetilde{\psi}_{ik}^{(\nu)}\rangle} \ge \|f\|^2(1-\varepsilon).$$

It follows from (2.39) and the Cauchy inequality that

$$\sqrt{\sum_{i=-\infty}^{\infty} \sum_{\nu=1}^{m-1} \sum_{k\in\mathbb{Z}^d} |\langle f, \psi_{ik}^{(\nu)}\rangle|^2} \ge \frac{1}{\sqrt{\widetilde{B}}\|f\|} \sum_{i=-\infty}^{j'} \sum_{\nu=1}^{m-1} \sum_{k\in\mathbb{Z}^d} \langle f, \psi_{ik}^{(\nu)}\rangle \overline{\langle f, \widetilde{\psi}_{ik}^{(\nu)}\rangle} \ge \frac{\|f\|}{\sqrt{\widetilde{B}}}(1-\varepsilon).$$

Passing to the limit as $\varepsilon \to 0$, we obtain that

$$\sum_{i=-\infty}^{\infty} \sum_{\nu=1}^{m-1} \sum_{k\in\mathbb{Z}^d} |\langle f, \psi_{ik}^{(\nu)}\rangle|^2 \ge \frac{\|f\|^2}{\widetilde{B}}$$

for any function $f \in L^2(\mathbb{R}^d)$. Hence, the system $\{\psi_{jk}^{(\nu)}\}_{j,k,\nu}$ is a frame. Similarly, $\{\widetilde{\psi}_{jk}^{(\nu)}\}_{j,k,\nu}$ is a frame. It follows from the Cauchy–Schwarz inequality that the series

$$\sum_{j,k,\nu} \langle f, \psi_{jk}^{(\nu)}\rangle \overline{\langle g, \widetilde{\psi}_{jk}^{(\nu)}\rangle}$$

converges absolutely and hence unconditionally. Thus, due to (2.37), these frames are dual, and (ii) is proved.◊

Combining the results of Theorems 2.4.3 and 2.4.4, taking into account Proposition 2.4.1, we see that if all assumptions of Theorem 2.4.3 are satisfied and $\varphi = \widetilde{\varphi}$, $m_\nu = \widetilde{m}_\nu$, $\nu = 1, \dots, m-1$, then $\{\psi_{jk}^{(\nu)}\}_{j,k,\nu}$ is a Parseval's frame in

$L_2(\mathbb{R}^d)$ and, hence, an orthonormal system. It follows that the system $\{\psi_{jk}^{(\nu)}\}_{j,k,\nu}$ is an MRA-based orthogonal basis for $L_2(\mathbb{R}^d)$.

The situation is different for the biorthogonal case. If all assumptions of Theorem 2.4.3 are satisfied, then the systems $\{\psi_{jk}^{(\nu)}\}_{j,k,\nu}$, $\{\widetilde{\psi}_{jk}^{(\nu)}\}_{j,k,\nu}$ are biorthonormal. It is not difficult to check also that these are MRA-based dual wavelet systems (see the proof in [2] for the case $d = 1$, $M = 2$). We skip the proof of this fact because if even it would be proved, we could not state that $\{\psi_{jk}^{(\nu)}\}_{j,k,\nu}$ and $\{\widetilde{\psi}_{jk}^{(\nu)}\}_{j,k,\nu}$ are dual biorthogonal bases for $L_2(\mathbb{R}^d)$, generally speaking. Indeed, if the systems $\{\psi_{jk}^{(\nu)}\}_{k,\nu}$ and $\{\widetilde{\psi}_{jk}^{(\nu)}\}_{k,\nu}$ form Riesz bases for W_j and $\widetilde{W}_j$, respectively, we can only state that for every $j \in \mathbb{Z}$, there exist constants B_j, $\widetilde{B}_j$ such that

$$\sum_{k,\nu} |\langle f, \psi_{jk}^{(\nu)} \rangle|^2 \le B_j \|f\|^2, \quad \sum_{k,\nu} |\langle f, \widetilde{\psi}_{jk}^{(\nu)} \rangle|^2 \le \widetilde{B}_j \|f\|^2,$$

for all $f \in L_2(\mathbb{R}^d)$. However, this does not imply conditions (2.38), (2.39) under which Theorem 2.4.4 states that each of the systems $\{\psi_{jk}^{(\nu)}\}_{j,k,\nu}$, $\{\widetilde{\psi}_{jk}^{(\nu)}\}_{j,k,\nu}$ is a frame. Later, in Sect. 3.1, we shall prove that (2.38), (2.39) are satisfied under some additional assumptions on φ, $\widetilde{\varphi}$.

Theorem 2.4.4 was intended to study MRA-based wavelet systems. However, the assertions of this theorem are true for a wider class of wavelet systems because only relation (2.26) was assumed instead of the biorthogonality for the systems $\{\varphi_{0k}\}_k$, $\{\widetilde{\varphi}_{0k}\}_k$. Analyzing the proof of Theorem 2.4.4, we see that the class of wavelet systems can be more extended. Actually, the assumptions that the functions φ, $\widetilde{\varphi}$ generate MRAs were not used. We used only that φ and $\widetilde{\varphi}$ are refinable functions and that the systems $\{\varphi_{0k}\}_k$, $\{\widetilde{\varphi}_{0k}\}_k$ are Bessel. Moreover, since we used only that the columns of the matrices are biorthonormal, the number of wavelet functions can be increased.

Let $\varphi, \widetilde{\varphi} \in L_2(\mathbb{R}^d)$ be refinable functions with masks $m_0, \widetilde{m}_0$, respectively, and let $\{q_0, \dots, q_{m-1}\}$ be an arbitrary collection of digits of the matrix M^*. Suppose there exist functions $m_\nu, \widetilde{m}_\nu \in L_2(\mathbb{T}^d)$, $\nu = 1, \dots, r$, $r \ge m - 1$, such that the columns of the $(r+1) \times m$ matrices

$$\{m_\nu(\xi + M^{*-1} q_k)\}_{\nu,k}, \quad \{\widetilde{m}_\nu(\xi + M^{*-1} q_k)\}_{\nu,k},$$

are biorthonormal for almost all $\xi \in \mathbb{R}^d$, and define wavelet functions $\psi^{(\nu)}$, $\widetilde{\psi}^{(\nu)}$, $\nu = 1, \dots, r$, by

$$\widehat{\psi^{(\nu)}}(\xi) = m_\nu(M^{*-1}\xi)\widehat{\varphi}(M^{*-1}\xi), \quad \widehat{\widetilde{\psi}^{(\nu)}}(\xi) = \widetilde{m}_\nu(M^{*-1}\xi)\widehat{\widetilde{\varphi}}(M^{*-1}\xi).$$

We shall say that the corresponding systems $\{\psi_{jk}^{(\nu)}\}_{j,k,\nu}$, $\{\widetilde{\psi}_{jk}^{(\nu)}\}_{j,k,\nu}$ are dual wavelet systems generated from φ, $\widetilde{\varphi}$ by *matrix extension principle* (MEP in the sequel). In the case $\varphi = \widetilde{\varphi}$ and $m_\nu = \widetilde{m}_\nu$, $\nu = 1, \dots, r$, we shall say that the wavelet system $\{\psi_{jk}^{(\nu)}\}_{j,k,\nu}$ is generated from φ by MEP. As above, $m_\nu, \widetilde{m}_\nu$, $\nu = 1, \dots, r$, are called wavelet masks.

Using this terminology and taking into account the above comments related to Theorem 2.4.4, we obtain the following statements.

Theorem 2.4.5 *Let $\varphi \in L_2(\mathbb{R}^d)$ be a refinable function such that $\widehat{\varphi}$ is continuous at the origin, $\widehat{\varphi}(\mathbf{0}) = 1$, and $\{\varphi_{0k}\}_k$ is a Bessel system, and let $\{\psi_{jk}^{(\nu)}\}_{j,k,\nu}$ be a wavelet system generated from φ by MEP. Then, $\{\psi_{jk}^{(\nu)}\}_{j,k,\nu}$ is a Parseval's frame in $L_2(\mathbb{R}^d)$.*

Theorem 2.4.6 *Let $\varphi, \widetilde{\varphi} \in L_2(\mathbb{R}^d)$ be refinable functions such that $\widehat{\varphi}, \widehat{\widetilde{\varphi}}$ are continuous at the origin, $\widehat{\varphi}(\mathbf{0}) \cdot \widehat{\widetilde{\varphi}}(\mathbf{0}) = 1$, and $\{\varphi_{0k}\}_k$, $\{\widetilde{\varphi}_{0k}\}_k$ are Bessel systems. Suppose $\{\psi_{jk}^{(\nu)}\}_{j,k,\nu}$, $\{\widetilde{\psi}_{jk}^{(\nu)}\}_{j,k,\nu}$ are dual wavelet systems generated from $\varphi, \widetilde{\varphi}$ by MEP. If these systems are Bessel, then they are dual frames in $L_2(\mathbb{R}^d)$.*

2.5 Matrix Extension Problem

The construction of MRA-based wavelet systems studied in Theorems 2.4.3, 2.4.4 is based on the existence of wavelet masks $m_\nu, \widetilde{m}_\nu$, $\nu = 1, \ldots, m-1$, for which the rows of the matrices (2.30) are biorthonormal. The first rows of the matrices are known, and their inner product equals 1 due to Proposition 2.4.1. All other entries depend on unknown functions $m_\nu, \widetilde{m}_\nu \in L_2(\mathbb{T}^d)$, $\nu = 1, \ldots, m-1$. Thus, matrix extension should be done properly to construct wavelet functions.

Now, we discuss the matrix extension problem. It is rather complicated to find appropriate wavelet masks. The following notion of polyphase components simplifies the problem to some extent.

Let t be a function in $L_2(\mathbb{T}^d)$, and $t(\xi) = \frac{1}{\sqrt{m}} \sum\limits_{k\in\mathbb{Z}^d} h_k e^{2\pi i(k,\xi)}$. Suppose $D(M) = \{s_0, \ldots, s_{m-1}\}$. The function defined by $\tau_k(\xi) = \sum\limits_{l\in\mathbb{Z}^d} h_{Ml+s_k} e^{2\pi i(l,\xi)}$ is called the *polyphase component* of t corresponding to the digit s_k, $k = 0, \ldots, m-1$. By definition, for every function $t \in L_2(\mathbb{T}^d)$, there exist its polyphase components τ_k, $k = 0, \ldots, m-1$, which are in $L_2(\mathbb{T}^d)$. The converse is also true, i.e., arbitrary functions $\tau_k \in L_2(\mathbb{T}^d)$, $k = 0, \ldots, m-1$, are the polyphase components of a function $t \in L_2(\mathbb{T}^d)$, where the function t is given by

$$t(\xi) := \frac{1}{\sqrt{m}} \sum_{k=0}^{m-1} e^{2\pi i(s_k,\xi)} \tau_k(M^*\xi).$$

Note that by Lemma 2.1.5, τ_k can be expressed as

$$\tau_k(\xi) = \frac{1}{\sqrt{m}} \sum_{l=0}^{m-1} e^{-2\pi i(M^{-1}s_k,\xi+q_l)} t(M^{*-1}(\xi + q_l)),$$

where $D(M^*) = \{q_0, \ldots, q_{m-1}\}$. For convenience, we will assume that $s_0 = q_0 = \mathbf{0}$.

Also, we introduce the following utility $m \times m$ matrix

$$U(\xi) = \left\{ \frac{1}{\sqrt{m}} e^{2\pi i (s_k, \xi + M^{*-1} q_l)} \right\}_{k,l=0,\dots,m-1} = \tag{2.40}$$

$$= \frac{1}{\sqrt{m}} \begin{pmatrix} 1 & 1 & \dots & 1 \\ e^{2\pi i (s_1,\xi)} & e^{2\pi i (s_1, \xi + M^{*-1} q_1)} & \dots & e^{2\pi i (s_1, \xi + M^{*-1} q_{m-1})} \\ \vdots & \vdots & \ddots & \vdots \\ e^{2\pi i (s_{m-1},\xi)} & e^{2\pi i (s_{m-1}, \xi + M^{*-1} q_1)} & \dots & e^{2\pi i (s_{m-1}, \xi + M^{*-1} q_{m-1})} \end{pmatrix},$$

for $\xi \in \mathbb{R}^d$. By Lemma 2.1.5, matrix $U(\mathbf{0})$ is unitary. Moreover, since the inner product of two rows of the matrix $U(\xi)$ is

$$\sum_{q \in D(M^*)} [U(\xi)]_{s,q} \overline{[U(\xi)]_{v,q}} = \frac{1}{m} \sum_{q \in D(M^*)} e^{2\pi i (s-v, \xi + M^{*-1} q)} = e^{2\pi i (s-v,\xi)} \delta_{s,v} \equiv \delta_{s,v},$$

for all $s, v \in D(M)$, matrix $U(\xi)$ is also a unitary matrix for every $\xi \in \mathbb{R}^d$, i.e., $U^* U \equiv I_m$ or $U U^* \equiv I_m$.

Now, assume that all assumptions of Theorem 2.4.3 are satisfied. Let $\mu_{\nu 0}, \dots, \mu_{\nu,m-1}$ be the polyphase components of mask m_ν and $\widetilde{\mu}_{\nu 0}, \dots, \widetilde{\mu}_{\nu,m-1}$ be the polyphase components of mask $\widetilde{m}_\nu$, $\nu = 0, \dots, m-1$.

Let us denote the matrices (2.30) by

$$\mathbf{M}(\xi) := \{m_\nu(\xi + M^{*-1} q_k)\}_{\nu,k=0}^{m-1}, \quad \widetilde{\mathbf{M}}(\xi) := \{\widetilde{m}_\nu(\xi + M^{*-1} q_k)\}_{\nu,k=0}^{m-1},$$

and the corresponding polyphase matrices are

$$\mathcal{M}(\xi) := \{\mu_{\nu k}(\xi)\}_{\nu,k=0}^{m-1}, \quad \widetilde{\mathcal{M}}(\xi) := \{\widetilde{\mu}_{\nu k}(\xi)\}_{\nu,k=0}^{m-1}. \tag{2.41}$$

In the next statement, we show the equivalence between the biorthonormality of the rows of matrices $\mathbf{M}(\xi)$ and $\widetilde{\mathbf{M}}(\xi)$ and the biorthonormality of the rows of polyphase matrices $\mathcal{M}(\xi)$ and $\widetilde{\mathcal{M}}(\xi)$.

Lemma 2.5.1 *Let $\mu_{\nu 0}, \dots, \mu_{\nu,m-1}$ be the polyphase components of mask m_ν and $\widetilde{\mu}_{\nu 0}, \dots, \widetilde{\mu}_{\nu,m-1}$ be the polyphase components of mask $\widetilde{m}_\nu$, $\nu = 0, \dots, m-1$. Then, $\mathbf{M}(\xi)\widetilde{\mathbf{M}}^*(\xi) = \mathcal{M}(M^*\xi)\widetilde{\mathcal{M}}^*(M^*\xi)$. Moreover, the equality $\mathbf{M}\widetilde{\mathbf{M}}^* \equiv I_m$ is equivalent to*

$$\mathcal{M}\widetilde{\mathcal{M}}^* \equiv I_m. \tag{2.42}$$

Proof Since

$$\mathcal{M}(M^*\xi) U(\xi) = \mathbf{M}(\xi), \quad \mathcal{M}(M^*\xi) = \mathbf{M}(\xi) U^*(\xi)$$

and

$$\widetilde{\mathcal{M}}(M^*\xi)U(\xi) = \widetilde{\mathbf{M}}(\xi), \quad \widetilde{\mathcal{M}}(M^*\xi) = \widetilde{\mathbf{M}}(\xi)U^*(\xi),$$

and matrix $U(\xi)$ is unitary, we get the required statement.◊

So, instead of the construction of matrices (2.30), one can construct the corresponding polyphase matrices. Hence, the problem is reduced to another extension problem. One has to construct two matrices $\mathcal{M}$ and $\widetilde{\mathcal{M}}$ satisfying (2.42) with entries in $L_2(\mathbb{T}^d)$, given their first rows $\{\mu_{0k}\}_{k=0}^{m-1}$ and $\{\widetilde{\mu}_{0k}\}_{k=0}^{m-1}$, respectively. Due to Lemma 2.5.1, we have

$$\sum_{k=0}^{m-1} \mu_{0k}(\xi)\overline{\widetilde{\mu}_{0k}(\xi)} = 1 \quad \text{a.e.} \tag{2.43}$$

Note that for $m = 2$, this problem has a trivial solution

$$\mathcal{M} := \begin{pmatrix} \mu_{00} & \mu_{01} \\ -\overline{\widetilde{\mu}_{01}} & \overline{\widetilde{\mu}_{00}} \end{pmatrix}, \quad \widetilde{\mathcal{M}} := \begin{pmatrix} \widetilde{\mu}_{00} & \widetilde{\mu}_{01} \\ -\overline{\mu_{01}} & \overline{\mu_{00}} \end{pmatrix}.$$

Further, we shall discuss a possibility of such extension for an arbitrary m in the assumptions of Theorem 2.4.3. First, consider the situation (corresponding to the construction of an orthogonal wavelet system), where $\mu_{0k} = \widetilde{\mu}_{0k}$, i.e., given functions $\mu_{0k} \in L_2(\mathbb{T}^d)$, $k = 0, \dots, m-1$, such that

$$\sum_{k=0}^{m-1} |\mu_{0k}(\xi)|^2 = 1 \quad \text{a.e.} \tag{2.44}$$

We need to extend the row $\{\mu_{0k}\}_{k=0}^{m-1}$ to a unitary matrix $\mathcal{M}$. This can be done by Householder's transform. Set

$$\mu_{k0} = \overline{\mu_{0k}}\frac{1-\mu_{00}}{1-\overline{\mu_{00}}}, \qquad \mu_{kl} = \delta_{kl} - \frac{\mu_{0l}\overline{\mu_{0k}}}{1-\overline{\mu_{00}}}$$

for $k, l = 1, \dots, m-1$. Or denoting by $P = (\mu_{01}, \dots, \mu_{0,m-1})$, the matrix extension can be written as follows:

$$\mathcal{M} = \left(\begin{array}{c|c} \mu_{00} & P \\ \hline P^* \frac{1-\mu_{00}}{1-\overline{\mu_{00}}} & I_{m-1} - \frac{PP^*}{1-\overline{\mu_{00}}} \end{array}\right).$$

Direct calculation shows that this matrix is unitary. Let us prove that the functions constructed above belong to the space $L_2(\mathbb{T}^d)$. Indeed, from (2.44), it follows that

$$|\mu_{0k}| \le \sqrt{1-|\mu_{00}|^2}$$

for all $k = 1, \dots, m-1$, which implies that $|\mu_{kl}| \le 2 + |\mu_{00}|$.

Let us now consider the biorthogonal case. Given functions are $\mu_{0k}, \widetilde{\mu}_{0k} \in L_2(\mathbb{T}^d)$, $k = 0, \dots, m-1$, satisfying (2.43). Since the integer shifts of the corresponding refinable function φ form a Riesz system, by Lemma 2.4.2 the sum

$$\sum_{q \in D(M^*)} |m_0(\xi + M^{*-1}q)|^2$$

is bounded away from zero almost everywhere. By Lemma 2.5.1, this yields

$$\sum_{k=0}^{m-1} |\mu_{0k}(\xi)|^2 \ge C > 0 \quad \text{a.e.}$$

It follows that the functions defined by

$$\mu'_{0k}(\xi) := \frac{\mu_{0k}(\xi)}{\left(\sum\limits_{l=0}^{m-1} |\mu_{0l}(\xi)|^2\right)^{1/2}}$$

belong to the space $L_2(\mathbb{T}^d)$. Using the row $\mu'_{00}, \dots, \mu'_{0(m-1)}$, we construct a unitary matrix with elements from $L_2(\mathbb{T}^d)$ (for instance, by Householder's transform described above). Denote by $\widetilde{\mathcal{M}}_1, \dots, \widetilde{\mathcal{M}}_{m-1}$ the rows of this matrix from the second to the last one and set

$$\mathcal{M}_0 = (\mu_{00}, \dots, \mu_{0,m-1}), \quad \widetilde{\mathcal{M}}_0 = (\widetilde{\mu}_{00}, \dots, \widetilde{\mu}_{0,m-1}).$$

Since the matrix is unitary, it follows from relation (2.43) that

$$\widetilde{\mathcal{M}}_n \widetilde{\mathcal{M}}_k^* = \delta_{nk}, \quad \mathcal{M}_0 \widetilde{\mathcal{M}}_l^* = \delta_{0l} \quad \text{a.e.}$$

for all $n, k = 1, \dots, m-1$, $l = 0, \dots, m-1$. Setting

$$\mathcal{M}_k = \widetilde{\mathcal{M}}_k - \widetilde{\mathcal{M}}_k \widetilde{\mathcal{M}}_0^* \mathcal{M}_0, \quad k = 1, \dots, m-1,$$

we obtain

$$\mathcal{M}_k \widetilde{\mathcal{M}}_l^* = \delta_{kl} \quad \text{a.e.}$$

for all $k, l = 0, \dots, m-1$. Thus, given the first rows $\{\mu_{0k}\}_{k=0}^{m-1}$, $\{\widetilde{\mu}_{0k}\}_{k=0}^{m-1}$, we have found the matrices $\mathcal{M}, \widetilde{\mathcal{M}}$ such that

$$\mathcal{M}\widetilde{\mathcal{M}}^* = I_m \quad \text{a.e.}$$

If the refinable masks $m_0, \widetilde{m}_0$ are smooth, then the wavelet masks constructed by the above algorithm are not necessarily smooth (or even continuous). Let us show that in the case $d < 2m - 1$, smoothness can be preserved. Suppose that the functions $\mu_{0k}, \widetilde{\mu}_{0k}$ have continuous partial derivatives of order r, $r \geq 1$. Obviously, the functions μ'_{0k} defined above are smooth. The smooth mapping $(\mu'_{00}, \dots, \mu'_{0,m-1})$ takes the d-dimensional torus $\mathbb{T}^d$ to a unit sphere S in the complex m-dimensional space $\mathbf{C}^{\mathbf{m}}$. Let us cover $\mathbb{T}^d$ with finitely many sets, whose diameters do not exceed certain $\varepsilon > 0$. Then, the diameters of the images of these sets do not exceed $C\varepsilon$, where C is a constant that depends only on the functions μ'_{0k}. Since the dimension of $\mathbb{T}^d$ is less than the dimension of S, it is clear that for sufficiently small ε, the union of the images of these sets does not cover S. Hence, there exists a point $\alpha \in S$ such that some of its neighborhood does not belong to the image of the torus. Suppose U is a unitary numerical matrix and U^T takes the vector α to the vector $(1, 0, \dots, 0)^T$. Set

$$(\mu''_{00}, \dots, \mu''_{0,m-1}) = (\mu'_{00}, \dots, \mu'_{0,m-1})U,$$
$$(\widetilde{\mu}''_{00}, \dots, \widetilde{\mu}''_{0,m-1}) = (\widetilde{\mu}_{00}, \dots, \widetilde{\mu}_{0,m-1})U.$$

Both new rows preserve smoothness of order r; the row $(\mu''_{00}, \dots, \mu''_{0,m-1})$ can be extended to a unitary matrix by Householder's transform, which also preserves smoothness since all values of the function μ''_{00} are bounded away from unity. The consequent steps of the construction are based on linear transformations. Thus, we can construct mutually inverse matrices $\{\mu''_{\nu k}(\xi)\}_{\nu,k=0}^{m-1}$, $\{\overline{\widetilde{\mu}''_{k\nu}(\xi)}\}_{\nu,k=0}^{m-1}$, consisting of functions with continuous derivatives of order r. To construct the required matrices, it remains to set

$$(\mu_{\nu 0}, \dots, \mu_{\nu,m-1}) = (\mu''_{\nu 0}, \dots, \mu''_{\nu,m-1})U^*,$$
$$(\widetilde{\mu}_{\nu 0}, \dots, \widetilde{\mu}_{\nu,m-1}) = (\widetilde{\mu}''_{\nu 0}, \dots, \widetilde{\mu}''_{\nu,m-1})U^*.$$

for $\nu = 1, \dots, m - 1$.

The case where all masks are trigonometric polynomials is of special interest for engineering applications. It will be shown in Sect. 2.6 that starting with an appropriate trigonometric polynomial as a refinable mask, one can construct a compactly supported scaling function. It is easy to see that if all wavelet masks are also trigonometric polynomials, then all wavelet functions are compactly supported. Let us now discuss the problem of the existence of compactly supported wavelet functions.

Let refinable masks $m_0, \widetilde{m}_0$ be trigonometric polynomials. Then, the corresponding polyphase functions $\mu_{0k}, \widetilde{\mu}_{0k}$ are also trigonometric polynomials. Our aim is to find matrices $\mathcal{M}, \widetilde{\mathcal{M}}$, whose entries are trigonometric polynomials such that $\mathcal{M}\widetilde{\mathcal{M}}^* = I_m$, given their first rows $\{\mu_{0k}\}_{k=0}^{m-1}$, $\{\widetilde{\mu}_{0k}\}_{k=0}^{m-1}$. This can be done in the following way. Set

$$\mathcal{M}_0 = (\mu_{00}, \dots, \mu_{0,m-1}), \quad \widetilde{\mathcal{M}}_0 = (\overline{\widetilde{\mu}_{00}}, \dots, \overline{\widetilde{\mu}_{0,m-1}}).$$

Suppose we have extended $\mathcal{M}_0$ to a *unimodular* matrix, i.e., a square matrix whose elements are trigonometric polynomials and the absolute value of the determinant is identically equal to 1 (this is essential and the most complicated part of the construction). Denote by $\mathcal{M}'_1, \ldots, \mathcal{M}'_{m-1}$ the rows of the matrix from the second to the last one. It is clear that the inverse matrix exists and its elements are also trigonometric polynomials. Denote by $\widetilde{\mathcal{M}}^*_1, \ldots, \widetilde{\mathcal{M}}^*_{m-1}$ the columns of the inverse matrix from the second to the last one. Consequently,

$$\mathcal{M}_0\widetilde{\mathcal{M}}^*_l = 0, \quad \mathcal{M}'_k\widetilde{\mathcal{M}}^*_l = \delta_{kl}$$

for $k, l = 1, \ldots, m-1$. Let us set

$$\mathcal{M}_k = \mathcal{M}'_k - \mathcal{M}'_k\widetilde{\mathcal{M}}^*_0\mathcal{M}_0, \quad k = 1, \ldots, m-1.$$

It is not hard to see that

$$\mathcal{M}_k\widetilde{\mathcal{M}}^*_l = \delta_{kl}.$$

for all $k, l = 0, \ldots, m-1$. Therefore, $\mathcal{M} = (\mathcal{M}_0, \mathcal{M}_1, \ldots, \mathcal{M}_{m-1})^T$, $\widetilde{\mathcal{M}} = (\widetilde{\mathcal{M}}_0, \widetilde{\mathcal{M}}_1, \ldots, \widetilde{\mathcal{M}}_{m-1})^T$ are the required matrices.

Let us now return to the first step of the construction and reformulate the problem in algebraic terms. Setting $z = (z_1, \ldots, z_d)$, $z_k = e^{2\pi i \xi_k}$, we shall consider the Laurent polynomials instead of trigonometric ones, using the same notation. We have a unimodular row of Laurent polynomials (i.e., a row $\mathcal{M}_0$, for which there exists a row $\widetilde{\mathcal{M}}_0$ such that $\mathcal{M}_0\widetilde{\mathcal{M}}^*_0 = 1$ for all $z \neq \mathbf{0}$). We need to extend the row $\mathcal{M}_0$ to a unimodular matrix (i.e., a square matrix, whose elements are Laurent polynomials and the determinant is equal to 1 in modulus for all $z \neq \mathbf{0}$). Problems of this sort have been of algebraical interest for a long time. In particular, the following problem (so-called Serre's problem) has been studied: Is it always possible to construct a unimodular matrix such that its elements are multivariate algebraical polynomials, given its first unimodular row. This problem was solved in 1976 by Quillen and Suslin independently (see [3]). Moreover, in 1977, Suslin [4] extended the result to a wider class of rings, in particular, to the ring of the Laurent polynomials.

Suslin's result is of theoretical character. Park and Woodburn [5] presented a constructive proof of this result, which formally gives an algorithm for constructing a unimodular matrix from a given row. This algorithm is based on the construction of Gröbner bases, which is rather complicated and thus cannot be used in practice except for some rare cases of polynomials of sufficiently small degree. Another version of the algorithm for the unimodular completion was suggested by Amidou and Yengui in [6].

The question of the possibility of extending any suitable row to a unitary matrix, whose elements are Laurent polynomials, still remains unsolved.

In spite of the considerable difficulties, which may occur in the implementation of the method of extending a row to a unimodular matrix, the process of constructing biorthogonal compactly supported wavelets described above is useful since in some particular cases one can sometimes guess a required unimodular matrix (naturally, this is more likely than to guess two mutually inverse matrices).

Example 2.5.2 Let $M = \begin{pmatrix} 2 & 0 \\ 0 & 2 \end{pmatrix}$. As a scaling function, we take a box spline B_3 (see Example 2.3.8) defined by the equality

$$\widehat{B_3}(\xi) = 2^2 \prod_{k=1}^{3} \frac{1 - e^{2\pi i(a_k,\xi)}}{2\pi i(a_k, \xi)},$$

where $a_1 = (1, 0)$, $a_2 = (0, 1)$, $a_3 = (1, 1)$. Its mask is the polynomial

$$m_0(\xi) = \prod_{k=1}^{3} \frac{1 + e^{2\pi i(a_k,\xi)}}{2}.$$

As a set of digits, consider the vectors

$$s_0 = (0, 0), s_1 = (0, 1), s_2 = (1, 0), s_3 = (1, 1)$$

and represent the mask in the form

$$m_0(\xi) = \frac{1}{2}\left(e^{2\pi i(s_0,\xi)} \frac{1 + e^{4\pi i(\xi_1+\xi_2)}}{4} + e^{2\pi i(s_1,\xi)} \frac{1 + e^{4\pi i\xi_1}}{4} + \right.$$
$$\left. e^{2\pi i(s_2,\xi)} \frac{1 + e^{4\pi i\xi_2}}{4} + \frac{1}{2} e^{2\pi i(s_3,\xi)}\right).$$

Setting $z_1 = e^{2\pi i\xi_1}$, $z_2 = e^{2\pi i\xi_2}$, we find the polyphase functions

$$\mu_{00}(\xi) = \frac{1 + z_1 z_2}{4}, \quad \mu_{01}(\xi) = \frac{1 + z_1}{4}, \quad \mu_{02}(\xi) = \frac{1 + z_2}{4}, \quad \mu_{03}(\xi) = \frac{1}{2}.$$

This row can easily be extended to the diagonal matrix

$$\begin{pmatrix} \frac{1+z_1z_2}{4} & \frac{1+z_1}{4} & \frac{1+z_2}{4} & \frac{1}{2} \\ 0 & 0 & 2 & 0 \\ 0 & 1 & 0 & 0 \\ 1 & 0 & 0 & 0, \end{pmatrix}$$

with the determinant equal to 1.◊

2.6 Refinable Functions

In the previous section, we described a general method for the construction of wavelet bases and frames. The starting point of the method is finding an appropriate refinable function or an appropriate pair of refinable functions. Now, we study refinable functions and discuss how they can be constructed.

Recall that a function $\varphi \in L_2(\mathbb{R}^d)$ is refinable if it satisfies the refinement equation (2.16), or equivalently, its Fourier transform $\widehat{\varphi}$ satisfies (2.17). Any refinable function defines its mask. The converse is also true. Taking an appropriate function $m_0 \in L_2(\mathbb{T}^d)$ as a mask, we can construct a function $\widehat{\varphi}$ that satisfies equation (2.17) with this mask. Indeed, the right-hand side of (2.17) depends on the function $\widehat{\varphi}$, to which we again can apply (2.17). Continuing this process, for any positive integer n, we get

$$\widehat{\varphi}(\xi) = \widehat{\varphi}(M^{*-n}\xi) \prod_{j=1}^{n} m_0(M^{*-j}\xi).$$

If the function $\widehat{\varphi}$ is continuous at the origin and $\widehat{\varphi}(\mathbf{0}) = 1$, then passing to the limit in this equality as $n \to \infty$ and taking into account (2.1), we get

$$\widehat{\varphi}(\xi) = \prod_{j=1}^{\infty} m_0(M^{*-j}\xi). \tag{2.45}$$

For the infinite product to converge at the origin, it is necessary to have

$$m_0(\mathbf{0}) = 1. \tag{2.46}$$

Certainly, the convergence at the origin is not sufficient. It makes sense to consider only functions m_0 that provide the convergence of the infinite product at least almost everywhere. The following assertion shows that the class of such masks is sufficiently wide.

Proposition 2.6.1 *Suppose a function* $m_0(\xi) = \sum_{k\in\mathbb{Z}^d} c_k e^{2\pi i(k,\xi)}$ *satisfies (2.46) and*

$$c_k = O(|k|^{-d-1-\varepsilon}), \quad \varepsilon > 0. \tag{2.47}$$

Then, the infinite product (2.45) converges[1] *absolutely and uniformly on any compact set.*

Proof From (2.46) and (2.47), we get

$$|m_0(\xi) - 1| \le 2 \sum_{k\in\mathbb{Z}^d} |c_k \sin \pi(k,\xi)| \le C \sum_{k\in\mathbb{Z}^d} |k|^{-d-\varepsilon}|\xi| =: C_1|\xi|. \tag{2.48}$$

[1] Unlike in the traditional terminology, the infinite product will be considered convergent even in the case when it is equal to zero.

This relation and (2.2) imply the uniform convergence of the series

$$\sum_{j=1}^{\infty} |m_0(M^{*-j}\xi) - 1|,$$

on any compact set. It follows that the infinite product (2.45) converges uniformly on any compact set.◊

In particular, condition (2.47) holds for trigonometric polynomials, which are of most interest for applications because the function φ defined by (2.45) is compactly supported in this case, what will be proved soon.

Corollary 2.6.2 *Under the conditions of Proposition 2.6.1, the function $\widehat{\varphi}$ defined by equality (2.45) is continuous on $\mathbb{R}^d$.*

The proof follows from the uniform convergence of partial products to $\widehat{\varphi}$ on compact sets and the continuity of the function m_0, which holds due to the uniform convergence of its Fourier series.

It appears that a refinement equation with polynomial mask always has a unique (up to a constant factor) solution in the space of tempered distributions $\mathcal{S}'$; moreover, this solution is compactly supported. In particular, if there exists a solution in $L_2(\mathbb{R}^d)$, it is unique (up to normalization) and compactly supported. To prove this, let us introduce a *transition operator* on the space of distributions:

$$[Tf](x) = \sqrt{m} \sum_{k \in \mathbb{Z}^d} h_k f(Mx + k). \tag{2.49}$$

This is a linear continuous operator in the space $\mathcal{S}'$. The refinable function is its eigenvector corresponding to the eigenvalue 1.

First, we prove a lemma.

Lemma 2.6.3 *Suppose M is a dilation matrix in $\mathbb{R}^d$ and A is a finite subset of $\mathbb{R}^d$. Then, there exists a convex compact set $K \subset \mathbb{R}^d$ centrally symmetric with respect to the origin, having a nonempty interior and such that*

$$M^{-1}(K - a) \subset K \quad \forall a \in A. \tag{2.50}$$

Proof Let U be a closed ball with the center in the origin containing A. Set

$$K = \sum_{j=1}^{\infty} M^{-j} U,$$

i.e., K consists of all vectors $x = \sum_{j=1}^{\infty} M^{-j} x_j,\ x_j \in U$. The latter series converges due to (2.2). It is easy to see that K is a compact set centrally symmetric with respect to the origin, having some nonempty interior. Since $K = M^{-1}(K + U)$ and $A \subset U$, we have (2.50).◊

Theorem 2.6.4 *A refinement equation with polynomial mask that satisfies (2.46) always has a unique (up to a constant factor) compactly supported solution $\varphi \in \mathcal{S}'$. This solution is given by formula (2.45). Moreover, for any compactly supported distribution $f \in \mathcal{S}'$, the sequence $f_n = T^n f$ converges to the function $c \cdot \varphi$ in the space $\mathcal{S}'$, where the normalizing factor c is $\widehat{f}(\mathbf{0})$.*

Proof We have to show that the infinite product $\prod_{k=1}^{\infty} m_0\left(M^{*-k}\xi\right)$ converges in the space $\mathcal{S}'$. It follows that the function φ defined as the inverse Fourier transform of this product belongs to $\mathcal{S}'$. This will also imply (if we use the inverse Fourier transform) convergence of $T^n f \to \varphi$ in the space $\mathcal{S}'$. The convergence takes place for any compactly supported function $f \in S'$ such that $\widehat{f}(\mathbf{0}) = 1$. In particular, it follows that the solution φ is unique. Applying Lemma 2.6.3 to the set $A = \text{spec}(m_0)$, we deduce that the operator T preserves the space $\{f \in \mathcal{S}' : \text{ supp } f \subset K\}$. Hence, if we choose f such that $\text{supp } f \subset K$, $\widehat{f}(\mathbf{0}) = 1$, then the limit $\lim\limits_{n\to\infty} T^n f$ has a support containing in the compact set K. Thus, it remains to prove the convergency of

$$\widehat{T^n f}(\xi) = \prod_{j=1}^{n} m_0\left(M^{*-j}\xi\right) \widehat{f}\left(M^{*-n}\xi\right) \to \widehat{f}(\mathbf{0}) \prod_{k=1}^{\infty} m_0\left(M^{*-k}\xi\right) \tag{2.51}$$

in the space $\mathcal{S}'$ for any compactly supported distribution $f \in \mathcal{S}'$. We have already proved convergency on each compact set in $\mathbb{R}^d$, which, however, is not sufficient for the convergency in $\mathcal{S}'$.

Let us estimate the rate of growth of the function $\widehat{T^n f}(\xi)$ as $\xi \to \infty$. Let $\rho = \rho(M^{*-1})$ be the spectral radius of the operator M^{*-1} (the maximal absolute value of its eigenvalues). Since M is a dilation matrix, the spectral radius of M^{-1} is less than 1, which yields $\rho < 1$. Fix an arbitrary $q \in (\rho, 1)$, and find a constant C_q such that $\|M^{*-k}\| \le C_q q^k$, $k \in \mathbb{N}$. As is known, there exists a constant C such that $|1 - m_0(\xi)| \le C|\xi|$ for all $\xi \in \mathbb{R}^d$. Now, take an arbitrary $\xi \in \mathbb{R}^d$, $|\xi| > \frac{1}{C_q C}$. Let p be the minimal integer such that $|M^{*-p}\xi| \le \frac{1}{C_q C}$. For each $n \ge p + 1$, we have

$$\left|\widehat{f}\left(M^{*-p}\xi\right) \prod_{k=1}^{n} m_0\left(M^{*-k}\xi\right)\right| \le C_1 \left|\prod_{k=1}^{p} m_0\left(M^{*-k}\xi\right)\right| \cdot \left|\prod_{j=1}^{n-p}\left(1 + q^j\right)\right| \tag{2.52}$$

$$\le C_1 C_2 \|m_0\|_\infty^p < C_1 C_2 \|m_0\|_\infty^{\log_q \frac{q}{C_q C\|\xi\|}} = C_1 C_2 \left(\frac{C_q C|\xi|}{q}\right)^{\log_{1/q} \|m_0\|_\infty},$$

where

$$C_1 = \max\left\{|\widehat{f}(\eta)|,\ |\eta| \le \frac{1}{C_q C}\right\}, \quad C_2 = \left|\prod_{j=1}^{\infty}\left(1 + q^j\right)\right|.$$

To estimate the number p, we used the relation $|M^{*-p+1}\xi| > \frac{1}{C_q C}$, whence it follows that

$$q^{p-1} > \frac{1}{C_q C|\xi|},$$

which implies that

$$p < \log_q \left(\frac{q}{C_q C|\xi|} \right).$$

Thus, for sufficiently large n, the rate of growth of the function $\widehat{T^n f}(\xi)$ cannot be more than that of a polynomial of degree $\log_{1/q} \|m_0\|_\infty$ as $\xi \to \infty$. Let us now take an arbitrary test function $h \in \mathcal{S}$. Since it decreases at infinity more rapidly than any power of $|\xi|$, it follows that for any $\varepsilon > 0$, there exists a sufficiently large $R = R(\varepsilon)$ such that $\left| \int\limits_{|\xi|>R} \widehat{T^n f}\, h\, d\xi \right| < \varepsilon$ for all $n \geq p+1$. On the other hand, in the ball $\{\xi \in \mathbb{R}^d : |\xi| \leq R\}$, the product (2.51) uniformly converges to $\widehat{f}(\mathbf{0})\widehat{\varphi}(\xi)$, and hence, $(\widehat{T^n f}, h) \to (\widehat{f}(\mathbf{0})\widehat{\varphi}, h)$ as $n \to \infty$, which completes the proof.◊

Thus, if a function $\varphi \in \mathcal{S}'$, whose Fourier transform is continuous at the origin and $\widehat{\varphi}(\mathbf{0}) = 1$, satisfies a refinement equation with polynomial mask, then it is compactly supported. Moreover, for any function $f \in \mathcal{S}'$, whose Fourier transform is continuous at the origin, the sequence $f_n = T^n f$ converges to $\widehat{f}(\mathbf{0}) \cdot \varphi$ in the space of distributions.

Corollary 2.6.5 *If a refinement equation with polynomial mask has a summable solution φ normalized by the condition $\int\limits_{-\infty}^{\infty} \varphi\, dx = 1$, then this solution is compactly supported and defined by formula (2.45). Moreover, for any summable function f, the sequence $f_n = T^n f$ converges to the function $\widehat{f}(\mathbf{0}) \cdot \varphi$ in the space $\mathcal{S}'$.*

Now, we know that each trigonometric polynomial m_0 satisfying (2.46) defines a compactly supported refinable function φ. Unfortunately, φ is a tempered distribution, generally speaking. To construct wavelet bases or frames using the method described in Sect. 2.4, we need φ to be in $L_2(\mathbb{R}^d)$. Thus, we are interested under what conditions on m_0 we have $\varphi \in L_2(\mathbb{R}^d)$.

Proposition 2.6.6 *Let m_0 be as in Proposition 2.6.1 and such that*

$$\sum_{s \in D(M^*)} |m_0(\xi + M^{*-1}s)|^2 \leq 1, \tag{2.53}$$

and let φ be defined by (2.45). Then, $\varphi \in L_2(\mathbb{R}^d)$, $\|\varphi\|_2 \leq 1$.

Proof For $k \in \mathbb{N}$, set

$$f_k(x) = \prod_{j=1}^{k} m_0(M^{*-j}x)\chi_{M^{*k}[-1/2,1/2]^d}(x).$$

Using Lemma 2.1.4 and 1-periodicity (on each variable) of m_0, we have

$$\|f_k\|_2^2 = \int\limits_{M^{*k}[-1/2,1/2]^d} \prod_{j=1}^{k} |m_0(M^{*-j}x)|^2\,dx = m^k \int\limits_{\mathbb{T}^d} \prod_{j=1}^{k} |m_0(M^{*k-j}x)|^2\,dx =$$

$$m^k \sum_{r\in D(M^*)} \int\limits_{M^{*-1}\mathbb{T}^d+M^{*-1}r} \prod_{j=1}^{k} |m_0(M^{*k-j}x)|^2\,dx =$$

$$m^k \sum_{r\in D(M^*)} \int\limits_{M^{*-1}\mathbb{T}^d} \prod_{j=1}^{k-1} |m_0(M^{*k-j}x)|^2 |m_0(x+M^{*-1}r)|^2\,dx \le$$

$$m^k \int\limits_{M^{*-1}\mathbb{T}^d} \prod_{j=1}^{k-1} |m_0(M^{*k-j}x)|^2\,dx = m^{k-1} \int\limits_{\mathbb{T}^d} \prod_{j=1}^{k-1} |m_0(M^{*k-1-j}x)|^2\,dx =$$

$$\int\limits_{M^{*k-1}[-1/2,1/2]^d} \prod_{j=1}^{k-1} |m_0(M^{*-j}x)|^2\,dx = \|f_{k-1}\|_2^2.$$

By Corollary 2.6.2, the infinite product converges at each point, which yields $f_k(x) \underset{k\to+\infty}{\longrightarrow} \widehat{\varphi}(x)$ for all $x \in \mathbb{R}^d$. It follows from Fatou's lemma and $\|f_1\|_2 \le 1$ that

$$\int\limits_{\mathbb{R}^d} |\widehat{\varphi}(\xi)|^2\,d\xi = \int\limits_{\mathbb{R}^d} \prod_{j=1}^{\infty} |m_0(M^{*-j}\xi)|^2\,d\xi \le \liminf_{k\to\infty} \int\limits_{\mathbb{R}^d} |f_k(\xi)|^2\,d\xi \le 1. \ \Diamond$$

Thus, under condition (2.53), the refinable function φ given by (2.45) is in $L_2(\mathbb{R}^d)$. In particular, $\varphi \in L_2(\mathbb{R}^d)$ whenever

$$\sum_{s\in D(M^*)} |m_0(x+M^{*-1}s)|^2 = 1,$$

what is required for the construction of orthogonal bases and tight frames. Additionally, we can establish the following statement.

Proposition 2.6.7 *Let m_0 be a trigonometric polynomial satisfying (2.46) and such that $m_0(\xi) \ge 0$ for all $\xi \in \mathbb{R}^d$. Suppose*

$$\sum_{q\in D(M^*)} m_0(\xi+M^{*-1}q) \le 1, \quad \xi \in \mathbb{R}^d.$$

Then, $\widehat{\varphi} \in L_1(\mathbb{R}^d)$ and $\varphi := \mathcal{F}^{-1}\widehat{\varphi}$ is a continuous compactly supported refinable function.

Proof Using arguments similar to the proof of Proposition 2.6.6, it can be proved that $\widehat{\varphi} \in L_1(\mathbb{R}^d)$, and therefore, φ is a continuous function. On the other hand, by Theorem 2.6.4, φ is compactly supported and refinable.$\Diamond$

Propositions 2.6.6 and 2.6.7 give sufficient conditions under which a compactly supported function φ defined by (2.45) is in $L_2(\mathbb{R}^d)$. Unfortunately, it is much more complicated to check belonging φ to $L_2(\mathbb{R}^d)$ in the general case. Algorithms that allow to reduce this problem to the computation of the eigenvalues for some finite-dimensional operators will be given in Sect. 6.8.

Next, we estimate the support of a refinable function with polynomial mask.

Theorem 2.6.8 *Let $\varphi \in L_2(\mathbb{R}^d)$ be given by (2.45), where m_0 is a trigonometric polynomial satisfying (2.46). Then,* $\operatorname{supp}\varphi \subset \{x \in \mathbb{R}^d : |x| \le R\}$, *where*

$$R \le \max_{k \in \operatorname{spec}(m_0)} |k| \sum_{j=1}^{\infty} \|M^{-j}\| < \infty.$$

Proof First of all, note that $\operatorname{supp}\varphi$ is a compact set because of Theorem 2.6.4. Let

$$m_0(\xi) = \frac{1}{\sqrt{m}} \sum_{k \in \mathbb{Z}^d} h_k e^{2\pi i (k,\xi)},$$

then

$$\varphi(x) = \sqrt{m} \sum_{k \in \mathbb{Z}^d} h_k \varphi(Mx + k) \quad \text{a.e.} \tag{2.54}$$

Let V denote the set of all vectors $x \in \mathbb{R}^d$ for which (2.54) does not hold, V_k, $k \in \operatorname{spec}(m_0)$, denote the set of all vectors $x \in \mathbb{R}^d$ such that $Mx + k \in V$, and W be the union of V and all V_k. Since $\mu W = 0$, it suffices to estimate the set $U := \{x \in \operatorname{supp}\varphi : \varphi(x) \ne 0\} \setminus W$.

Let $x_0 \in U$. Then, due to (2.54), there exists $k_1 \in \operatorname{spec}(m_0)$ such that $x_1 := Mx_0 + k \in U$, which yields $x_0 = M^{-1}(x_1 - k_1)$. Similarly, there exist $k_2 \in \operatorname{spec}(m_0)$ and $x_2 \in U$ such that $x_1 = M^{-1}(x_2 - k_2)$, and hence, $x_0 = M^{-2}x_2 - M^{-1}k_1 - M^{-2}k_2$. We can repeat the process, and then, for every $N \in \mathbb{N}$

$$x_0 = M^{-N}x_N - \sum_{j=1}^{N} M^{-j}k_j, \quad x_n \in U, \quad k_j \in \operatorname{spec}(m_0).$$

Using (2.1), we can pass to the limit as $N \to +\infty$ and get

$$x_0 = -\sum_{j=1}^{\infty} M^{-j} k_j.$$

It follows that

$$|x_0| \le \sum_{j=1}^{\infty} \|M^{-j}\| |k_j| \le \max_{j\in\mathbb{N}} |k_j| \sum_{j=1}^{\infty} \|M^{-j}\|,$$

where the latter series is convergent due to (2.2), what was to be proved.◊

Finally, we prove that any compactly supported refinable function from $L_2(\mathbb{R}^d)$ has some Sobolev smoothness. Recall that the Sobolev space $W_2^{\nu}(\mathbb{R}^d)$ consists of functions $f \in L_2(\mathbb{R}^d)$ such that

$$\int_{\mathbb{R}^d} |\widehat{f}(\xi)|^2 (1 + |\xi|^2)^s d\xi < \infty.$$

Theorem 2.6.9 *Let φ be a compactly supported refinable function with a polynomial mask m_0. If φ is in $L_2(\mathbb{R}^d)$, then there exists $\nu > 0$, such that $\varphi \in W_2^{\nu}(\mathbb{R}^d)$. Moreover, $(1 + |\cdot|)^{\nu}\widehat{\varphi} \in L_{\infty}(\mathbb{R}^d)$.*

Proof Let $n \in \mathbb{N}$. By the refinement equation (2.17), we have

$$\widehat{\varphi}(M^{*n}\xi) = m^{(n)}(\xi)\widehat{\varphi}(\xi), \quad \text{with} \quad m^{(n)}(\xi) := \prod_{j=1}^{n} m_0(M^{*n-j}\xi).$$

Since $m^{(n)}$ is 1-periodic (with respect to each variable), due to item (ii) in Lemma 2.2.1, we have

$$[\widehat{\varphi}(M^{*n}\cdot), \widehat{\varphi}(M^{*n}\cdot)](\xi) = |m^{(n)}(\xi)|^2 [\widehat{\varphi}, \widehat{\varphi}](\xi). \tag{2.55}$$

Suppose the standard basis for $\mathbb{R}^d$ is denoted by $e_1 \dots, e_d$. Let us define function u by

$$u(\xi) := [\widehat{\varphi}, \widehat{\varphi}](\xi) \sum_{j=1}^{d} |1 - e^{2\pi i (e_j, \xi)}|^2 = [\widehat{\varphi}, \widehat{\varphi}](\xi) \sum_{j=1}^{d} |1 - e^{2\pi i \xi_j}|^2, \tag{2.56}$$

where $\xi = (\xi_1, \dots, \xi_d) \in \mathbb{R}^d$. It follows from item (vi) in Lemma 2.2.1 that $[\widehat{\varphi}, \widehat{\varphi}](\xi)$ is equivalent to the trigonometric polynomial

$$\sum_{k\in\mathbb{Z}^d} e^{2\pi i (k,\xi)} \int_{\mathbb{R}^d} \varphi(x+k)\overline{\varphi(x)} dx.$$

Therefore, without loss of generality, we can consider that u is a trigonometric polynomial.

Now, we take a closer look at $\|\varphi - \varphi(\cdot + M^{-n}e_j)\|_2$, $j = 1, \ldots, d$. Using the above considerations and item (iii) in Lemma 2.2.1, we obtain

$$\sum_{j=1}^{d} \int_{\mathbb{R}^d} |\varphi(x) - \varphi(x + M^{-n}e_j)|^2 dx = \sum_{j=1}^{d} \int_{\mathbb{R}^d} |\widehat{\varphi}(\xi) - e^{2\pi i (M^{-n}e_j, \xi)} \widehat{\varphi}(\xi)|^2 d\xi =$$

$$\sum_{j=1}^{d} \int_{\mathbb{R}^d} |1 - e^{2\pi i (M^{-n}e_j, \xi)}|^2 |\widehat{\varphi}(\xi)|^2 d\xi = m^n \sum_{j=1}^{d} \int_{\mathbb{R}^d} |1 - e^{2\pi i (e_j, \xi)}|^2 |\widehat{\varphi}(M^{*n}\xi)|^2 d\xi =$$

$$m^n \sum_{j=1}^{d} \int_{\mathbb{T}^d} |1 - e^{2\pi i (e_j, \xi)}|^2 [\widehat{\varphi}(M^{*n}\cdot), \widehat{\varphi}(M^{*n}\cdot)](\xi) d\xi =$$

$$= m^n \sum_{j=1}^{d} \int_{\mathbb{T}^d} |1 - e^{2\pi i (e_j, \xi)}|^2 |m^{(n)}(\xi)|^2 [\widehat{\varphi}, \widehat{\varphi}](\xi) d\xi = \int_{\mathbb{T}^d} |m^{(n)}(\xi)|^2 u(\xi) d\xi.$$

Let us define the transition operator T as follows:

$$Tv(\xi) := \sum_{q \in D(M^*)} |m_0(M^{*-1}(\xi + q))|^2 v(M^{*-1}(\xi + q)),$$

where v is a trigonometric polynomial. It is not hard to check that Tv is also a trigonometric polynomial. Now, we show by induction that for any $n \in \mathbb{N}$

$$\int_{\mathbb{T}^d} |m^{(n)}(\xi)|^2 u(\xi) d\xi = m^{-n} \int_{\mathbb{T}^d} T^n u(\xi) d\xi. \tag{2.57}$$

Let $n = 1$. Since $m^{(1)}(\xi) = m_0(\xi)$, Lemma 2.1.4 yields that

$$\int_{\mathbb{T}^d} |m_0(\xi)|^2 u(\xi) d\xi = \sum_{q \in D(M^*)} \int_{M^{*-1}(\mathbb{T}^d + q)} |m_0(\xi)|^2 u(\xi) d\xi =$$

$$\frac{1}{m} \sum_{q \in D(N)} \int_{\mathbb{T}^d + q} |m_0(M^{*-1}\xi)|^2 u(M^{*-1}\xi) d\xi =$$

$$\frac{1}{m} \int_{\mathbb{T}^d} \sum_{q \in D(M^*)} |m_0(M^{*-1}(\xi + q))|^2 u(M^{*-1}(\xi + q)) d\xi = \frac{1}{m} \int_{\mathbb{T}^d} Tu(\xi) d\xi.$$

Assume now that

$$\int_{\mathbb{T}^d} |m^{(n-1)}(\xi)|^2 u(\xi) = m^{-n+1} \int_{\mathbb{T}^d} T^{n-1} u(\xi) d\xi$$

is valid. Note that $m^{(n)}(\xi) = m^{(n-1)}(M^*\xi) m_0(\xi)$. Again using Lemma 2.1.4, we get

$$\int_{\mathbb{T}^d} |m^{(n)}(\xi)|^2 u(\xi) d\xi = \int_{\mathbb{T}^d} |m^{(n-1)}(M^*\xi)|^2 |m_0(\xi)|^2 u(\xi) d\xi =$$

$$\sum_{q\in D(N)} \int_{M^{*-1}(\mathbb{T}^d+q)} |m^{(n-1)}(M^*\xi)|^2 |m_0(\xi)|^2 u(\xi) d\xi =$$

$$\frac{1}{m} \sum_{q\in D(M^*)} \int_{\mathbb{T}^d+q} |m^{(n-1)}(\xi)|^2 |m_0(M^{*-1}\xi)|^2 u(M^{*-1}\xi) d\xi =$$

$$\frac{1}{m} \int_{\mathbb{T}^d} |m^{(n-1)}(\xi)|^2 \sum_{q\in D(M^*)} |m_0(M^{*-1}(\xi+q))|^2 u(M^{*-1}(\xi+q)) d\xi =$$

$$\frac{1}{m} \int_{\mathbb{T}^d} |m^{(n-1)}(\xi)|^2 T u(\xi) d\xi = m^{-n} \int_{\mathbb{T}^d} T^n u(\xi) d\xi,$$

which completes the proof of (2.57).

Note that $u(\xi) \geq 0$, and therefore, $T^n u(\xi) \geq 0$ for all $n \in \mathbb{N}$ and $\xi \in \mathbb{R}^d$.

Now, we can connect the norm of $\varphi - \varphi(\cdot + M^{-n}e_j)$, $j = 1, \dots, d$, with the transition operator T as follows:

$$\sum_{j=1}^{d} \|\varphi - \varphi(\cdot + M^{-n}e_j)\|_2^2 = \sum_{j=1}^{d} \int_{\mathbb{R}^d} |\varphi(x) - \varphi(x + M^{-n}e_j)|^2 dx =$$

$$m^n \int_{\mathbb{T}^d} |m^{(n)}(\xi)|^2 u(\xi) d\xi = \int_{\mathbb{T}^d} |T^n u(\xi)| d\xi.$$

Since M is a dilation matrix, $\lim_{n\to+\infty} \|\varphi - \varphi(\cdot + M^{-n}e_j)\|_2^2 = 0$ for every $j = 1, \dots, d$. Hence,

$$\lim_{n\to+\infty} \int_{\mathbb{T}^d} |T^n u(\xi)| d\xi = \lim_{n\to+\infty} \sum_{j=1}^{d} \|\varphi - \varphi(\cdot + M^{-n}e_j)\|_2^2 = 0.$$

Next, we estimate the rate of decay of $\lim_{n\to+\infty} \int_{\mathbb{T}^d} |T^n u(\xi)| d\xi$. Let

$$V := \text{span}\,\{T^n\; u, n \in \mathbb{N}\}.$$

Note that $V \subset L(\mathbb{T}^d)$, since u and $T^n u$ are trigonometric polynomials. Let us show that V is a finite-dimensional space. Denote $w := |m_0|^2$, and let w_k be the coefficients of trigonometric polynomial w, namely $w(\xi) = \sum_{k\in\mathbb{Z}^d} w_k e^{2\pi i(k,\xi)}$, and u_k be the coefficients of trigonometric polynomial u, namely $u(\xi) = \sum_{k\in\mathbb{Z}^d} u_k e^{2\pi i(k,\xi)}$. Then

$$
\begin{aligned}
Tu(\xi) :=& \sum_{q\in D(M^*)} w(M^{*-1}(\xi+q))u(M^{*-1}(\xi+q)) = \\
& \sum_{q\in D(M^*)} \sum_{l\in\mathbb{Z}^d} w_l e^{2\pi i(l, M^{*-1}(\xi+q))} \sum_{k\in\mathbb{Z}^d} u_k e^{2\pi i(k, M^{*-1}(\xi+q))} = \\
& \sum_{l\in\mathbb{Z}^d}\sum_{k\in\mathbb{Z}^d} u_k w_l e^{2\pi i(l+k, M^{*-1}\xi)} \sum_{q\in D(M^*)} e^{2\pi i(l+k, M^{*-1}q)}.
\end{aligned}
$$

By Lemma 2.1.5,

$$
\sum_{q\in D(M^*)} e^{2\pi i(l+k, M^{*-1}q)} = \begin{cases} m, & \text{if } l+k \in M\mathbb{Z}^d, \\ 0, & \text{if } l+k \notin M\mathbb{Z}^d, \end{cases}
$$

Thus, the spectrum of the trigonometric polynomial Tu is inside of the set $M^{-1}(U+W)$, where $U :=$ spec (u), $W :=$ spec (w). (Here, spec (u) denotes the spectrum of u.) Therefore, the spectrum of T^2u is inside $M^{-2}(U+W)+M^{-1}U$. So, the spectrum of $T^n u$ is

$$
\text{spec } (T^n u) \subset M^{-n}W + \sum_{i=1}^{n} M^{-i}U.
$$

Assume that $N_u \in \mathbb{R}$ is such that for all $k \in$ spec (u), $|k| \le N_u$ and N_w is such that for all $k \in$ spec (w), $|k| \le N_w$. If now $k \in$ spec $(T^n u)$, then $|k| \le AN_w + BN_u$, where $A = \sup_{n>0} \|M^{-n}\|$, $B = \sup_{n>0} \sum_{k=1}^{n} \|M^{-n}\|$, and A, B are finite due to (2.2). This proves that V is a finite-dimensional space and the degrees of trigonometric polynomials $T^n u$ are uniformly bounded.

Next, let us consider the action of operator T on the invariant space V. We assume that the space V is equipped with the usual norm $\|v\|_{L_1} = \int_{\mathbb{T}^d} |v(\xi)| d\xi$, $v \in V$. Let $\rho(T\big|_V)$ be the spectral radius of the operator T on V. Since V is a finite-dimensional space and $\|T^n v\|_{L_1} \to 0$ as $n \to +\infty$ for all $v \in V$, we have $\|T^n\| \to 0$. This yields $\rho(T\big|_V) = \inf_{n\ge1} \|T^n\|^{1/n} < 1$. Since $\rho(T\big|_V) = \lim_{n\to+\infty} \|T^n\|^{1/n}$, there exists $\rho < 1$ such that $\|T^n\| < \rho^n$ for all big enough n. It follows that

$$
\int_{\mathbb{T}^d} |T^n u(\xi)| d\xi \le C\rho^n, \quad \forall n \in \mathbb{N}. \tag{2.58}
$$

Let us prove now that $\varphi \in W_2^{\nu}(\mathbb{R}^d)$, where

$$
0 < \nu < \frac{1}{2} \ln(\rho^{-1}) / \ln(1 + d\|M^*\|).
$$

Since M is a dilation matrix, there exists an open set Q such that $\mathbf{0} \in Q \subseteq [-1/3, 1/3]^d$ and $M^{*-1}Q \subseteq Q$. Indeed, the set Q can be constructed as follows: Let $\widetilde{Q}$ denote the set of all sums $x_0 + M^{*-1}x_1 + \cdots + M^{*-r}x_r$, where $|x_i| < 1$, $i = 1, \ldots, r$. It is easy to see that $\mathbf{0} \in \widetilde{Q}$, $\widetilde{Q}$ is an open bounded set and $M^{*-1}\widetilde{Q} \subseteq \widetilde{Q}$.

The set $\widetilde{Q}$ can be appropriately scaled such that $Q := \widetilde{q}\widetilde{Q} \subseteq [-1/3, 1/3]^d$ with some $\widetilde{q} \in \mathbb{R}$. Setting $U := Q \setminus M^{*-1}Q$, we have $\mathbb{R}^d = Q \cup \cup_{n=1}^{\infty} M^{*n}U$. Since $\mathbf{0} \notin U \subseteq [-1/3, 1/3]^d$, we obtain

$$C_1 := \min_{\xi\in U} \sum_{j=1}^{d} |1 - e^{2\pi i \xi_j}|^2 > 0.$$

It follows from (2.55) and (2.56) that

$$|\widehat{\varphi}(M^{*n}\xi)|^2 \leq [\widehat{\varphi}(M^{*n}\cdot), \widehat{\varphi}(M^{*n}\cdot)](\xi) \leq$$

$$C_1^{-1}[\widehat{\varphi}(M^{*n}\cdot), \widehat{\varphi}(M^{*n}\cdot)](\xi) \sum_{j=1}^{d} |1 - e^{2\pi i \xi_j}|^2 = C_1^{-1}|m^{(n)}(\xi)|^2 u(\xi) \quad \forall \xi \in U.$$

Therefore, using (2.57) and (2.58), we have

$$\int_U |\widehat{\varphi}(M^{*n}\xi)|^2 d\xi \leq C_1^{-1} \int_{\mathbb{T}^d} |m^{(n)}(\xi)|^2 u(\xi) d\xi =$$

$$C_1^{-1} m^{-n} \int_{\mathbb{T}^d} |T^n u(\xi)| d\xi \leq C_1^{-1} C \rho^n m^{-n} \quad \forall n \in \mathbb{N}.$$

Taking into account that $\max_{\xi\in M^{*n}U} |\xi| \leq \|M^*\|^n d^{1/2}$ because $U \subseteq [-1/3, 1/3]^d$, and

$$\int_U |\widehat{\varphi}(M^{*n}\xi)|^2 d\xi = m^{-n} \int_{M^{*n}U} |\widehat{\varphi}(\xi)|^2 d\xi,$$

we obtain

$$\int_{M^{*n}U} |\widehat{\varphi}(\xi)|^2 (1 + |\xi|^2)^{\nu} d\xi \leq (1 + \|M^*\|^{2n} d)^{\nu} \int_{M^{*n}U} |\widehat{\varphi}(\xi)|^2 d\xi =$$

$$(1 + d\|M^*\|)^{2\nu n} m^n \int_U |\widehat{\varphi}(M^{*n}\xi)|^2 d\xi \leq C_1^{-1} C((1 + d\|M^*\|)^{2\nu}\rho)^n.$$

By the choice of ν, $\rho_1 := (1 + d\|M^*\|)^{2\nu}\rho < 1$. Since $\mathbb{R}^d = Q \cup \cup_{n=1}^{\infty} M^{*n}U$, we can state that

$$\int_{\mathbb{R}^d} |\widehat{\varphi}(\xi)|^2(1+|\xi|^2)^\nu d\xi \le$$

$$\int_Q |\widehat{\varphi}(\xi)|^2(1+|\xi|^2)^\nu d\xi + \sum_{n=1}^{\infty} \int_{M^{*n}U} |\widehat{\varphi}(\xi)|^2(1+|\xi|^2)^\nu d\xi \le$$

$$(1+d)^\nu \|\widehat{\varphi}\|_\infty + CC_1^{-1} \sum_{n=1}^{\infty} \rho_1^n < \infty.$$

Here, $\|\widehat{\varphi}\|_\infty < \infty$ since φ is compactly supported. Thus, $\varphi \in W_2^\nu(\mathbb{R}^d)$.

To complete the proof, it remains to show that if $f \in W_2^\nu(\mathbb{R}^d)$ for some $0 < \nu \le 1$ and f is compactly supported, then $(1+|\cdot|)^\nu \widehat{f}(\cdot) \in L_\infty(\mathbb{R}^d)$. Note that for $0 < \nu \le 1$, there exists a constant $C_1 > 0$ such that $|\sin \pi x| < C_1|x|^\nu$ for all $x \in \mathbb{R}^d$. Therefore, for every $y \in \mathbb{R}^d$, we obtain

$$\|f - f(\cdot + y)\|_2^2 = \|\widehat{f}(1 - e^{2\pi i(y,\cdot)})\|_2^2 = 16 \int_{\mathbb{R}^d} |\widehat{f}(\xi)|^2 \,|\sin \pi(y,\xi)|^2\, d\xi \le$$

$$16C_1^2 \int_{\mathbb{R}^d} |\widehat{f}(\xi)|^2 |(y,\xi)|^{2\nu} d\xi \le 16C_1^2|y|^{2\nu} \int_{\mathbb{R}^d} |\widehat{f}(\xi)|^2|\xi|^{2\nu} d\xi = C_2|y|^{2\nu}, \quad (2.59)$$

where $C_2 = 4C_1^2 \int_{\mathbb{R}^d} |\widehat{f}(\xi)|^2|\xi|^{2\nu}d\xi < \infty$.

Using the appropriate change of variable, it is easy to see that

$$\widehat{f}(\xi) = -\int_{\mathbb{R}^d} f\left(x - \frac{\xi}{2|\xi|^2}\right) e^{-2\pi i(x,\xi)} dx, \quad \xi \neq 0.$$

Since f is a compactly supported function, there exists a constant C_3 depending only on the support of f such that $\|f - f(\cdot + y)\|_1 \le C_3\|f - f(\cdot + y)\|_2$ whenever $|y| \le 1/2$. It follows from (2.59) that

$$|2\widehat{f}(\xi)| = \left| \int_{\mathbb{R}^d} \left[f(x) - f\left(x - \frac{\xi}{2|\xi|^2}\right)\right] e^{-2\pi i(x,\xi)} d\xi \right| \le \left\| f - f\left(\cdot - \frac{\xi}{2|\xi|^2}\right)\right\|_1 \le$$

$$C_3 \left\| f - f\left(\cdot - \frac{\xi}{2|\xi|^2}\right)\right\|_2 \le \sqrt{C_2}C_3 \left|\frac{\xi}{2|\xi|^2}\right|^\nu = \sqrt{C_2}C_3 2^{-\nu}|\xi|^{-\nu}, \quad \forall |\xi| \ge 1.$$

This completes the proof.◇

2.7 Conditions of Biorthogonality

In this section, we consider refinable functions $\varphi, \widetilde{\varphi}$ with masks

$$m_0(\xi) := m^{-1/2} \sum_{k\in\mathbb{Z}^d} h_k e^{2\pi i(k,\xi)}, \quad \widetilde{m}_0(\xi) := m^{-1/2} \sum_{k\in\mathbb{Z}^d} \widetilde{h}_k e^{2\pi i(k,\xi)}, \tag{2.60}$$

respectively. We want to know when the integer shifts of functions $\varphi, \widetilde{\varphi}$ form a biorthonormal system in $L_2(\mathbb{R}^d)$ (in particular, when the integer shifts of one function φ form an orthonormal system), and what is required for the construction of biorthogonal (in particular, orthogonal) wavelet bases for $L_2(\mathbb{R}^d)$ (see Sect. 2.4).

By Proposition 2.4.1, condition (2.26) is necessary for the integer shifts of $\varphi, \widetilde{\varphi}$ to be biorthonormal. Note that (2.26) can be given in the equivalent form in terms of the Fourier coefficients of the functions $m_0, \widetilde{m}_0$, due to the following statement.

Proposition 2.7.1 *Let there be given* 1*-periodic functions*

$$m_0(\xi) = m^{-1/2} \sum_{k\in\mathbb{Z}^d} h_k e^{2\pi i(k,\xi)}, \quad \widetilde{m}_0(\xi) = m^{-1/2} \sum_{k\in\mathbb{Z}^d} \widetilde{h}_k e^{2\pi i(k,\xi)},$$

with absolutely convergent Fourier series. Then, condition (2.26) is equivalent to the fact that the equality

$$\sum_{k\in\mathbb{Z}^d} h_k \overline{\widetilde{h}_{k-Mn}} = \delta_{n0} \tag{2.61}$$

holds for any $n \in \mathbb{Z}^d$.

Proof We have the equality

$$\sum_{q\in D(M^*)} m_0(\xi + M^{*-1}q)\overline{\widetilde{m}_0(\xi + M^{*-1}q)} =$$

$$m^{-1} \sum_{q\in D(M^*)} \sum_{k\in\mathbb{Z}^d} \sum_{l\in\mathbb{Z}^d} h_k \overline{\widetilde{h}_l} e^{2\pi i(k-l,\xi)} e^{2\pi i(k-l,M^{*-1}q)} =$$

$$m^{-1} \sum_{n\in\mathbb{Z}^d} \sum_{k\in\mathbb{Z}^d} e^{2\pi i(n,\xi)} h_k \overline{\widetilde{h}_{k-n}} \sum_{q\in D(M^*)} e^{2\pi i(M^{-1}n,q)},$$

where the change of the order of summation is justified by absolute convergence over the $2d$-dimensional index (n, k) of the series on the right-hand side, while the sum over s is finite. From Lemma 2.1.5, it follows that

$$\sum_{q\in D(M^*)} m_0(\xi + M^{*-1}q)\overline{\widetilde{m}_0(\xi + M^{*-1}q)} = \sum_{n\in\mathbb{Z}^d} e^{2\pi i(Mn,\xi)} \sum_{k\in\mathbb{Z}^d} h_k \overline{\widetilde{h}_{k-Mn}} \tag{2.62}$$

for all $\xi \in \mathbb{R}^d$. Thus, condition (2.26) is equivalent to the fact that

$$1 = \sum_{n \in \mathbb{Z}^d} e^{2\pi i (n,\omega)} \left(\sum_{k \in \mathbb{Z}^d} h_k \overline{\widetilde{h}_{k-Mn}} \right), \quad \omega = M^* \xi,$$

for all $\omega \in \mathbb{R}^d$. Hence, we have the Fourier expansion of unity. Taking the Fourier coefficients of this series, we get the required equalities.◊

Due to (2.61), it is very easy to construct polynomial masks $m_0, \widetilde{m}_0$ satisfying (2.26) as well as to provide conditions $m_0(\mathbf{0}) = \widetilde{m}_0(\mathbf{0}) = 1$ additionally. However, all this is not enough for the biorthonormality of systems $\{\varphi_{0k}\}_k$, $\{\widetilde{\varphi}_{0k}\}_k$, which is demonstrated by the following example.

Example 2.7.2 Let $d = 1$, and $\varphi = \widetilde{\varphi} = \chi_{[0,3]}$. Obviously, the function φ is refinable for $M = 2$ and its Fourier transform is equal to $\widehat{\varphi}(\xi) = \frac{e^{2\pi i 3\xi} - 1}{2\pi i \xi}$. For its mask $m_0(\xi) = \frac{1}{2}\left(1 + e^{6\pi i \xi}\right)$, the following identity holds

$$|m_0(\xi)|^2 + |m_0(\xi + 1/2)|^2 = 1.$$

However, it is clear that all integer shifts of the function φ are not orthogonal to each other. Moreover, these integer shifts do not form a Riesz system. Indeed, direct calculation gives that

$$\sum_{l \in \mathbb{Z}} |\widehat{\varphi}(\xi + l)|^2 = 3 + 4 \cos 2\pi \xi + 2 \cos 4\pi \xi.$$

We see that the right-hand side of this equality vanishes for $\xi = 1/3$.◊

Evidently, similar examples exist for any $d > 1$.

Unfortunately, no sufficient conditions that would be simply formulated and easily checked in practice have been found so far. The sufficient conditions described below in various terms do not seem very good in this sense. Still, they can often be verified in some specific cases.

First, we shall prove some auxiliary results. Recall (see Sect. 2.1) that a set $K \subset \mathbb{R}^d$ is congruent modulo $\mathbb{Z}^d$ to a set $L \subset \mathbb{R}^d$ if the set K can be partitioned into a finite number of mutually disjoint measurable subsets K_n, $n = 1, \dots, N$, in such a way that there exist integer vectors $l_1, ..., l_N$ such that $L = \bigcup_{n=1}^{N} (K_n + l_n)$, where $(K_n + l_n) \cap (K_{n_1} + l_{n_1}) = \emptyset$ for $n \neq n_1$.

Lemma 2.7.3 *Let m_0 be a continuous 1-periodic (with respect to each variable) function, $m_0(\mathbf{0}) = 1$ and*

$$F(\xi) := \prod_{j=1}^{\infty} m_0(M^{*-j}\xi).$$

If the function F is continuous and

$$\sum_{l\in\mathbb{Z}^d} |F(\xi+l)| \neq 0 \tag{2.63}$$

for all $\xi \in \mathbb{R}^d$, then there exists a set K congruent to the cube $[-\frac{1}{2}, \frac{1}{2}]^d$ modulo $\mathbb{Z}^d$, containing a neighborhood of the origin and such that

$$\inf_{k\in\mathbb{N}} \inf_{\xi\in K} |m_0(M^{*-k}\xi)| \neq 0. \tag{2.64}$$

Proof Set

$$C_\xi = \sum_{l\in\mathbb{Z}^d} |F(\xi+l)|$$

if the series on the right hand side converges. Otherwise, set $C_\xi = 1$. It follows from (2.63) that for any $\xi \in [-1/2, 1/2]^d$, there exists $l_\xi \in \mathbb{N}$ such that

$$\sum_{\|l\|_\infty \le l_\xi} |F(\xi+l)| \ge \frac{1}{2} C_\xi.$$

The function $|F|$ is continuous, so the finite sum $\sum_{\|l\|_\infty \le l_\xi} |F(\cdot+l)|$ is also continuous. Hence, for any ξ, there exists a neighborhood $N_\xi = \{\theta \in \mathbb{R}^d : |\xi - \theta| < R_\xi\}$ such that

$$\sum_{\|l\|_\infty \le l_\xi} |F(\theta+l)| \ge \frac{1}{4} C_\xi$$

for all $\theta \in N_\xi$. Since $[-1/2, 1/2]^d$ is a compact set and $\{N_\xi\}_{\xi\in\mathbb{R}^d}$ is its open covering, one can select in it a finite subcovering $\{N_\xi\}_{\xi\in\Omega}$. Denote by l_0 the maximal element among l_ξ, $\xi \in \Omega$ and by $4C$ the minimal element among C_ξ, $\xi \in \Omega$. Clearly, for any $\xi \in [-1/2, 1/2]^d$, there exists $l \in \mathbb{Z}^d$, $\|l\|_\infty \le l_0$ such that

$$|F(\xi+l)| \ge \min\left\{1, \frac{C}{(2l_0+1)^d}\right\} =: C_1.$$

Let us renumber the vectors $l \in \mathbb{Z}^d$, $\|l\|_\infty \le l_0$ in such a way that $l(0) = \mathbf{0}$. Thus, we obtain the sequence $l(n)$, $n = 0, ..., (2l_0+1)^d =: L$. Define sets S_n as follows:

$$S_0 = \{\xi \in [-1/2, 1/2]^d : F(\xi) \ge C_1\},$$

$$S_n = \{\xi \in [-1/2, 1/2]^d \setminus \bigcup_{k=0}^{n-1} S_k : |F(\xi + l(n))| \ge C_1\}.$$

It is clear that the sets S_n, $n = 0, ..., L$, are measurable and form a partition of $[-1/2, 1/2]^d$. The function F is continuous and $F(0) = 1$; hence, S_0 contains a neighborhood of the origin. Set

$$K = \bigcup_{n=0}^{L} (S_n + l(n)). \tag{2.65}$$

It is not hard to see that the set K is congruent to the cube $[-1/2, 1/2]^d$. By construction, $F(\xi) \geq C_1$ on K and K contains a neighborhood of the origin.

Let us show that K satisfies (2.64). Let $K \subset [-R, R]^d$. The function m_0 is continuous; hence, relations (2.46), (2.1) imply that

$$|m_0(M^{*-k}\xi)| > \frac{1}{2}$$

for all $\xi \in [-R, R]^d$ for $k > k_0$. Thus, it suffices to show that

$$\inf_{\xi \in K} |m_0(M^{*-k}\xi)| \neq 0$$

for a finite number of k, $1 \leq k \leq k_0$. For $\xi \in K$, we have

$$F(\xi) = \left(\prod_{k=1}^{k_0} m_0(M^{*-k}\xi) \right) F(M^{*-k_0}\xi).$$

The left-hand side of this equality is bounded away from zero, and the second factor on the right-hand side is bounded since the function F is continuous and thus is bounded on the compact set $\overline{K}$. It follows that

$$\prod_{k=1}^{k_0} |m_0(M^{*-k}\xi)| \geq C_2 > 0$$

for any $\xi \in K$. Since the function m_0 is continuous, each factor in this product is bounded away from zero, i.e.,

$$|m_0(M^{*-k}\xi)| \geq C_3 > 0$$

for all $k = 1, \ldots, k_0$ and any $\xi \in K$, which completes the proof.◊

Lemma 2.7.4 *Suppose m_0 satisfies the conditions of Proposition 2.6.1 and there exists a set K congruent to the cube $[-1/2, 1/2]^d$ modulo $\mathbb{Z}^d$ containing a neighborhood of the origin and such that*

$$C := \inf_{k \in \mathbb{N}} \inf_{\xi \in K} |m_0(M^{*-k}\xi)| \neq 0. \tag{2.66}$$

Let

$$F_n(\xi) = \prod_{j=1}^{n} m_0(M^{*-j}\xi)\chi_K(M^{*-n}\xi), \quad F(\xi) = \prod_{j=1}^{\infty} m_0(M^{*-j}\xi).$$

If $F \in L(\mathbb{R}^d)$, then F_n converges to F in $L(\mathbb{R}^d)$, and if $F \in L_2(\mathbb{R}^d)$, then F_n converges to F in $L_2(\mathbb{R}^d)$.

Proof First of all, we note that by Proposition 2.6.1, the infinite product defining the function F converges at every point. Using the inequality $e^{-2x} \le 1 - x$, which is true for $0 \le x \le 1/2$, and relations (2.66), (2.48), for sufficiently large N (which we fix) and any $\xi \in K$, we get

$$|F(\xi)| \ge C^N \prod_{j=N+1}^{\infty} \left(1 - |1 - m_0(M^{*-j}\xi)|\right) \ge$$

$$C^N \exp\left(-2C_1 \max_{\xi \in K} |\xi| \sum_{k=N+1}^{\infty} \|M^{-k}\| \right) =: C_2 > 0.$$

This can be reformulated as follows:

$$\chi_K(\xi) \le |F(\xi)|/C_2 \tag{2.67}$$

for all $\xi \in \mathbb{R}^d$. The compact set K contains a neighborhood of the origin; hence, (2.1) implies that $\chi_K(M^{*-n}\xi) \to 1$ as $n \to +\infty$ for any $\xi \in \mathbb{R}^d$. Combining this with Proposition (2.6.1), we conclude that $F_n \to F$ converges pointwise on $\mathbb{R}^d$. From inequality (2.67), we get

$$|F_n(\xi)| \le \frac{1}{C_2} \prod_{j=1}^{n} |m_0(M^{*-j}\xi)||F(M^{*-n}\xi)| = \frac{1}{C_2}|F(\xi)|$$

for all $\xi \in \mathbb{R}^d$. Thus, if $F \in L(\mathbb{R}^d)$, the functions F_n have a summable majorant. By Lebesgue's dominated convergence theorem, it follows that F_n converges to F in $L(\mathbb{R}^d)$. If $F \in L_2(\mathbb{R}^d)$, then we get

$$|F_n(\xi) - F(\xi)|^2 \le |F_n(\xi)|^2 + 2|F_n(\xi)|\,|F(\xi)| + |F(\xi)|^2 \le$$

$$\left(\frac{1}{C_2^2} + \frac{2}{C_2} + 1\right) |F(\xi)|^2 =: C_4|F(\xi)|^2.$$

The function $|F_n - F|^2$ converges pointwise to zero and has the summable majorant $C_4|F(\xi)|^2$; therefore, $F_n \to F$ in $L_2(\mathbb{R}^d)$.$\Diamond$

Theorem 2.7.5 (Cohen's criterion) *Let $\varphi, \widetilde{\varphi} \in L_2(\mathbb{R}^d)$ be refinable functions with masks $m_0, \widetilde{m}_0$, respectively, and let these masks satisfy the conditions of*

Proposition 2.6.1 and (2.26). A sufficient condition for the biorthonormality of integer shifts of the functions φ, $\widetilde{\varphi}$ is the existence of a set K congruent to the cube $[-\frac{1}{2}, \frac{1}{2}]^d$ modulo $\mathbb{Z}^d$, containing a neighborhood of the origin and such that

$$\inf_{k\in\mathbb{N}}\inf_{\xi\in K}|m_0(M^{*-k}\xi)| \neq 0, \quad \inf_{k\in\mathbb{N}}\inf_{\xi\in K}|\widetilde{m}_0(M^{*-k}\xi)| \neq 0. \tag{2.68}$$

If, in addition,

$$\varphi(x), \widetilde{\varphi}(x) = O\left(|x|^{-(d+\varepsilon)}\right), \quad \varepsilon > 0, \quad |x| \to \infty, \tag{2.69}$$

then this condition is necessary and sufficient.

Remark 2.7.6 In practice, to verify that condition (2.68) holds, it suffices to find the minimal value of a finite number of functions on $\overline{K}$. Indeed, the compactness of $\overline{K}$, continuity of m_0 and $\widetilde{m}_0$, equalities $m_0(\mathbf{0}) = \widetilde{m}_0(\mathbf{0}) = 1$, and condition (2.1) imply that $|m_0(M^{*-k}\xi)\widetilde{m}_0(M^{*-k}\xi)| > 1/2$ for all $\xi \in [-R, R]^d$ for $k > k_0$. Therefore, it remains to show that $2k_0$ functions

$$m_0(M^{*-1}\xi),\ \widetilde{m}_0(M^{*-1}\xi), ..., m_0(M^{*-k_0}\xi),\ \widetilde{m}_0(M^{*-k_0}\xi)$$

have no zeros on $\overline{K}$. The set $\overline{K}$ is called Cohen's compact set.

Proof of Theorem 2.7.5. Sufficiency. Suppose that

$$m_0(\xi) = m^{-1/2}\sum_{k\in\mathbb{Z}^d} h_k e^{2\pi i(k,\xi)}, \quad \widetilde{m}_0(\xi) = m^{-1/2}\sum_{k\in\mathbb{Z}^d} \widetilde{h}_k e^{2\pi i(k,\xi)}$$

with $h_k = O(|k|^{-d-1-\varepsilon}), \widetilde{h}_k = O(|k|^{-d-1-\varepsilon}), \varepsilon > 0$. Denote $m_0^*(\xi) := m_0(\xi)\overline{\widetilde{m}_0(\xi)}$. The rate of decay of the Fourier coefficients of m_0^* is the same as for m_0 and $\widetilde{m}_0$. Indeed, since

$$m_0^*(\xi) = m^{-1}\sum_{k\in\mathbb{Z}^d}\sum_{l\in\mathbb{Z}^d} h_k\overline{\widetilde{h}_l}e^{2\pi i(k-l,\xi)} =$$

$$m^{-1}\sum_{n\in\mathbb{Z}^d}\sum_{k\in\mathbb{Z}^d} h_k\overline{\widetilde{h}_{k-n}}e^{2\pi i(n,\xi)} = m^{-1}\sum_{n\in\mathbb{Z}^d}(h*\widetilde{h})_n e^{2\pi i(n,\xi)},$$

where $(h*\widetilde{h})_n$ is the nth element of the convolution sequence $h*\widetilde{h}$, $h = \{h_k\}_{k\in\mathbb{Z}^d}$, $\widetilde{h} = \{\widetilde{h}_k\}_{k\in\mathbb{Z}^d}$, and then, the rate of decay of $(h*\widetilde{h})_n$ as $n\to\infty$ can be computed as follows. Let us consider $|n| > 2R$ for some $R > 0$. Then, either $|k| > R$ or $|n-k| > R$. Therefore,

$$(h*\widetilde{h})_n = \sum_{|k|>R} h_k\overline{\widetilde{h}_{k-n}} + \sum_{|n-k|>R} h_k\overline{\widetilde{h}_{k-n}} \leq CR^{-d-1-\varepsilon}.$$

Thus, we can apply Lemma 2.7.4 to the function $m_0^*(\xi)$ and deduce that the sequence

$$\mu_k(\xi) := \prod_{j=1}^{k} m_0(M^{*-j}\xi)\overline{\widetilde{m}_0(M^{*-j}\xi)}\, \chi_K(M^{*-k}\xi), \quad k \in \mathbb{N}, \tag{2.70}$$

converges to $\widehat{\varphi}\cdot\overline{\widehat{\widetilde{\varphi}}}$ in $L(\mathbb{R}^d)$. Since the set K is congruent to the cube $[-\frac{1}{2}, \frac{1}{2}]^d$ modulo $\mathbb{Z}^d$, we get

$$\int_K f = \int_{[-1/2,1/2]^d} f = \int_{[0,1]^d} f$$

for any 1-periodic in each variable function f. In particular, using (2.70), for any $n \in \mathbb{Z}^d$, we obtain

$$\int_{\mathbb{R}^d} \mu_k(\xi)e^{2\pi i(n,\xi)}d\xi = m^k \int_{\mathbb{R}^d} \mu_k(M^{*k}\theta)e^{2\pi i(n,M^{*k}\theta)}d\theta =$$

$$= m^k \int_K \prod_{j=1}^{k} m_0(M^{*k-j}\theta)\overline{\widetilde{m}_0(M^{*k-j}\theta)}e^{2\pi i(n,M^{*k}\theta)}d\theta =$$

$$= m^k \int_{[0,1]^d} \prod_{l=1}^{k-1} m_0(M^{*l}\theta)\overline{\widetilde{m}_0(M^{*l}\theta)}e^{2\pi i(n,M^{*k}\theta)}m_0(\theta)\overline{\widetilde{m}_0(\theta)}d\theta.$$

Under the integral sign on the right-hand side, we have a function 1-periodic in each variable; hence, using Lemma 2.1.4 for $A = M^*$, we represent this integral in the form

$$m^k \sum_{r\in D(M^*)} \int_{Q_1+M^{*-1}r} \prod_{l=1}^{k-1} m_0(M^{*l}\theta)\overline{\widetilde{m}_0(M^{*l}\theta)}e^{2\pi i(n,M^{*k}\theta)}m_0(\theta)\overline{\widetilde{m}_0(\theta)}d\theta,$$

where $Q_1 = M^{*-1}[0,1)^d$. Making the change of variables $\theta = \xi + M^{*-1}r$ in each summand, putting the sum under the integral sign, and using (2.26), we get

$$m^k \int_{Q_1} \prod_{l=1}^{k-1} m_0(M^{*l}\xi)\overline{\widetilde{m}_0(M^{*l}\xi)}e^{2\pi i(n,M^{*k}\xi)}d\xi.$$

The change of variables $\theta = M^*\xi$ in the integral and the change of index $l-1 \to l$ in the product give

$$m^{k-1} \int_{[0,1)^d} \prod_{l=0}^{k-2} m_0(M^{*l}\theta)\overline{\widetilde{m}_0(M^{*l}\theta)}e^{2\pi i(n,M^{*k-1}\theta)}d\theta =$$

$$= \int_{\mathbb{R}^d} \mu_{k-1}(\theta) e^{2\pi i (n,\theta)} d\theta.$$

Thus, for any $k,\ n \in \mathbb{N}$, we have

$$\int_{\mathbb{R}^d} \mu_k(\theta) e^{2\pi i (n,\theta)} d\theta = \int_{\mathbb{R}^d} \mu_{k-1}(\theta) e^{2\pi i (n,\theta)} d\theta.$$

Constructing an analogous chain of equalities for $k = 1$, we obtain

$$\int_{\mathbb{R}^d} \mu_1(\xi) e^{2\pi i (n,\xi)} d\xi = \dots = \int_{[0,1)^d} e^{2\pi i (n,\theta)} d\theta = \delta_{n0} \ \forall n \in \mathbb{N},$$

and therefore,

$$\int_{\mathbb{R}^d} \widehat{\varphi}(\xi)\, \overline{\widehat{\widetilde{\varphi}}(\xi)} e^{2\pi i (n,\xi)} d\xi = \lim_{k\to\infty} \int_{\mathbb{R}^d} \mu_k(\xi) e^{2\pi i (n,\xi)} d\xi = \delta_{n0},$$

which, by the Plancherel theorem, yields the equality

$$\int_{\mathbb{R}^d} \varphi(x)\, \overline{\widetilde{\varphi}(x-n)}\, dx = \delta_{n0}.$$

Necessity. Assume that (2.69) holds and prove that the function

$$\sum_{l\in\mathbb{Z}^d} \widehat{\varphi}(\xi+l) \overline{\widehat{\widetilde{\varphi}}(\xi+l)} =: \Phi(\xi) \tag{2.71}$$

is continuous on $\mathbb{R}^d$. Set

$$G(t) = \sum_{k\in\mathbb{Z}^d} \varphi(t+k) e^{-2\pi i (t+k,\xi)}, \quad \widetilde{G}(t) = \sum_{k\in\mathbb{Z}^d} \widetilde{\varphi}(t+k) e^{-2\pi i (t+k,\xi)}.$$

By the Poisson's summation formula, we have

$$\widehat{G}(l) = \widehat{\varphi}(\xi+l), \quad \widehat{\widetilde{G}}(l) = \widehat{\widetilde{\varphi}}(\xi+l)$$

for any $l \in \mathbb{Z}^d$. Using the generalized Parseval's equality, we get

$$\Phi(\xi) = \sum_{l\in\mathbb{Z}^d} \widehat{G}(l)\overline{\widehat{\widetilde{G}}(l)} = \int_{[0,1]^d} G(t)\overline{\widetilde{G}(t)}\,dt =$$

$$\int_{[0,1]^d} \sum_{k\in\mathbb{Z}^d} \varphi(t+k)e^{-2\pi i(t+k,\xi)}\overline{\widetilde{G}(t)}\,dt = \int_{\mathbb{R}^d} \varphi(t)e^{-2\pi i(t,\xi)}\overline{\widetilde{G}(t)}\,dt =$$

$$\int_{\mathbb{R}^d} \varphi(t) \sum_{k\in\mathbb{Z}^d} \overline{\widetilde{\varphi}(t+k)}e^{-2\pi i(k,\xi)}\,dt.$$

Since in view of condition (2.69),

$$\sum_{k\in\mathbb{Z}^d}\int_{\mathbb{R}^d} |\varphi(t)\widetilde{\varphi}(t+k)|\,dt = \int_{\mathbb{R}^d} |\varphi(t)| \sum_{k\in\mathbb{Z}^d} |\widetilde{\varphi}(t+k)|\,dt < \infty,$$

interchanging summation and integration represents $\Phi(\xi)$ as the sum of absolutely convergent trigonometric series, which is continuous.

The continuity of the function $[\widehat{\varphi}, \widehat{\widetilde{\varphi}}]$ and Proposition 2.2.2 implies that the equality $[\widehat{\varphi}, \widehat{\widetilde{\varphi}}](\xi) = 1$ holds for all $\xi \in \mathbb{R}^d$, so $\widehat{\varphi}\cdot\widehat{\widetilde{\varphi}}$ satisfies all the conditions of Lemma 2.7.3. Thus, $\inf\limits_{k\in\mathbb{N}} \inf\limits_{\xi\in K} |m_0(M^{*-k}\xi)||\widetilde{m}_0(M^{*-k}\xi)| \neq 0$, which is equivalent to (2.68). ♢

Corollary 2.7.7 *Suppose $\varphi, \widetilde{\varphi} \in L_2(\mathbb{R}^d)$ are refinable functions with masks m_0, $\widetilde{m}_0$, respectively, satisfying (2.26) and the conditions of Proposition 2.6.1. If m_0 and $\widetilde{m}_0$ do not vanish on the set $\bigcup\limits_{k=1}^{\infty} M^{*-k}[-1/2, 1/2]^d$, then the integer shifts of the functions φ, $\widetilde{\varphi}$ form a biorthonormal system.*

The assumptions of this corollary are a special case of the assumptions of Theorem 2.7.5, where the cube $[-1/2, 1/2]^d$ is taken as a set K.

In view of (2.1), there exists k_0 such that the relation $M^{*-k}[-1/2, 1/2]^d \subset M^{*-1}[-1/2, 1/2]^d$ holds for any $k > k_0$; hence, under the conditions of Corollary 2.7.7, it suffices to check that the functions m_0 and $\widetilde{m}_0$ have no zeros on the sets $M^{*-k}[-1/2, 1/2]^d$, $k = 1, \dots, k_0$. If $M^{*-k}[-1/2, 1/2]^d \subset M^{*-k+1}[-1/2, 1/2]^d$, then it is sufficient that m_0 and $\widetilde{m}_0$ have no zeros on $M^{*-1}[-1/2, 1/2]^d$.

Example 2.7.8 Let $d = 2$, $M = \begin{pmatrix} 1 & 1 \\ 1 & -1 \end{pmatrix}$,

$$m_0(\xi) = \frac{1}{2} + \frac{1}{4}\cos 2\pi\xi_1 + \frac{1}{4}\cos 2\pi\xi_2,$$

$$\widetilde{m}_0(\xi) = 1 + \frac{1}{4}(\cos 2\pi\xi_1 + \cos 2\pi\xi_2) -$$

$$\frac{1}{8}(\cos^2 2\pi\xi_1 + \cos^2 2\pi\xi_2) - \frac{1}{4}\cos 2\pi\xi_1 \cos 2\pi\xi_2.$$

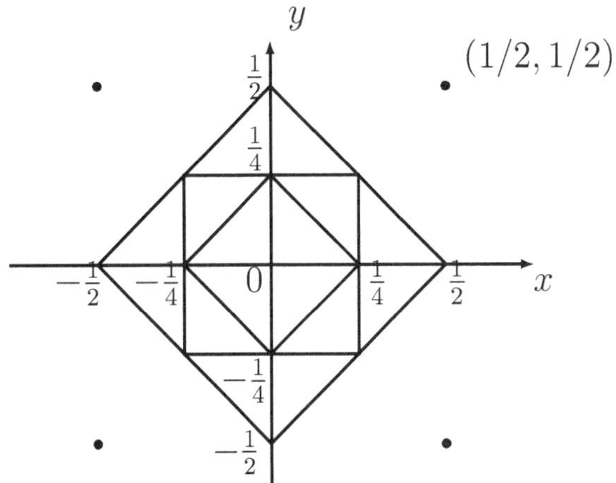

Fig. 2.2 Sets $M^{*-k}[-1/2, 1/2]^2$ for $k = 1, 2, 3$

The coefficients of the polynomials m_0 and $\widetilde{m}_0$ in the matrix form are the following:

$$\begin{pmatrix} 0 & 1/8 & 0 \\ 1/8 & 1/2 & 1/8 \\ 0 & 1/8 & 0 \end{pmatrix}, \quad \begin{pmatrix} 0 & 0 & -1/32 & 0 & 0 \\ 0 & -1/16 & 1/8 & -1/16 & 0 \\ -1/32 & 1/8 & 7/8 & 1/8 & -1/32 \\ 0 & -1/16 & 1/8 & -1/16 & 0 \\ 0 & 0 & -1/32 & 0 & 0 \end{pmatrix}$$

It is not hard to verify that (2.46) and (2.61) (and thus (2.26)) hold. Let us show that these masks satisfy the sufficient condition of biorthogonality of Corollary 2.7.7. Indeed, the polynomials m_0 and $\widetilde{m}_0$ vanish only at the point $(1/2, 1/2)^T$ on $\mathbb{T}^d$, since the equalities $m_0(\xi) = 0, \widetilde{m}_0(\xi) = 0$ hold only if all cosines in them are equal to -1. Hence, no zero of the functions $m_0, \widetilde{m}_0$ belongs to the set $M^{-1}[-1/2, 1/2]^d$, and it is easily seen that the inclusion $M^{-k}[-1/2, 1/2]^d \subset M^{-1}[-1/2, 1/2]^d$ takes place for all $k \geq 2$ (see Fig. 2.2).$\Diamond$

Consider two refinable functions $\varphi, \widetilde{\varphi} \in L(\mathbb{R}^d)$ and set

$$\varphi^*(x) := \int_{\mathbb{R}^d} \varphi(t)\overline{\widetilde{\varphi}(x-t)}dt, \quad x \in \mathbb{R}^d, \tag{2.72}$$

$$h_k^* := \sum_{n \in \mathbb{Z}^d} h_n \overline{\widetilde{h}_{n-k}}, \quad k \in \mathbb{Z}^d. \tag{2.73}$$

The Plancherel theorem implies the following equality:

$$\varphi^*(x) = \int_{\mathbb{R}^d} \widehat{\varphi}(\xi)\overline{\widehat{\widetilde{\varphi}}(\xi)}e^{-2\pi i(\xi,x)}\,d\xi,$$

from which it is easy to see that the function φ^* is continuous on $\mathbb{R}^d$. Moreover, by the condition $\varphi, \widetilde{\varphi} \in L_2(\mathbb{R}^d)$, the function φ^* is bounded, and $\varphi^* \in L(\mathbb{R}^d)$ by the condition $\varphi, \widetilde{\varphi} \in L(\mathbb{R}^d)$. Hence, $\varphi^* \in L_2(\mathbb{R}^d)$. Let us find the Fourier transform of this function. Since the function under the integral sign in

$$\int\limits_{\mathbb{R}^d}\int\limits_{\mathbb{R}^d} \varphi(t)\overline{\widetilde{\varphi}(x-t)}e^{-2\pi i(\xi,x)}\,dt\,dx$$

is summable on $\mathbb{R}^d \times \mathbb{R}^d$, changing the order of integration, we get

$$\widehat{\varphi^*}(\xi) = \widehat{\varphi}(\xi)\overline{\widehat{\widetilde{\varphi}}(-\xi)}.$$

Applying equality (2.17) to both factors on the right-hand side, we obtain

$$\widehat{\varphi^*}(\xi) = m_0(M^{*-1}\xi)\overline{\widetilde{m}_0(-M^{*-1}\xi)}\widehat{\varphi}(M^{*-1}\xi)\overline{\widehat{\widetilde{\varphi}}(-M^{*-1}\xi)} =$$
$$m_0(M^{*-1}\xi)\overline{\widetilde{m}_0(-M^{*-1}\xi)}\widehat{\varphi^*}(M^{*-1}\xi).$$

Thus, φ^* is a refinable function with mask

$$m_0(\xi)\overline{\widetilde{m}_0(-\xi)} =: m_0^*(\xi).$$

It is easy to verify that

$$m_0^*(\xi) = m^{-1}\sum_{k\in\mathbb{Z}^d} h_k^* e^{2\pi i(k,\xi)}.$$

Lemma 2.7.9 *Suppose $\varphi, \widetilde{\varphi} \in L_2(\mathbb{R}^d)$ are refinable functions with masks $m_0, \widetilde{m_0}$, respectively, the Fourier series of these masks are absolutely convergent, and the function φ^* is given by equality (2.72). Then, for any $\xi \in \mathbb{R}^d$, we have the following equality:*

$$\varphi^*(\xi) = \sum_{n\in\mathbb{Z}^d} h_n^*\varphi^*(M\xi + n). \tag{2.74}$$

Proof The absolute convergency of the Fourier series of the functions $m_0, \widetilde{m}_0$ and relation (2.73) implies that

$$\sum_{n\in\mathbb{Z}^d} |h_n^*| < \infty. \tag{2.75}$$

Since the function φ^* is bounded, we can state that the series on the right-hand side of (2.74) converges uniformly. Hence, in view of the continuity of φ^*, its sum is continuous. On the other hand, this series converges to the function φ^* in $L_2(\mathbb{R}^d)$ since relation (2.74) is a refinement equation for φ^*, which implies that (2.74) holds at each point. ◇

Introduce the operator W_{φ^*}, which takes $\ell_1(\mathbb{Z}^d)$ to $\ell_1(\mathbb{Z}^d)$:

$$(W_{\varphi^*}c)_k := \sum_{n\in\mathbb{Z}^d} h_n^* c_{n+Mk}, \quad c = \{c_n\}_{n\in\mathbb{Z}^d} \in \ell_1(\mathbb{Z}^d). \tag{2.76}$$

Theorem 2.7.10 (Lawton's criterion) *Let $\varphi, \widetilde{\varphi} \in L_2(\mathbb{R}^d)$ be refinable functions with masks $m_0, \widetilde{m}_0$, respectively. Suppose the Fourier series of these masks converge absolutely and satisfy (2.46), and the function φ^* is defined by equality (2.72). A necessary condition for integer shifts of the functions φ, $\widetilde{\varphi}$ to be biorthonormal is that the sequence $\delta := \{\delta_{k0}\}_{k\in\mathbb{Z}^d}$ be the only eigenvector of the operator W_{φ^*} corresponding to eigenvalue 1 in the space $\ell_1(\mathbb{Z}^d)$. If, in addition,*

$$\varphi(x), \widetilde{\varphi}(x) = O\left(|x|^{-d-\varepsilon}\right), \quad \varepsilon > 0, \quad |x| \to \infty, \tag{2.77}$$

then the above necessary condition is sufficient.

Proof Sufficiency. Let δ be the only eigenvector of the operator W_{φ^*} corresponding to the eigenvalue 1 in the space $\ell_1(\mathbb{Z}^d)$. By Lemma 2.7.9,

$$\varphi^*(k) = \sum_{n\in\mathbb{Z}^d} h_n^* \varphi^*(Mk+n)$$

for any $k \in \mathbb{Z}^d$, i.e., the sequence $\{\varphi^*(k)\}_{k\in\mathbb{Z}^d}$ is an eigenvector of the operator W_{φ^*} corresponding to the eigenvalue 1. Let us show that

$$\sum_{n\in\mathbb{Z}^d} |\varphi^*(n)| = \sum_{n\in\mathbb{Z}^d} \left| \int_{\mathbb{R}^d} \varphi(x)\overline{\widetilde{\varphi}}(x+n)\,dx \right| < \infty. \tag{2.78}$$

To do this, it suffices to verify the convergence of the series

$$\sum_{k\in\mathbb{Z}^d} \sum_{n\in\mathbb{Z}^d} \int_{\mathbb{T}^d} |\varphi(x+k)\widetilde{\varphi}(x+k+n)|\,dx. \tag{2.79}$$

From (2.77), it follows that there exist $C > 0$ and $R > 1$ such that

$$|\varphi(x)|, |\widetilde{\varphi}(x)| \le \frac{C}{|x|^{d+\varepsilon}}$$

for $|x| \ge R$. Therefore,

$$\sum_{|k|\ge R+1} \sum_{n\in\mathbb{Z}^d} \int_{\mathbb{T}^d} |\varphi(x+k)\widetilde{\varphi}(x+k+n)|\,dx \le \sum_{|k|\ge R+1} \frac{2^{d+\varepsilon}C}{|k|^{d+\varepsilon}} \|\widetilde{\varphi}\|_1 < \infty.$$

To prove the convergency of series (2.79), it remains to note that for any fixed $k \in \mathbb{Z}^d$,

$$\sum_{|n+k|\ge R+1} \int_{\mathbb{T}^d} |\varphi(x+k)\widetilde{\varphi}(x+k+n)|\,dx \le \sum_{|n+k|\ge R+1} \frac{2^{d+\varepsilon}C}{|n+k|^{d+\varepsilon}} \|\varphi\|_1 < \infty.$$

Relation (2.78) means that $\{\varphi^*(k)\}_{k\in\mathbb{Z}^d} \in \ell_1(\mathbb{Z}^d)$. Consequently, $\varphi^*(k) = \delta_{k0}$ for any $k \in \mathbb{Z}^d$, which means that integer shifts of the functions φ, $\widetilde{\varphi}$ are biorthonormal. Necessity. Suppose $\varphi^*(k) = \delta_{k0}$ for any $k \in \mathbb{Z}^d$ and $c = \{c_k\}_{k\in\mathbb{Z}^d} \in \ell_1(\mathbb{Z}^d)$ is an eigenvector of the operator W_{φ^*} corresponding to the eigenvalue 1. Let us show that the equality

$$\sum_{k\in\mathbb{Z}^d} c_{-k}\varphi^*(r + M^{-N}k) = c_r. \tag{2.80}$$

holds for any positive integer N and any $r \in \mathbb{Z}^d$. The proof is by induction. First, take $N = 1$. Apply (2.74) to the left-hand side of (2.80), which gives

$$\sum_{k\in\mathbb{Z}^d} c_{-k} \sum_{n\in\mathbb{Z}^d} h_n^*\varphi^*(Mr + k + n) = \sum_{k\in\mathbb{Z}^d} c_{-k}h^*_{-k-Mr} = (W_{\varphi^*}c)_r = c_r.$$

Now, take $N+1$ in place of N in (2.80). Applying (2.74), we get

$$\sum_{k\in\mathbb{Z}^d} c_{-k}\varphi^*(r + M^{-N-1}k) = \sum_{k\in\mathbb{Z}^d} c_{-k} \sum_{n\in\mathbb{Z}^d} h_n^*\varphi^*(Mr + M^{-N}k + n).$$

Changing the order of summation on the right-hand side (which is possible since the double series converges absolutely in view of relation (2.75) and boundedness of φ^*), and using the induction hypothesis, we obtain

$$\sum_{k\in\mathbb{Z}^d} c_{-k}\varphi^*(r + M^{-N-1}k) = \sum_{n\in\mathbb{Z}^d} h_n^* c_{Mr+n} = (W_{\varphi^*}c)_r = c_r.$$

It follows from the absolute convergence of the series $\sum_{k\in\mathbb{Z}^d} c_k$ and boundedness of φ^* that

$$\lim_{N\to+\infty} \sum_{k\in\mathbb{Z}^d} c_{-k}\varphi^*(r + M^{-N-1}k) = \sum_{k\in\mathbb{Z}^d} c_{-k} \lim_{N\to+\infty} \varphi^*(r + M^{-N-1}k).$$

Passing to the limit in (2.80) and taking into account (2.1), we get

$$c_r = \varphi^*(r)\sum_{k\in\mathbb{Z}^d} c_{-k} = \delta_{r0}\sum_{k\in\mathbb{Z}^d} c_k,$$

which obviously implies that $c_r = \delta_{r0}$ for any $r \in \mathbb{Z}^d$. It remains to note that Propositions 2.4.1, 2.7.1 and equality (2.73) imply that δ is an eigenvector of the operator W_{φ^*} corresponding to the eigenvalue 1. ◊

In practice, Theorem 2.7.10 is better suited for disproving the orthogonality or biorthogonality conditions (though even by this theorem, it is usually rather hard to do). As an example, we check the absence of orthogonality for the mask considered in Example 2.7.2.

Example 2.7.11 Let $d = 1$, $M = 2$, $m_0(\xi) = \frac{1}{2}\left(1 + e^{6\pi i\xi}\right)$. Let us show that this mask does not generate a refinable function φ, whose integer shifts form an orthonormal system. The mask m^* of the function φ^* has three nonzero elements $h_0^* = 1$, $h_3^* = 1/2$, $h_{-3}^* = 1/2$; hence, the eigenvectors of the operator W_{φ^*} corresponding to eigenvalue 1 can be found from the system

$$c_k = c_{2k} + \frac{1}{2}c_{2k+3} + \frac{1}{2}c_{2k-3}, \quad k \in \mathbb{Z}.$$

It is clear that δ is a solution of this system. Consider the vector c with only four nonzero components $c_1 = c_{-1} = 1$, $c_2 = c_{-2} = 1/2$. These components appear only in the equations

$$\begin{cases} c_1 &= c_2 + \frac{1}{2}c_{-1} + \frac{1}{2}c_5 \\ c_2 &= c_4 + \frac{1}{2}c_1 + \frac{1}{2}c_7 \\ c_{-1} &= c_{-2} + \frac{1}{2}c_1 + \frac{1}{2}c_{-5} \\ c_{-2} &= c_{-4} + \frac{1}{2}c_{-1} + \frac{1}{2}c_{-7} \end{cases}$$

which turn into equalities. Thus, we have found another eigenvector of the operator W_{φ^*} corresponding to the eigenvalue 1 in $\ell_1(\mathbb{Z}^d)$. ◊

References

1. Sadovnichii, V.A.: Theory of operators. Roger Cooke Publisher, New York, Consultants Bureau (1991)
2. Novikov, I.Y., Protasov, V.Y., Skopina, M.A.: Wavelet Theory, vol. 239. AMS, Providence, RI, Translations Mathematical Monographs (2011)
3. Lam, T.Y.: Serre's conjecture, I in "Lecture Notes in Mathematics", vol. 635. Springer-Verlag, New York (1978)
4. Suslin, A.A.: On the structure of the special linear group over polynomial rings. Math. USSR—Izvestija **11**(2), 221–238 (1977)
5. Park, H., Woodburn, C.: An algorithmic proof of Suslin's stability theorem for polynomial rings. J. Algebra **178**, 277–298 (1995)
6. Amidou, M., Yengui, I.: An algorithm for unimodular completion over Laurent polynomial rings. Linear Algebra Appl. **429**(7), 1687–1698 (2008)

Chapter 3
Construction of Wavelet Frames Generated by MEP

Abstract Sufficient conditions for a dual wavelet system to be a dual wavelet frame are studied. Algorithmic methods for the construction of tight and dual compactly supported wavelet frames, providing an arbitrary approximation order and other important features, are discussed.

3.1 Sufficient Frame Conditions

In this section, we consider dual wavelet systems generated from refinable functions $\varphi, \widetilde{\varphi}$ by MEP. Let $m_0, \widetilde{m}_0$ be the masks of $\varphi, \widetilde{\varphi}$, respectively. We assume that $\widehat{\varphi}, \widehat{\widetilde{\varphi}}$ are continuous at the origin and $\widehat{\varphi}(\mathbf{0}) \cdot \widehat{\widetilde{\varphi}}(\mathbf{0}) = 1$. Recall (see Sect. 2.4) that wavelet systems $\{\psi_{jk}^{(\nu)}\}_{j,k,\nu}, \{\widetilde{\psi}_{jk}^{(\nu)}\}_{j,k,\nu}$ are generated from $\varphi, \widetilde{\varphi}$ by MEP if there exist functions (wavelet masks) $m_\nu, \widetilde{m}_\nu \in L_2(\mathbb{T}^d)$, $\nu = 1, \dots, r$, $r \geq m - 1$, such that the columns of the $(r+1) \times m$ matrices

$$\{m_\nu(\xi + M^{*-1}q_k)\}_{\nu,k}, \quad \{\widetilde{m}_\nu(\xi + M^{*-1}q_k)\}_{\nu,k}$$

are biorthonormal (here $\{q_0, \dots, q_{m-1}\}$ is an arbitrary collection of digits of the matrix M^*). The wavelet functions $\psi^{(\nu)}, \widetilde{\psi}^{(\nu)}$, $\nu = 1, \dots, r$, are defined by

$$\widehat{\psi^{(\nu)}}(\xi) = m_\nu(M^{*-1}\xi)\widehat{\varphi}(M^{*-1}\xi), \quad \widehat{\widetilde{\psi}^{(\nu)}}(\xi) = \widetilde{m}_\nu(M^{*-1}\xi)\widehat{\widetilde{\varphi}}(M^{*-1}\xi). \tag{3.1}$$

Theorem 2.4.6 states that each of the systems $\{\psi_{jk}^{(\nu)}\}_{j,k,\nu}, \{\widetilde{\psi}_{jk}^{(\nu)}\}_{j,k,\nu}$ is a frame whenever these systems and the systems $\{\varphi_{0k}\}_k, \{\widetilde{\varphi}_{0k}\}_k$ are Bessel. Now we are interested under what conditions on $\varphi, \widetilde{\varphi}$ and wavelet masks $m_\nu, \widetilde{m}_\nu$ this is satisfied.

Lemma 3.1.1 *Let the Fourier transform of a function $\psi \in L_2(\mathbb{R}^d)$ satisfies the conditions*

$$|\widehat{\psi}(\xi)| \leq C_0|\xi|, \tag{3.2}$$
$$|\widehat{\psi}(\xi)| \leq C_1(1 + |\xi|)^{-d/2-\varepsilon}, \quad \varepsilon > 0, \tag{3.3}$$

for almost all $\xi \in \mathbb{R}^d$. Then there exists $C > 0$ such that

A. Krivoshein et al., *Multivariate Wavelet Frames*,
Industrial and Applied Mathematics, DOI 10.1007/978-981-10-3205-9_3

$$\sum_{j\in\mathbb{Z}}\sum_{k\in\mathbb{Z}^d} |\langle f, \psi_{jk}\rangle|^2 \le C\|f\|^2 \tag{3.4}$$

for any function f in $L_2(\mathbb{R}^d)$.

Proof Let us fix $\varepsilon > 0$ and $\delta \in (0, 2\varepsilon/(d+2\varepsilon)$. It follows from (3.3) that $[\widehat{\psi}, \widehat{\psi}]$ is essentially bounded on $\mathbb{R}^d$. Using Lemma 2.2.4 and the Cauchy–Schwarz inequality, we obtain

$$\sum_{k\in\mathbb{Z}^d} \left|\langle f, \psi_{jk}\rangle\right|^2 = m^j \int_{\mathbb{T}^d} \left|[\widehat{f}(M^{*j}\cdot), \widehat{\psi}](\xi)\right|^2 d\xi =$$

$$\int_{M^{*j}\mathbb{T}^d} \left|\sum_{l\in\mathbb{Z}^d} \widehat{f}(\xi + M^{*j}l)\overline{\widehat{\psi}(M^{*-j}\xi + l)}\right|^2 d\xi \le$$

$$\le \int_{M^{*j}\mathbb{T}^d} \left(\sum_{l\in\mathbb{Z}^d} |\widehat{f}(\xi + M^{*j}l)|^2 |\widehat{\psi}(M^{*-j}\xi + l)|^{2\delta}\right) \times$$

$$\left(\sum_{k\in\mathbb{Z}^d} |\widehat{\psi}(M^{*-j}\xi + k)|^{2(1-\delta)}\right) d\xi =$$

$$\int_{\mathbb{R}^d} |\widehat{f}(\xi)|^2 |\widehat{\psi}(M^{*-j}\xi)|^{2\delta} \sum_{k\in\mathbb{Z}^d} |\widehat{\psi}(M^{*-j}\xi + k)|^{2(1-\delta)} d\xi.$$

Changing the order of summation and integration is justified by the monotone convergence theorem if the last integral is finite. Indeed, $\widehat{\psi}$ is essentially bounded on $\mathbb{R}^d$ and, since $2(1-\delta)(d/2+\varepsilon) > d$, it follows from (3.3) that $\sum\limits_k |\widehat{\psi}(\xi + k)|^{2(1-\delta)}$ is also essentially bounded on $\mathbb{R}^d$. Hence,

$$\sum_{j\in\mathbb{Z}}\sum_{k\in\mathbb{Z}^d} |\langle f, \psi_{jk}\rangle|^2 \le C_2 \int_{\mathbb{R}^d} |\widehat{f}(\xi)|^2 \sum_{j\in\mathbb{Z}} |\widehat{\psi}(M^{*-j}\xi)|^{2\delta} d\xi \le$$

$$C_2 \operatorname*{vraisup}_{\xi\in\mathbb{R}^d} \sum_{j\in\mathbb{Z}} |\widehat{\psi}(M^{*-j}\xi)|^{2\delta} \|\widehat{f}\|^2 = C_2 \operatorname*{vraisup}_{\xi\in\mathbb{R}^d} \sum_{j\in\mathbb{Z}} |\widehat{\psi}(M^{*-j}\xi)|^{2\delta} \|f\|^2.$$

Here, changing the order of summation and integration is justified by the Lebesgue's dominated convergence theorem if the right-hand side is finite. To prove this, set $\alpha(\xi) = \sum_{j\in\mathbb{Z}} |\widehat{\psi}(M^{*-j}\xi)|^{2\delta}$. Since $\alpha(\xi) = \alpha(M^{*n}\xi)$ for any $n \in \mathbb{Z}$, the essential supremum over $\mathbb{R}^d$ can be replaced by the essential supremum over some bounded set which is separated from zero, for instance, over the set

$$\Omega := M^{*k_0}[-1, 1]^d \backslash (-1, 1)^d,$$

where k_0 is the minimal integer, for which the inclusion $[-1,1]^d \subset M^{*k_0}[-1,1]^d$ holds. Indeed, if $\xi \in (-1,1)^d$, $\xi \neq \mathbf{0}$, then the vector $M^{*k_0}\xi$ is contained either in Ω or in $(-1,1)^d$. Consider the vectors $M^{*2k_0}\xi$, $M^{*3k_0}\xi$, etc. For some number n, the vector $M^{*nk_0}\xi$ belongs to Ω, since M is a dilation matrix. Therefore,

$$\alpha(\xi) = \alpha(M^{*nk_0}\xi) < \operatorname*{vraisup}_{\xi\in\Omega} \alpha(\xi)$$

for almost all $\xi \in (-1,1)^d$. If $\xi \in \mathbb{R}^d \setminus M^{*k_0}[-1,1]^d$, then there exists positive integer j_0 such that $M^{*-j_0}\xi \in (-1,1)^d$, i.e., we have the case considered above. Thus,

$$\sum_{j\in\mathbb{Z}} \sum_{k\in\mathbb{Z}^d} |\langle f, \psi_{jk} \rangle|^2 \leq C_2 \operatorname*{vraisup}_{\xi\in\Omega} \sum_{j\in\mathbb{Z}} |\widehat{\psi}(M^{*-j}\xi)|^{2\delta} \|\widehat{f}\|^2.$$

Using (3.2) and (2.2), for almost all $\xi \in \Omega$, we have

$$\sum_{j=0}^{\infty} |\widehat{\psi}(M^{*-j}\xi)|^{2\delta} \leq C_0^{2\delta} \sum_{j=0}^{\infty} |M^{*-j}\xi|^{2\delta} \leq$$

$$C_0^{2\delta} \sup_{\Omega} |\xi|^{2\delta} \sum_{j=0}^{\infty} \|M^{*-j}\|^{2\delta} := C_3.$$

Using now (3.3), we estimate the second part of the sum for $\xi \in \Omega$

$$\sum_{j=-\infty}^{-1} |\widehat{\psi}(M^{*-j}\xi)|^{2\delta} \leq C_1^{2\delta} \sum_{j=1}^{\infty} (1 + |M^{*j}\xi|)^{2\delta(-d/2-\varepsilon)}$$

Since $2\delta(-d/2 - \varepsilon) = -\delta(d + 2\varepsilon) \leq -\delta$, it follows that

$$\sum_{j=-\infty}^{-1} |\widehat{\psi}(M^{*-j}\xi)|^{2\delta} \leq C_1^{2\delta} \sum_{j=1}^{\infty} (1 + |M^{*j}\xi|)^{-\delta} \leq C_1^{2\delta} \sum_{j=1}^{\infty} |M^{*j}\xi|^{-\delta}.$$

Note that $1/|M^{*j}\xi| \leq \|M^{*-j}\|/|\xi|$; hence, continuing this chain of inequalities and using (2.2), we obtain

$$\sum_{j=-\infty}^{-1} |\widehat{\psi}(M^{*-j}\xi)|^{2\delta} \leq C_1^{2\delta} \sup_{\Omega} \frac{1}{|\xi|^{\delta}} \sum_{j=1}^{\infty} \|M^{*-j}\|^{\delta} := C_4.$$

Combining these estimates, we get (3.4).◊

Theorem 3.1.2 *Let $\varphi, \widetilde{\varphi}$ be refinable functions such that $\widehat{\varphi}, \widehat{\widetilde{\varphi}}$ are continuously differentiable in a neighborhood of the origin, $\widehat{\varphi}(\mathbf{0}) \cdot \widehat{\widetilde{\varphi}}(\mathbf{0}) = 1$ and*

$$|\widehat{\varphi}(\xi)|,\ |\widehat{\widetilde{\varphi}}(\xi)| \le C(1+|\xi|)^{-d/2-\varepsilon}. \tag{3.5}$$

Suppose $\{\psi_{jk}^{(\nu)}\}_{j,k,\nu}$, $\{\widetilde{\psi}_{jk}^{(\nu)}\}_{j,k,\nu}$ *are dual wavelet systems generated from* φ $\widetilde{\varphi}$ *by MEP, where the wavelet masks* m_ν, $\widetilde{m}_\nu$ *are bounded, continuously differentiable in a neighborhood of the origin and*

$$m_\nu(\mathbf{0}) = \widetilde{m}_\nu(\mathbf{0}) = 0, \quad \nu = 1, \dots, r. \tag{3.6}$$

Then the systems $\{\psi_{jk}^{(\nu)}\}_{j,k,\nu}$, $\{\widetilde{\psi}_{jk}^{(\nu)}\}_{j,k,\nu}$ *are dual frames in* $L_2(\mathbb{R}^d)$.

Proof The systems $\{\varphi_{0k}\}_k$, $\{\widetilde{\varphi}_{0k}\}_k$ are Bessel, due to (3.5), since $[\widehat{\varphi}, \widehat{\varphi}]$ and $[\widehat{\widetilde{\varphi}}, \widehat{\widetilde{\varphi}}]$ are essentially bounded on $\mathbb{R}^d$. Let us show that the functions $\psi^{(\nu)}$, $\nu = 1, \dots, r$, satisfy the conditions of Lemma 3.1.1. Indeed, (3.3) follows from (3.5), (3.1), and boundedness of the wavelet masks. Since the function $\widehat{\varphi}$ and the masks $m_1, \dots, m_r$ are bounded and have bounded partial derivatives in the neighborhood of the origin, (3.2) follows from (3.6) and (3.1). Thus, by Lemma 3.1.1, the system $\{\psi_{jk}^{(\nu)}\}_{j,k}$ is Bessel for each ν, which yields that $\{\psi_{jk}^{(\nu)}\}_{j,k,\nu}$ is a Bessel system. Similarly, $\{\widetilde{\psi}_{jk}^{(\nu)}\}_{j,k,\nu}$ is a Bessel system. It remains to apply Theorem 2.4.6.◊

Now we pay our attention to the case where dual wavelet systems are generated from compactly supported refinable functions. First we prove some auxiliary statements.

Lemma 3.1.3 *Suppose* $\theta \in L_2(\mathbb{R}^d)$ *is a compactly supported function such that* $\widehat{\theta}(\mathbf{0}) = 0$. *If* $\psi \in W_2^\nu(\mathbb{R}^d)$ *for some* $\nu \in (0, 1)$, *then*

$$\sum_{k\in\mathbb{Z}^d} |\langle \psi_{-j,0}, \theta_{0,k}\rangle|^2 \le C\|M^{*-j}\|^{2\nu}, \quad \forall j \in \mathbb{Z}_+,$$

where

$$C := \left\| \sum_{k\in\mathbb{Z}^d} |\cdot + k|^{-2\nu} |\widehat{\theta}(\cdot + k)|^2 \right\|_\infty \int_{\mathbb{R}^d} |\xi|^{2\nu} |\widehat{\psi}(\xi)|^2 d\xi < \infty.$$

Proof Note that $\widehat{\psi_{-j,0}}(\xi) = m^{j/2}\widehat{\psi}(M^{*j}\xi)$. By item (v) in Lemma 2.2.1, $\langle \psi_{-j,0}, \theta_{0,k}\rangle$ is the k-th Fourier coefficient of $[\widehat{\psi_{-j,0}}, \widehat{\theta}](\xi)$. Let us show that $[\widehat{\psi_{-j,0}}, \widehat{\theta}] \in L_2(\mathbb{T}^d)$. Indeed, using the Cauchy–Schwarz inequality, we obtain

$$
\begin{aligned}
\int\limits_{\mathbb{T}^d} \left|\widehat{[\psi_{-j,0}, \widehat{\theta}]}(\xi)\right|^2 d\xi &= \int\limits_{\mathbb{T}^d} \left|\sum_{k\in\mathbb{Z}^d} \widehat{\psi_{-j,0}}(\xi+k)\overline{\widehat{\theta}(\xi+k)}\right|^2 d\xi \\
&= \int\limits_{\mathbb{T}^d} \left|\sum_{k\in\mathbb{Z}^d} \widehat{\psi_{-j,0}}(\xi+k)|\xi+k|^{\nu}\overline{\widehat{\theta}(\xi+k)}|\xi+k|^{-\nu}\right|^2 d\xi \\
\le C_1 \int\limits_{\mathbb{T}^d} \sum_{k\in\mathbb{Z}^d} \left|\widehat{\psi_{-j,0}}(\xi+k)\right|^2 |\xi+k|^{2\nu} d\xi &\le C_1 \int\limits_{\mathbb{R}^d} \left|\widehat{\psi_{-j,0}}(\xi)\right|^2 |\xi|^{2\nu} d\xi \\
= C_1 m^j \int\limits_{\mathbb{R}^d} \left|\widehat{\psi}(M^{*j}\xi)\right|^2 |\xi|^{2\nu} d\xi &= C_1 \int\limits_{\mathbb{R}^d} |\widehat{\psi}(\xi)|^2 |M^{*-j}\xi|^{2\nu} d\xi \\
&\le C_1 \|M^{*-j}\|^{2\nu} \int\limits_{\mathbb{R}^d} |\widehat{\psi}(\xi)|^2 |\xi|^{2\nu} d\xi,
\end{aligned}
$$

where $C_1 := \left\|\sum_{k\in\mathbb{Z}^d} |\cdot+k|^{-2\nu}|\widehat{\theta}(\cdot+k)|^2\right\|_\infty$, and changing the order of summation and integration is valid by Lebesgue's dominated convergence theorem. Let us show that $C_1 < \infty$. By item (vi) in Lemma 2.2.1, $[\widehat{\theta}, \widehat{\theta}] \in L_\infty(\mathbb{R}^d)$, which yields

$$
\sum_{k\in\mathbb{Z}^d, k\neq\mathbf{0}} |\xi+k|^{-2\nu}|\widehat{\theta}(\xi+k)|^2 < \infty
$$

for almost all $\xi \in [-1/2, 1/2]^d$. On the other hand, since $\widehat{\theta}(\mathbf{0}) = 0$ and $\widehat{\theta}$ is continuously differentiable, $\widehat{\theta}(\xi)|\xi|^{-\nu}$ is bounded near the origin. So, C_1 is finite.◊

Next we need to introduce additional notations. For $i \in \mathbb{Z}_+$ and a sequence $u = \{u_n\}_{n\in\mathbb{Z}^d} \in \ell_2(\mathbb{Z}^d)$, we define

$$
(D_{M^i}u)_k := \begin{cases} u_{M^{-i}k} & \text{if } k \in M^i\mathbb{Z}^d, \\ 0 & \text{otherwise.} \end{cases}
$$

and define $(D_{M^{-i}}u)_k = u_{M^ik}$, $k \in \mathbb{Z}^d$.

For two sequences $u \in \ell_0(\mathbb{Z}^d)$, $v \in \ell_2(\mathbb{Z}^d)$, their convolution is defined by

$$
(u * v)_n := \sum_{k\in\mathbb{Z}^d} u_{n-k}v_k = \sum_{k\in\mathbb{Z}^d} v_{n-k}u_k, \quad n \in \mathbb{Z}^d.
$$

Lemma 3.1.4 *Let $L \in \mathbb{N}$, $i \in \mathbb{Z}_+$. For any sequence $u \in \ell_0(\mathbb{Z}^d)$ which is supported in $M^i\{x \in \mathbb{R}^d : |x| \le L\}$, and $v \in \ell_2(\mathbb{Z}^d)$,*

$$
\|(D_{M^i}v) * u\|_{\ell_2} \le L'\|v\|_{\ell_2}\|u\|_{\ell_2} \tag{3.7}
$$

and

$$\|D_{M^{-i}}(v * u)\|_{\ell_2} \leq L'\|v\|_{\ell_2}\|u\|_{\ell_2}, \tag{3.8}$$

where L' depends only on L and d.

Proof For any integer $l \in \mathbb{Z}^d$, define a sequence $u^{(l)}$ by

$$u_k^{(l)} = \begin{cases} u_k & \text{if } k \in \mathbb{Z}^d \cap M^i(l + [0,1)^d), \\ 0 & \text{otherwise.} \end{cases}$$

In other words,

$$u_k^{(l)} = \begin{cases} u_k & \text{if } k = M^i l + q, q \in D(M^i) \\ 0 & \text{otherwise.} \end{cases}$$

Here $D(M^i)$ is the set of digits of M^i defined by $D(M^i) = \mathbb{Z}^d \cap M^i[0,1)^d$. It is easy to see that $u^{(l)}$ are disjoint and

$$u = \sum_{l \in \mathbb{Z}^d, |l| \leq L + \sqrt{d}} u^{(l)}, \quad \|u\|_{\ell_2}^2 = \sum_{l \in \mathbb{Z}^d, |l| \leq L + \sqrt{d}} \|u^{(l)}\|_{\ell_2}^2.$$

Using the fact that $u^{(l)}$ is supported on $\mathbb{Z}^d \cap M^i(l + [0,1)^d)$, it can be shown that $\|(D_{M^i} v) * u^{(l)}\|_{\ell_2} = \|v\|_{\ell_2}\|u^{(l)}\|_{\ell_2}$ for all $l \in \mathbb{Z}^d$. Indeed, note that

$$((D_{M^i} v) * u)_n = \sum_{k \in \mathbb{Z}^d} u_{n-k}(D_{M^i} v)_k = \sum_{k \in M^i \mathbb{Z}^d} u_{n-k} v_{M^{-i}k} = \sum_{k \in \mathbb{Z}^d} u_{n - M^i k} v_k.$$

Thus,

$$\|(D_{M^i} v) * u^{(l)}\|_{\ell_2}^2 = \sum_{n \in \mathbb{Z}^d} \left| \sum_{k \in \mathbb{Z}^d} u_{n - M^i k}^{(l)} v_k \right|^2 = \sum_{p \in \mathbb{Z}^d} \sum_{s \in D(M^i)} \left| \sum_{k \in \mathbb{Z}^d} u_{M^i p + s - M^i k}^{(l)} v_k \right|^2.$$

For a fixed p, there exist a unique k such that $M^i(p-k)+s = M^i l + s$ and $k = p - l$. Therefore,

$$\|(D_{M^i} v) * u^{(l)}\|_{\ell_2}^2 = \sum_{p \in \mathbb{Z}^d} \sum_{s \in D(M^i)} |u_{M^i l + s}^{(l)} v_{p-l}|^2 =$$

$$\sum_{p \in \mathbb{Z}^d} |v_{p-l}|^2 \sum_{s \in D(M^i)} |u_{M^i l + s}^{(l)}|^2 = \|v\|_{\ell_2}\|u^{(l)}\|_{\ell_2}.$$

Consequently, inequality (3.7) can be obtained as follows:

$$\|(D_{M^i} v) * u\|_{\ell_2}^2 \leq \sum_{l \in \mathbb{Z}^d, |l| \leq L + \sqrt{d}} \|(D_{M^i} v) * u^{(l)}\|_{\ell_2}^2 = \|v\|_{\ell_2} \sum_{l \in \mathbb{Z}^d, |l| \leq L + \sqrt{d}} \|u^{(l)}\|_{\ell_2} \leq$$

$$\|v\|_{\ell_2}\left(\sum_{l\in\mathbb{Z}^d,|l|\le L+\sqrt{d}} 1\right)^{1/2}\left(\sum_{l\in\mathbb{Z}^d,|l|\le L+\sqrt{d}}\|u^{(l)}\|^2_{\ell_2}\right)^{1/2}\le L'\|v\|_{\ell_2}\|u\|_{\ell_2}.$$

To prove (3.8) we note that $D_{M^{-i}}(v*u)$ is a subsequence of $v*u$, which yields

$$\|D_{M^{-i}}(v*u)\|_{\ell_2}\le\|v*u\|_{\ell_2}=\|((D_{M^0}v)*u)\|_{\ell_2}.$$

Thus (3.8) follows from (3.7).◊

Lemma 3.1.5 *For any dilation matrix M there exist compactly supported functions $\theta^{(1)},\dots,\theta^{(m-1)}$ such that $\{\theta^{(\nu)}_{jk}\}_{jk\nu}$ is a tight wavelet frame in $L_2(\mathbb{R}^d)$.*

Proof Let $A_{\nu,k}\in\mathbb{R}$, $k,\nu=0,\dots,m-1$, be entries of a unitary matrix such that $A_{0,k}=m^{-1/2}$, $k=0,\dots,m-1$. Set

$$m_\nu(\xi):=\frac{1}{\sqrt{m}}\sum_{k=0}^{m-1}A_{\nu,k}e^{2\pi i(s_k,\xi)},$$

where $\{s_0,\dots,s_{m-1}\}=D(M)$, $s_0=\mathbf{0}$. Obviously. every m_ν is a trigonometric polynomial, $m_0(\mathbf{0})=1$, and the functions $\mu_{\nu,k}\equiv A_{\nu,k}$ are the polyphase components of m_ν. It follows from Theorem 2.6.4 that m_0 is a mask of a compactly supported refinable function φ, and $\varphi\in L_2(\mathbb{R}^d)$ due to Lemma 2.42 and Proposition 2.6.6 because $\sum_{k=0}^{m-1}\mu_{0k}\equiv 1$. Define functions $\theta^{(\nu)}$, $\nu=1,\dots,m-1$, by $\widehat{\theta^{(r)}}(M^*\xi)=m_\nu(\xi)(\xi)\widehat{\varphi}(\xi)$. It follows from Theorem 2.4.5 and Proposition 1.1.8 that $\{\theta^{(\nu)}_{jk}\}_{jk\nu}$ is a tight frame.◊

Now we can state sufficient conditions for the system $\{\psi_{jk}\}_{j,k}$ to be a Bessel.

Theorem 3.1.6 *If $\psi\in L_2(\mathbb{R}^d)$ is a compactly supported function such that $\widehat{\psi}(\mathbf{0})=0$ and $\psi\in W^\nu_2(\mathbb{R}^d)$ for some $\nu>0$, then $\{\psi_{jk}\}_{j,k}$ is a Bessel system.*

Proof Let $\theta^{(1)},\dots,\theta^{(m-1)}$ from Lemma 3.1.5, $f\in L_2(\mathbb{R}^d)$. Then

$$f=\sum_{\nu=1}^{m-1}\sum_{j\in\mathbb{Z}}\sum_{k\in\mathbb{Z}^d}u^{(\nu,j)}_k\theta^{(\nu)}_{j,k},\tag{3.9}$$

where $u^{(\nu,j)}_k:=\langle f,\theta^{(\nu)}_{j,k}\rangle$. Let $c^j_k:=\langle f,\psi_{j,k}\rangle$, $c^j:=\{c^j_k\}_{k=1}^\infty$ and $u^{(\nu,j)}:=\{u^{(\nu,j)}_k\}_{k=1}^\infty$.
First we prove that

$$c^j=\sum_{\nu=1}^{m-1}\sum_{n=0}^{\infty}D_{M^{-n}}(u^{(\nu,n+j)}*v^{(\nu,n)})+\sum_{\nu=1}^{m-1}\sum_{n=1}^{\infty}D_{M^n}(u^{(\nu,j-n)})*w^{(\nu,n)},$$

where the sequences $v^{(\nu,n)}$ and $w^{(\nu,n)}$ are defined by $v^{(\nu,n)}_k=\langle\theta^{(\nu)}_{0,-k},\psi_{-n,0}\rangle$ and $w^{(\nu,n)}_k=\langle\theta^{(\nu)}_{-n,0},\psi_{0,k}\rangle$. It follows from (3.9) that

$$c_l^i = \sum_{\nu=1}^{m-1} \sum_{j\in\mathbb{Z}} \sum_{k\in\mathbb{Z}^d} \langle f, \theta_{j,k}^{(\nu)}\rangle \langle \theta_{j,k}^{(\nu)}, \psi_{i,l}\rangle =$$

$$\sum_{\nu=1}^{m-1} \sum_{j\geq i} \sum_{k\in\mathbb{Z}^d} \langle f, \theta_{j,k}^{(\nu)}\rangle \langle \theta_{j,k}^{(\nu)}, \psi_{i,l}\rangle + \sum_{\nu=1}^{m-1} \sum_{j< i} \sum_{k\in\mathbb{Z}^d} \langle f, \theta_{j,k}^{(\nu)}\rangle \langle \theta_{j,k}^{(\nu)}, \psi_{i,l}\rangle. \quad (3.10)$$

Direct computations yield

$$\langle \theta_{j,k}^{(\nu)}, \psi_{i,l}\rangle = \langle \theta_{0,k-M^{j-i}l}^{(\nu)}, \psi_{i-j,0}\rangle$$

Therefore, the first summand in (3.10) can be modified as follows.

$$\sum_{\nu=1}^{m-1} \sum_{j-i\geq 0} \sum_{k\in\mathbb{Z}^d} \langle f, \theta_{j,k}^{(\nu)}\rangle \langle \theta_{j,k}^{(\nu)}, \psi_{i,l}\rangle = \sum_{\nu=1}^{m-1} \sum_{j-i\geq 0} \sum_{k\in\mathbb{Z}^d} \langle f, \theta_{j,k}^{(\nu)}\rangle \langle \theta_{0,k-M^{j-i}l}^{(\nu)}, \psi_{i-j,0}\rangle =$$

$$\sum_{\nu=1}^{m-1} \sum_{n\geq 0} \sum_{k\in\mathbb{Z}^d} \langle f, \theta_{n+i,k}^{(\nu)}\rangle \langle \theta_{0,k-M^{n}l}^{(\nu)}, \psi_{-n,0}\rangle =$$

$$\sum_{\nu=1}^{m-1} \sum_{n\geq 0} \sum_{k\in\mathbb{Z}^d} u_k^{(\nu,n+i)} v_{M^n l-k}^{(\nu,n)} = \sum_{\nu=2}^{m} \sum_{n\geq 0} (D_{M^{-n}}(u^{(\nu,n+i)} * v^{(\nu,n)}))_l.$$

Next, using

$$\langle \theta_{j,k}^{(\nu)}, \psi_{i,l}\rangle = \langle \theta_{-(i-j),0}^{(\nu)}, \psi_{0,l-M^{i-j}k)}\rangle$$

we modify the second summand in (3.10).

$$\sum_{\nu=1}^{m-1} \sum_{i-j>0} \sum_{k\in\mathbb{Z}^d} \langle f, \theta_{j,k}^{(\nu)}\rangle \langle \theta_{j,k}^{(\nu)}, \psi_{i,l}\rangle = \sum_{\nu=1}^{m-1} \sum_{i-j>0} \sum_{k\in\mathbb{Z}^d} \langle f, \theta_{j,k}^{(\nu)}\rangle \langle \theta_{-(i-j),0}^{(\nu)}, \psi_{0,l-M^{i-j}k}\rangle =$$

$$\sum_{\nu=1}^{m-1} \sum_{n>0} \sum_{k\in\mathbb{Z}^d} \langle f, \theta_{i-n,k}^{(\nu)}\rangle \langle \theta_{-n,0}^{(\nu)}, \psi_{0,l-M^{n}k}\rangle =$$

$$\sum_{\nu=1}^{m-1} \sum_{n>0} \sum_{k\in\mathbb{Z}^d} u_k^{(\nu,i-n)} w_{l-M^n k}^{(\nu,n)} = \sum_{\nu=2}^{m} \sum_{n\geq 0} (D_{M^{n}}(u^{(\nu,i-n)}) * w^{(\nu,n)})_l.$$

Due to Theorem 2.6.9, we can state that the functions $\psi, \theta^{(1)}, \dots, \theta^{(m-1)}$ are in $W_2^{\alpha}(\mathbb{R}^d)$ for some $0 < \alpha < 1$. By Lemma 3.1.3, there is a constant $C_1 > 0$ such that

$$\max\{\|v^{(\nu,n)}\|_{\ell_2}, \|w^{(\nu,n)}\|_{\ell_2}\} \leq C_1 \|M^{*-n}\|^{\alpha} \quad \forall n \in \mathbb{Z}_+, \nu = 1, \dots, m-1. \quad (3.11)$$

Since ψ and $\theta^{(\nu)}$, $\nu = 1, \dots, m-1$ are compactly supported, we can assume that their supports are contained in the set $\{x \in \mathbb{R}^d : |x| \le L\}$ for some $L \in \mathbb{N}$. Let us show that all the sequences $v^{(\nu,n)}$, $w^{(\nu,n)}$ are supported in $M^n\{x \in \mathbb{R}^d : |x| \le L_1\}$, where $L_1 := L(1 + \sup_{n\in\mathbb{Z}_+} \|M^{-n}\|)$. Indeed, since $v_k^{(\nu,n)} = \langle \eta_{0,-k}^{(\nu)}, \psi_{-n,0}\rangle$ and $w_k^{(\nu,n)} = \langle \theta_{-n,0}^{(\nu)}, \psi_{0,k}\rangle$, we have

$$v_k^{(\nu,n)} = \int_{\mathbb{R}^d} \theta^{(\nu)}(x-k) m^{-n/2} \psi(M^{-n}x)dx =$$

$$\int_{\mathbb{R}^d} \theta^{(\nu)}(M^n x - k) m^{n/2} \psi(x)dx = m^{n/2} \int_{|x|\le L} \theta^{(\nu)}(M^n x - k)\psi(x)dx.$$

Hence, if $v_k^{(\nu,n)} \ne 0$, then there exists $|x| \le L$ such that $|M^n x - k| \le L$, which yields $|M^{-n}k| \le |x| + |x - M^{-n}k| \le L + \|M^{-n}\|L$. So, for any $n \in \mathbb{Z}_+$ the sequence $v^{(\nu,n)}$ is supported in $M^n\{x \in \mathbb{R}^d : |x| \le L_1\}$. Analogously, the sequence $w^{(\nu,n)}$ is supported in the same set.

Now, using (3.11) and Lemma 3.1.4, we can estimate ℓ_2-norm of the sequence c^i.

$$\|c^i\|_{\ell_2} \le \sum_{\nu=1}^{m-1}\sum_{n=0}^{\infty} \|D_{M^{-n}}(u^{(\nu,n+i)} * v^{(\nu,n)})\|_{\ell_2} + \sum_{\nu=1}^{m-1}\sum_{n=1}^{\infty} \|D_{M^n}(u^{(\nu,i-n)}) * w^{(\nu,n)}\|_{\ell_2} \le$$

$$C_2 \sum_{\nu=1}^{m-1}\left(\sum_{n=0}^{\infty} \|u^{(\nu,n+i)}\|_{\ell_2}\|v^{(\nu,n)}\|_{\ell_2} + \sum_{n=1}^{\infty} \|u^{(\nu,i-n)}\|_{\ell_2}\|w^{(\nu,n)}\|_{\ell_2}\right) \le$$

$$C_1C_2 \sum_{\nu=1}^{m-1}\left(\sum_{n=0}^{\infty} \|u^{(\nu,n+i)}\|_{\ell_2}\|M^{*-n}\|^{\alpha} + \sum_{n=1}^{\infty} \|u^{(\nu,i-n)}\|_{\ell_2}\|M^{*-n}\|^{\alpha}\right) \le$$

$$C_1C_2 \sum_{\nu=1}^{m-1}\sum_{n\in\mathbb{Z}} \|u^{(\nu,i-n)}\|_{\ell_2}\|M^{*-|n|}\|^{\alpha},$$

where $C_2 = L'$, $L' = L'(L_1, d)$ is a constant from Lemma 3.1.4. Recall Young's inequality for convolutions of discrete sequences, which states that for $a \in \ell_1(\mathbb{Z}^d)$ and $b \in \ell_2(\mathbb{Z}^d)$, we have $\|a * b\|_{\ell_2} \le \|a\|_{\ell_1}\|b\|_{\ell_2}$. Therefore,

$$\sum_{i\in\mathbb{Z}^d} \|c^i\|_{\ell_2}^2 \le C_1^2C_2^2 \sum_{i\in\mathbb{Z}^d}\left(\sum_{n\in\mathbb{Z}} \|M^{*-|n|}\|^{\alpha} \sum_{\nu=1}^{m-1} \|u^{(\nu,i-n)}\|_{\ell_2}\right)^2$$

$$C_1^2C_2^2\left(\sum_{n\in\mathbb{Z}} \|M^{*-|n|}\|^{\alpha}\right)^2 \sum_{\nu=1}^{m-1}\sum_{n\in\mathbb{Z}} \|u^{(\nu,n)}\|_{\ell_2}^2.$$

Since $\sum_{\nu=1}^{m-1}\sum_{n\in\mathbb{Z}}\|u^{(\nu,n)}\|_{\ell_2}^2=\|f\|_2^2$ and $\sum_{i\in\mathbb{Z}^d}\|c^i\|_{\ell_2}^2=\sum_{j\in\mathbb{Z}}\sum_{k\in\mathbb{Z}^d}|\langle f,\psi_{j,k}\rangle|^2$, then we obtain

$$\sum_{j\in\mathbb{Z}}\sum_{k\in\mathbb{Z}^d}|\langle f,\psi_{j,k}\rangle|^2\le C_3\|f\|_2^2,\quad \forall f\in L_2(\mathbb{R}^d),$$

where $C_3:=C_1^2C_2^2\left(\sum_{n\in\mathbb{Z}}\|M^{*-|n|}\|^{\alpha}\right)^2<\infty$. Therefore, $\{\psi_{jk}\}_{j,k}$ is a Bessel system.◊

Corollary 3.1.7 *Let $\psi\in L_2(\mathbb{R}^d)$ be defined by $\widehat{\psi}=t\widehat{\varphi}$, where φ is a compactly supported refinable function in $L_2(\mathbb{R}^d)$ with a polynomial mask, and t is a trigonometric polynomial satisfying $t(\mathbf{0})=0$. Then $\{\psi_{jk}\}_{j,k}$ is a Bessel system.*

The proof follows from Theorems 2.6.9 and 3.1.6.

Theorem 3.1.6 allows to state sufficient conditions for a pair of dual wavelet system to be dual frames.

Theorem 3.1.8 *Let $\varphi,\ \widetilde{\varphi}\in L_2(\mathbb{R}^d)$ be compactly supported refinable functions such that $\widehat{\varphi}(\mathbf{0})\cdot\widehat{\widetilde{\varphi}}(\mathbf{0})=1$. Suppose dual wavelet systems $\{\psi_{j,k}^{(\nu)}\}_{j,k,\nu}$, $\{\widetilde{\psi}_{j,k}^{(\nu)}\}_{j,k,\nu}$ are generated from $\varphi,\ \widetilde{\varphi}$ by MEP, where all wavelet masks $m_\nu,\ \widetilde{m}_\nu,\ \nu=1,\dots,r$ are trigonometric polynomials satisfying $m_\nu(\mathbf{0})=\widetilde{m}_\nu(\mathbf{0})=0$. Then the systems $\{\psi_{j,k}^{(\nu)}\}_{j,k,\nu}$, $\{\widetilde{\psi}_{j,k}^{(\nu)}\}_{j,k,\nu}$ are dual frames in $L_2(\mathbb{R}^d)$.*

The proof follows from Theorem 2.4.6 and Proposition 1.1.8, because each of the systems $\{\psi_{j,k}^{(\nu)}\}_{j,k}$, $\{\widetilde{\psi}_{j,k}^{(\nu)}\}_{j,k}$, $\nu=1,\dots,r$, is Bessel by Corollary 3.1.7

3.2 Approximation Order and Vanishing Moments

In the previous section, we proved that dual wavelet systems generated from two appropriate refinable functions by MEP are frames, provided the Fourier transform of each wavelet function vanishes at the origin. The latter assumption is necessary due to the following simple statement.

Theorem 3.2.1 *Let functions $\psi^{(\nu)}\in L_2(\mathbb{R}^d)$, $\nu=1,\dots,r$, be such that $\widehat{\psi^{(\nu)}}$ is continuous at the origin. If the corresponding wavelet system $\{\psi_{jk}^{(\nu)}\}_{j,k,\nu}$ is a frame, then $\widehat{\psi^{(\nu)}}(\mathbf{0})=0$ for each ν.*

Proof Firstly, we show that the system $\{\psi_{0k}^{(\nu)}\}_k$ is Bessel for all $j\in\mathbb{Z}^d$ and $\nu=1,\dots,r$. Let B be an upper frame bound, $c=\{c_n\}_{n\in\mathbb{Z}^d}\in\ell_2(\mathbb{Z}^d)$, and Ω be a finite subset of $\mathbb{Z}^d$. Since

$$\left\|\sum_{n\in\Omega}c_n\psi_{0n}^{(\nu)}\right\|=\left|\left\langle\sum_{n\in\Omega}c_n\psi_{0n}^{(\nu)},g_\Omega\right\rangle\right|=\left|\sum_{n\in\Omega}c_n\langle\psi_{0n}^{(\nu)},g_\Omega\rangle\right|,$$

where $\|g_\Omega\| \le 1$, it follows from the Cauchy–Schwarz inequality that

$$\left\|\sum_{n\in\Omega} c_n \psi_{0n}^{(\nu)}\right\| \le \sum_{n\in\Omega} |c_n|^2 \sum_{n\in\mathbb{Z}^d} \left|\langle \psi_{0n}^{(\nu)}, g_\Omega\rangle\right|^2 \le B\sum_{n\in\Omega} |c_n|^2 \|g_\Omega\|^2 \le B\sum_{n\in\Omega} |c_n|^2.$$

This yields that the series $\sum_{n\in\mathbb{Z}^d} c_n\psi_{0n}^{(\nu)}$ converges in $L_2(\mathbb{R}^d)$, and in a similar way we have

$$\left\|\sum_{n\in\mathbb{Z}^d} c_n \psi_{0n}^{(\nu)}\right\| \le B\sum_{n\in\mathbb{Z}^d} |c_n|^2.$$

Thus $\{\psi_{0k}^{(\nu)}\}_k$ is a Bessel system. Using Theorem 2.2.6 and taking into account Remark 1.1.7, we have

$$\sum_{k\in\mathbb{Z}^d} |\langle f, \psi_{jk}^{(\nu)}\rangle|^2 \underset{j\to+\infty}{\longrightarrow} |\widehat{\psi}^{(\nu)}(\mathbf{0})|^2\|f\|^2.$$

On the other hand, since $\{\psi_{jk}^{(\nu)}\}_{j,k,\nu}$ is a frame, the series

$$\sum_{j\in\mathbb{Z}}\sum_{k\in\mathbb{Z}^d} |\langle f, \psi_{jk}^{(\nu)}\rangle|^2$$

converges. Hence, $\widehat{\psi}^{(\nu)}(\mathbf{0}) = 0.\Diamond$

If $f \in L(\mathbb{R}^d)$, then the condition $\widehat{f}(\mathbf{0}) = 0$ is equivalent to

$$\int_{\mathbb{R}^d} f(x)\, dx = 0,$$

which is usually referred as vanishing moment of f. Similarly, the moment of order $\alpha \in \mathbb{Z}_+^d$, that is the integral

$$\int_{\mathbb{R}^d} x^\alpha f(x)\, dx = \int_{\mathbb{R}^d} x_1^{\alpha_1}\dots x_d^{\alpha_d} f(x)\, dx$$

vanishes at the origin if and only if $D^\alpha \widehat{f}(\mathbf{0}) = 0$. This approves the following terminology.

Definition 3.2.2 We say that a wavelet system $\{\psi_{jk}^{(\nu)}\}_{j,k,\nu}$ has *vanishing moments* up to order α, $\alpha \in \mathbb{Z}_+^d$, (has VM_α *property* in the sequel), if for every ν the equality $D^\beta\widehat{\psi}^{(\nu)}(\mathbf{0}) = 0$ holds for each $\beta \in \mathbb{Z}_+^d$, $\beta \le \alpha$.

Usually it is more useful to control univariate order of vanishing moment property (for example, to apply Taylor formula).

Definition 3.2.3 We say that a wavelet system $\{\psi_{jk}^{(\nu)}\}_{j,k,\nu}$ has vanishing moments up to order n, $n \in \mathbb{Z}_+$, (has VM^n *property* in the sequel), if for every ν the equality $D^\beta \widehat{\psi^{(\nu)}}(\mathbf{0}) = 0$ holds for each $\beta \in \mathbb{Z}_+^d$, $[\beta] < n$.

It will be clear soon why the high order of vanishing moments is useful for wavelet functions.

Let $\{\psi_{jk}^{(\nu)}\}_{j,k,\nu}$, $\{\widetilde{\psi}_{jk}^{(\nu)}\}_{j,k,\nu}$ be dual frames. Due to Corollary 1.2.5, every function $f \in L_2(\mathbb{R}^d)$ can be decomposed as

$$f = \sum_{j\in\mathbb{Z}^d} \sum_{k\in\mathbb{Z}^d} \sum_{\nu=1}^{r} \langle f, \widetilde{\psi}_{ik}^{(\nu)} \rangle \psi_{ik}^{(\nu)}. \tag{3.12}$$

Decompositions (3.12) is said to have *approximation order n* if

$$\left\| f - \sum_{i<j} \sum_{k\in\mathbb{Z}^d} \sum_{\nu=1}^{r} \langle f, \widetilde{\psi}_{ik}^{(\nu)} \rangle \psi_{ik}^{(\nu)} \right\|_2 \le C \|f\|_{W_2^n} \sum_{i=j}^{\infty} \|M^{-i}\|^n \quad \forall f \in W_2^n,$$

where W_2^n is the Sobolev space. In particular, if $M = \lambda I_d$, then

$$\left\| f - \sum_{i<j} \sum_{k\in\mathbb{Z}^d} \sum_{\nu=1}^{r} \langle f, \widetilde{\psi}_{ik}^{(\nu)} \rangle \psi_{ik}^{(\nu)} \right\|_2 \le C' \frac{\|f\|_{W_2^n}}{|\lambda|^{jn}};$$

if M is an arbitrary dilation matrix, then for any small enough $\varepsilon > 0$

$$\left\| f - \sum_{i<j} \sum_{k\in\mathbb{Z}^d} \sum_{\nu=1}^{r} \langle f, \widetilde{\psi}_{ik}^{(\nu)} \rangle \psi_{ik}^{(\nu)} \right\|_2 \le C(\varepsilon) \frac{\|f\|_{W_2^n}}{(|\lambda| - \varepsilon)^{jn}},$$

where λ is a minimal (in modulus) eigenvalue of M (see comments to (2.2) in Sect. 2.1).

Theorem 3.2.4 *Let* $\{\psi_{jk}^{(\nu)}\}_{j,k,\nu}$, $\{\widetilde{\psi}_{jk}^{(\nu)}\}_{j,k,\nu}$ *be dual wavelet frames,*

$$|\psi^{(\nu)}(x)| \le \frac{C}{(1+|x|)^\gamma}, \quad |\widetilde{\psi}^{(\nu)}(x)| \le \frac{C}{(1+|x|)^\gamma}, \quad \gamma > n + d,$$

for all $\nu = 1, \dots, r$, *and almost all* $x \in \mathbb{R}^d$. *If* $\{\widetilde{\psi}_{jk}^{(\nu)}\}_{j,k,\nu}$ *has* VM^n *property, then decomposition (3.12) has approximation order n.*

We need the following auxiliary statement for the proof.

Lemma 3.2.5 *Let* η *be a positive bounded function decreasing on* $[0, \infty)$ *so that* $\eta(|x|)$ *is summable on* $\mathbb{R}^d$. *Then there exists a constant K depending on* η *and d such that*

$$\sum_{k\in\mathbb{Z}^d} \eta(|x+k|)\eta(|y+k|) \le K\eta\left(\frac{|x-y|}{8}\right) \tag{3.13}$$

for all $x, y \in \mathbb{R}^d$.

Proof First of all note that there exist constants K_1, K_2 such that

$$\eta(|t|) \le K_1, \quad \sum_{k\in\mathbb{Z}^d} \eta(|t+k|) \le K_2 \;\; \forall t \in \mathbb{R}^d.$$

Since both the left- and the right-hand sides of (3.13) are invariant with respect to the operation $(x, y) \to (x+l, y+l), l \in \mathbb{Z}^d$, we can assume that $x \in [-\frac{1}{2}, \frac{1}{2}]^d$ and, therefore, $|x| \le \frac{\sqrt{d}}{2}$. If $|y| \le 2\sqrt{d}$, then

$$\sum_{k\in\mathbb{Z}^d} \eta(|x+k|)\eta(|y+k|) \le K_1K_2 \le \frac{K_1K_2}{\eta(3\sqrt{d})}\eta(|x-y|).$$

Now let $|y| \ge 2\sqrt{d}$. Since $|x-y| \le 2|y|$ and $|y| \ge 4|x|$, we have

$$\sum_{|k|\le\frac{|y|}{2}} \eta(|x+k|)\eta(|y+k|) \le \sum_{|k|\le\frac{|y|}{2}} \eta(|x+k|)\eta\left(\frac{|y|}{2}\right) \le K_2\eta\left(\frac{|y|}{2}\right) \le K_2\eta\left(\frac{|x-y|}{4}\right),$$

$$\sum_{|k|>\frac{|y|}{2}} \eta(|x+k|)\eta(|y+k|) \le \sum_{|k|>\frac{|y|}{2}} \eta\left(\frac{|k|}{2}\right)\eta(|y+k|) \le K_2\eta\left(\frac{|y|}{4}\right) \le K_2\eta\left(\frac{|x-y|}{8}\right).\Diamond$$

Proof of Theorem 3.2.4. Let $f \in W_2^n(\mathbb{R}^d)$, $j \in \mathbb{Z}_+$. It follows from (3.12) that

$$\left\| f - \sum_{i<j}\sum_{k\in\mathbb{Z}^d}\sum_{\nu=1}^{r} \langle f, \widetilde{\psi}_{ik}^{(\nu)}\rangle \psi_{ik}^{(\nu)} \right\|_2 \le \sum_{i\ge j}\sum_{\nu=1}^{r} \left\| \sum_{k\in\mathbb{Z}^d} \langle f, \widetilde{\psi}_{ik}^{(\nu)}\rangle \psi_{ik}^{(\nu)} \right\|_2. \tag{3.14}$$

Since VM^n property is equivalent to

$$\int_{\mathbb{R}^d} y^\alpha \widetilde{\psi}_{ik}^{(\nu)}(y)\,dy = 0, \quad \nu = 1, \ldots, r,\; i \in \mathbb{Z},\; k \in \mathbb{Z}^d, \quad \forall \alpha \in \mathbb{Z}_+^d,\; [\alpha] < n,$$

using Taylor formula with the integral remainder, we have

$$\left|\langle f, \widetilde{\psi}_{ik}^{(\nu)}\rangle\right| = \left|\int_{\mathbb{R}^d} f(y)\widetilde{\psi}_{ik}^{(\nu)}(y)\,dy\right| =$$

$$\left|\int_{\mathbb{R}^d} dy\,\widetilde{\psi}_{ik}^{(\nu)}(y)\left(\sum_{l=0}^{n-1}\frac{1}{l!}\left((y-x)_1\frac{\partial}{\partial x_1}+\cdots+(y-x)_d\frac{\partial}{\partial x_d}\right)^l f(x)+\right.\right.$$

$$\left.\left.\int_0^1 \frac{(1-t)^{n-1}}{(n-1)!}\left((y-x)_1\frac{\partial}{\partial x_1}+\cdots+(y-x)_d\frac{\partial}{\partial x_d}\right)^n f(x+t(y-x))\,dt\right)\right| \le$$

$$C_2\sum_{\substack{\alpha\in\mathbb{Z}_+^d\\ [\alpha]=n}}\int_{\mathbb{R}^d} dy\,|x-y|^n|\widetilde{\psi}_{ik}^{(\nu)}(y)|\int_0^1 |D^\alpha f(x+t(y-x))|\,dt.$$

From this, due to Lemma 3.2.5 and Cauchy–Schwarz inequality, we obtain

$$\left\|\sum_{k\in\mathbb{Z}^d}\langle f, \widetilde{\psi}_{ik}^{(\nu)}\rangle\psi_{ik}^{(\nu)}\right\|_2^2 \le$$

$$C_1^2\int_{\mathbb{R}^d} dx\left(\int_{\mathbb{R}^d} dy|x-y|^n\sum_{k\in\mathbb{Z}^d}|\widetilde{\psi}_{ik}^{(\nu)}(y)\psi_{ik}^{(\nu)}(x)|\int_0^1\sum_{\substack{\alpha\in\mathbb{Z}_+^d\\ [\alpha]=n}}|D^\alpha f(x+t(y-x))|\,dt\right)^2 \le$$

$$C_2\int_{\mathbb{R}^d} dx\left(\int_{\mathbb{R}^d} dy\int_0^1 dt\frac{m^i|x-y|^n}{(1+|M^i(x-y)|)^\gamma}\left(\sum_{\substack{\alpha\in\mathbb{Z}_+^d\\ [\alpha]=n}}|D^\alpha f(x+t(y-x))|^2\right)^{\frac{1}{2}}\right)^2 =$$

$$C_2\int_{\mathbb{R}^d} dx\left(\int_{\mathbb{R}^d} du\int_0^1 dt\frac{m^i|u|^n}{(1+|M^i u|)^\gamma}\left(\sum_{\substack{\alpha\in\mathbb{Z}_+^d\\ [\alpha]=n}}|D^\alpha f(x+tu)|^2\right)^{\frac{1}{2}}\right)^2 \le$$

$$C_2\int_{\mathbb{R}^d} dx\left(\int_{\mathbb{R}^d} du\int_0^1 dt\frac{m^i|u|^n}{(1+|M^i u|)^\gamma}\sum_{\substack{\alpha\in\mathbb{Z}_+^d\\ [\alpha]=n}}|D^\alpha f(x+tu)|^2\times\right.$$

$$\left.\int_{\mathbb{R}^d} du\int_0^1 dt\frac{m^i|u|^n}{(1+|M^i u|)^\gamma}\right) \le C_2\left(\|f\|_{W_2^n}\int_{\mathbb{R}^d}\frac{m^i|u|^n\,du}{(1+|M^i u|)^\gamma}\right)^2 \le$$

$$C_2\left(\|f\|_{W_p^2}\|M^{-i}\|^n\int_{\mathbb{R}^d}\frac{|v|^n\,dv}{(1+|v|)^\gamma}\right)^2 \le$$

$$C_2\left(\|f\|_{W_2^n}\|M^{-i}\|^n\int_{\mathbb{R}^d}\frac{dv}{(1+|v|)^{\gamma-n}}\right)^2 = C_3\|f\|_{W_2^n}^2\|M^{-i}\|^{2n}.$$

It remains to combine this estimation with (3.14).◊

3.3 Polyphase Characterization of Vanishing Moments

It was proved in the previous section that frame expansion (3.12) has approximation order n whenever wavelet functions have enough decay and VM^n properties are satisfied for $\{\widetilde{\psi}_{jk}^{(\nu)}\}_{j,k,\nu}$. Now we consider dual wavelet systems generated by MEP and discuss how to provide VM^n property for one of these systems.

Let $\varphi, \widetilde{\varphi} \in L_2(\mathbb{R}^d)$ be refinable functions with masks $m_0, \widetilde{m}_0$, respectively, $\widehat{\varphi}, \widehat{\widetilde{\varphi}}$ be continuous at the origin and $\widehat{\varphi}(\mathbf{0}) = \widehat{\widetilde{\varphi}}(\mathbf{0}) = 1$. We assume that dual wavelet systems $\{\psi_{jk}^{(\nu)}\}_{j,k,\nu}$, $\{\widetilde{\psi}_{jk}^{(\nu)}\}_{j,k,\nu}$ are generated from $\varphi, \widetilde{\varphi}$ by MEP. The systems depend on $\varphi, \widetilde{\varphi}$, and hence, on the refinable masks $m_0, \widetilde{m}_0$, as well as on a matrix extension, i.e., on a choice of wavelet masks $m_\nu, \widetilde{m}_\nu$, $\nu = 1, \dots, r$. We shall study under what conditions on the refinable and wavelet masks one of the systems has the vanishing moment property.

Assume that the functions $\widehat{\widetilde{\varphi}}$ and $\widetilde{m}_1(M^{*-1}\cdot), \dots, \widetilde{m}_r(M^{*-1}\cdot)$ have derivatives up to order $\alpha \in \mathbb{Z}_+^d$ at the origin. It easily follows from (3.1) and the Leibniz formula that VM_α property holds for $\{\widetilde{\psi}_{jk}^{(\nu)}\}_{j,k,\nu}$ if and only if

$$D^\beta(\widetilde{m}_\nu(M^{*-1}\xi))\Big|_{\xi=\mathbf{0}} = 0, \quad \nu = 1, \dots, r, \quad \forall \beta \in \mathbb{Z}_+^d, \quad \beta \le \alpha. \tag{3.15}$$

We have already saw in Sect. 2.5 that even without additional requirements, finding wavelet masks is a quite complicated problem, especially if we want to have polynomial masks which are needed to construct compactly supported wavelet functions. The requirement to provide (3.15) complicates the problem. But the polyphase representations of masks will help us.

Let $\mu_{\nu,k}, \widetilde{\mu}_{\nu,k}$, $k = 0, \dots, m-1$, be the polyphase components of the masks $m_\nu, \widetilde{m}_\nu$, respectively, for every $\nu = 0, \dots, r$, and let $D(M) = \{s_0, \dots, s_{m-1}\}$, $D(M^*) = \{q_0, \dots, q_{m-1}\}$, where $s_0 = q_0 = \mathbf{0}$.

By the definition of the polyphase components (see Sect. 2.5), we have

$$m_\nu(\xi) = \frac{1}{\sqrt{m}} \sum_{k=0}^{m-1} e^{2\pi i (s_k, \xi)} \mu_{\nu k}(M^*\xi). \tag{3.16}$$

Set

$$U(\xi) = \left\{ \frac{1}{\sqrt{m}} e^{2\pi i (s_k, \xi + M^{*-1} q_l)} \right\}_{k,l=0,\dots,m-1}.$$

and recall that $U(\xi)$ is a unitary matrix (see Sect. 2.5). It easily follows that $\mu_{\nu k}$ can be expressed as

$$\mu_{\nu k}(\xi) = \frac{1}{\sqrt{m}} \sum_{l=0}^{m-1} e^{-2\pi i (M^{-1} s_k, \xi + q_l)} m_\nu(M^{*-1}(\xi + q_l)). \tag{3.17}$$

It is clear from these formulas that a function m_ν is differentiable (up to order $\alpha \in \mathbb{Z}_+^d$) on the set $M^{*-1}(D(M^*))$ if and only if its polyphase components $\mu_{\nu k}$, $k = 0, \dots, m-1$, are differentiable (up to order $\alpha \in \mathbb{Z}_+^d$) at the origin. Also note that m_ν is a trigonometric polynomial if and only if its polyphase components are trigonometric polynomials.

Consider the following $(r+1) \times m$ matrices

$$\mathbf{M}(\xi) := \{m_\nu(\xi + M^{*-1}q_l)\}, \quad \widetilde{\mathbf{M}}(\xi) := \{\widetilde{m}_\nu(\xi + M^{*-1}q_l)\},$$

where $\nu = 0, \dots, m-1$, $l = 0, \dots, r$, and the corresponding polyphase matrices

$$\mathcal{M} \quad := \quad \begin{pmatrix} \mu_{00} & \dots & \mu_{0,m-1} \\ \vdots & \ddots & \vdots \\ \mu_{r,0} & \dots & \mu_{r,m-1} \end{pmatrix}, \qquad \widetilde{\mathcal{M}} \quad := \quad \begin{pmatrix} \widetilde{\mu}_{00} & \dots & \widetilde{\mu}_{0,m-1} \\ \vdots & \ddots & \vdots \\ \widetilde{\mu}_{r,0} & \dots & \widetilde{\mu}_{r,m-1} \end{pmatrix}, \tag{3.18}$$

Proposition 3.3.1 *The equality* $\mathbf{M}^*\widetilde{\mathbf{M}} = I_m$ *is equivalent to*

$$\mathcal{M}^*\widetilde{\mathcal{M}} \equiv I_m. \tag{3.19}$$

Proof Since

$$\mathcal{M}(M^*\xi)U(\xi) = \mathbf{M}(\xi), \quad \mathcal{M}(M^*\xi) = \mathbf{M}(\xi)U^*(\xi),$$

$$\widetilde{\mathcal{M}}(M^*\xi)U(\xi) = \widetilde{\mathbf{M}}(\xi), \quad \widetilde{\mathcal{M}}(M^*\xi) = \widetilde{\mathbf{M}}(\xi)U^*(\xi)$$

and the matrix $U(\xi)$ is unitary, we get the required statement.◊

Due to Proposition 3.3.1, Matrix Extension Principle can be reformulated in terms of polyphase functions as follows.

Let $\varphi, \widetilde{\varphi} \in L_2(\mathbb{R}^d)$ be refinable functions and $\mu_{0k}, \widetilde{\mu}_{0,k}$, $k = 0, \dots, m-1$, be the polyphase components of their masks $m_0, \widetilde{m}_0$, respectively. Suppose there exist functions $\mu_{\nu,k}, \widetilde{\mu}_{\nu,k} \in L_2(\mathbb{T}^d)$, $k = 0, \dots, m-1$, $\nu = 1, \dots, r$, such that matrices (3.18) satisfy $\mathcal{M}^*\widetilde{\mathcal{M}} \equiv I_m$ and define wavelet functions $\psi^{(\nu)}, \widetilde{\psi}^{(\nu)}$, $\nu = 1, \dots, r$, by

$$\widehat{\psi^{(\nu)}}(\xi) = m_\nu(M^{*-1}\xi)\widehat{\varphi}(M^{*-1}\xi), \quad \widehat{\widetilde{\psi}^{(\nu)}}(\xi) = \widetilde{m}_\nu(M^{*-1}\xi)\widehat{\widetilde{\varphi}}(M^{*-1}\xi),$$

where $m_\nu, \widetilde{m}_\nu$ are functions whose polyphase components are $\mu_{\nu,k}, \widetilde{\mu}_{\nu,k}$, $k = 0, \dots, m-1$, respectively. The wavelet systems $\{\psi_{jk}^{(\nu)}\}_{j,k,\nu}$, $\{\widetilde{\psi}_{jk}^{(\nu)}\}_{j,k,\nu}$ is said to be generated from $\varphi, \widetilde{\varphi}$ (or from $m_0, \widetilde{m}_0$) by MEP.

The following auxiliary assertion will be useful for us.

Proposition 3.3.2 *Let $\alpha \in \mathbb{Z}_+^d$, functions $\mu_{0k} \in L_2(\mathbb{T}^d)$, $k = 0, \dots, m-1$, have continuous derivatives up to order α at the origin, and let m_0 be the function whose polyphase components are μ_{0k}, $k = 0, \dots, m-1$. If there exist complex numbers λ_γ, $\gamma \in \mathbb{Z}_+^d$, $\gamma \le \alpha$, such that*

$$D^\beta \mu_{0k}(\mathbf{0}) = \frac{(2\pi i)^{[\beta]}}{\sqrt{m}} \sum_{\mathbf{0} \le \gamma \le \beta} \lambda_\gamma \binom{\beta}{\gamma} (-M^{-1}s_k)^{\beta-\gamma} \tag{3.20}$$

for all $\beta \in \mathbb{Z}_+^d$, $\beta \le \alpha$, and $k = 0, \dots, m-1$, then

$$\lambda_\beta = \frac{1}{(2\pi i)^{[\beta]}} D^\beta \left(m_0(M^{*-1}\xi)\right)\Big|_{\xi=\mathbf{0}} \quad \forall \beta \in \mathbb{Z}_+^d, \beta \le \alpha, \tag{3.21}$$

and

$$D^\beta \left(m_0(M^{*-1}\xi)\right)\Big|_{\xi=q} = 0 \quad \forall \beta \in \mathbb{Z}_+^d, \beta \le \alpha. \tag{3.22}$$

for every $q \in D(M^*)$, $q \ne \mathbf{0}$. Conversely, if (3.22) is satisfied for m_0 and numbers λ_γ are given by (3.21), then relation (3.20) holds.

Proof Let (3.20) holds for some complex numbers λ_γ and $q \in D(M^*)$. Setting $r_k = M^{-1}s_k$, by Leibniz formula and (3.20), we have

$$D^\beta \left(e^{2\pi i(r_k,\xi)}\mu_{0k}(\xi)\right)\Big|_{\xi=q} = \sum_{\mathbf{0}\le\gamma\le\beta} \binom{\beta}{\gamma} (2\pi i r_k)^\gamma e^{2\pi i(r_k,q)} D^{\beta-\gamma}\mu_{0k}(\mathbf{0}) =$$

$$\frac{(2\pi i)^{[\beta]}}{\sqrt{m}} e^{2\pi i(r_k,q)} \sum_{\mathbf{0}\le\gamma\le\beta} \binom{\beta}{\gamma} r_k{}^\gamma \sum_{\mathbf{0}\le\varepsilon\le\beta-\gamma} \lambda_\varepsilon \binom{\beta-\gamma}{\varepsilon} (-r_k)^{\beta-\gamma-\varepsilon} =$$

$$\frac{(2\pi i)^{[\beta]}}{\sqrt{m}} e^{2\pi i(r_k,q)} \sum_{\mathbf{0}\le\gamma\le\beta} \sum_{\mathbf{0}\le\varepsilon\le\beta-\gamma} \lambda_\varepsilon \binom{\beta-\gamma}{\varepsilon} \binom{\beta}{\gamma} (-r_k)^{\beta-\varepsilon}(-1)^{[\gamma]} =$$

$$\frac{(2\pi i)^{[\beta]}}{\sqrt{m}} e^{2\pi i(r_k,q)} \sum_{\mathbf{0}\le\varepsilon\le\beta} \lambda_\varepsilon (-M^{-1}s_k)^{\beta-\varepsilon} \binom{\beta}{\varepsilon} \sum_{\mathbf{0}\le\gamma\le\beta-\varepsilon} \binom{\beta-\varepsilon}{\gamma} (-1)^{[\gamma]}.$$

Since

$$\sum_{\mathbf{0}\le\gamma\le\beta-\varepsilon} \binom{\beta-\varepsilon}{\gamma} (-1)^{[\gamma]} = \prod_{j=1}^d (1-1)^{\beta_j-\varepsilon_j} = \begin{cases} 0, & \beta \ne \varepsilon, \\ 1, & \beta = \varepsilon, \end{cases}$$

we obtain

$$D^\beta \left(e^{2\pi i(M^{-1}s_k,\xi)}\mu_{0k}(\xi)\right)\Big|_{\xi=q} = \frac{(2\pi i)^{[\beta]}\lambda_\beta}{\sqrt{m}} e^{2\pi i(r_k,q)}, \quad k = 0, \dots, m-1.$$

It follows from (3.16) and Lemma 2.1.5 that

$$D^{\beta}(m_0(M^{*-1}\xi)\Big|_{\xi=q} = \frac{1}{\sqrt{m}}\sum_{k=0}^{m-1} D^{\beta}\left(e^{2\pi i(M^{-1}s_k,\xi)}\mu_{0k}(x)\right)\Big|_{\xi=q} =$$

$$\frac{(2\pi i)^{[\beta]}\lambda_\beta}{m}\sum_{k=0}^{m-1} e^{2\pi i(r_k,q)} = \begin{cases} (2\pi i)^{[\beta]}\lambda_\beta, & q = \mathbf{0}, \\ 0, & q \neq \mathbf{0}, \end{cases}$$

which completes the proof of the first claim.

Next we will prove the second claim by induction on $n := [\alpha]$. The base for $n = 0$ follows immediately from (3.17). To prove the inductive step $n \to n+1$, we assume that our second claim holds true for all $\alpha \in \mathbb{Z}^d$ such that $[\alpha] < n+1$. Now we suppose that $\alpha \in \mathbb{Z}^d$, $[\alpha] = n+1$, and (3.22) is satisfied. Let numbers λ_γ be given by (3.21) and $q \in D(M^*)$. Using (3.16), and Leibniz formula, we obtain

$$D^{\alpha}(m_0(M^{*-1}x)\Big|_{x=q} = \frac{1}{\sqrt{m}}\sum_{k=0}^{m-1}\sum_{\mathbf{0}\le\beta\le\alpha}\binom{\alpha}{\beta} D^{\alpha-\beta}\left(e^{2\pi i(r_k,x)}\right)\Big|_{x=q} D^{\beta}\mu_{0k}(\mathbf{0}) =$$

$$\frac{1}{\sqrt{m}}\sum_{k=0}^{m-1} e^{2\pi i(r_k,x)}\left(D^{\alpha}\mu_{0k}(0) + \sum_{\mathbf{0}\le\beta<\alpha}\binom{\alpha}{\beta} D^{\alpha-\beta}(2\pi i r_k)^{\alpha-\beta} D^{\beta}\mu_{0k}(\mathbf{0})\right). \quad (3.23)$$

Since if (3.22) is satisfied for α, then it is also satisfied for any $\alpha' \in \mathbb{Z}_+^d$, $\alpha' < \alpha$, because of the inductive hypotheses, we have

$$\sum_{\mathbf{0}\le\beta<\alpha}\binom{\alpha}{\beta}(2\pi i r_k)^{\alpha-\beta} D^{\beta}\mu_{0k}(\mathbf{0}) =$$

$$\frac{(2\pi i)^{[\alpha]}}{\sqrt{m}}\sum_{\mathbf{0}\le\beta<\alpha}\sum_{\mathbf{0}\le\gamma\le\beta}\lambda_\gamma\binom{\alpha}{\beta}\binom{\beta}{\gamma} r_k^{\alpha-\beta}(-r_k)^{\beta-\gamma} =$$

$$\frac{(2\pi i)^{[\alpha]}}{\sqrt{m}}\sum_{\mathbf{0}\le\gamma<\alpha}\lambda_\gamma r_k^{\alpha-\gamma}\sum_{\gamma\le\beta<\alpha}\binom{\alpha}{\beta}\binom{\beta}{\gamma}\prod_{j=1}^{d}(-1)^{\beta_j-\gamma_j} =$$

$$\frac{(2\pi i)^{[\alpha]}}{\sqrt{m}}\sum_{\mathbf{0}\le\gamma<\alpha}\binom{\alpha}{\gamma}\lambda_\gamma {r_k}^{\alpha-\gamma}\sum_{\mathbf{0}\le\delta<\alpha-\gamma}\binom{\alpha-\gamma}{\delta}\prod_{j=1}^{d}(-1)^{\delta_j} =$$

$$\frac{(2\pi i)^{[\alpha]}}{\sqrt{m}}\sum_{\mathbf{0}\le\gamma<\alpha}\binom{\alpha}{\gamma}\lambda_\gamma {r_k}^{\alpha-\gamma}\left(\prod_{j=1}^{d}(1-1)^{\alpha_j-\gamma_j} - \prod_{j=1}^{d}(-1)^{\alpha_j-\gamma_j}\right) =$$

$$-\frac{(2\pi i)^{[\alpha]}}{\sqrt{m}}\sum_{\mathbf{0}\le\gamma<\alpha}\binom{\alpha}{\gamma}\lambda_\gamma(-M^{-1}s_k)^{\alpha-\gamma}.$$

Combining this with (3.23), using (3.21) and (3.22) we obtain

$$\sum_{k=0}^{m-1} e^{2\pi i(r_k,q)}\left(D^\alpha \mu_{0k}(0) - \frac{(2\pi i)^{[\alpha]}}{\sqrt{m}} \sum_{0\le\gamma<\alpha} \binom{\alpha}{\gamma} \lambda_\gamma (-M^{-1}s_k)^{\alpha-\gamma}\right) =$$

$$\sqrt{m} D^\alpha (m_0(M^{*-1}x))\Big|_{x=q} = \begin{cases} \sqrt{m}(2\pi i)^{[\alpha]}\lambda_\alpha, & q = \mathbf{0}, \\ 0, & q \ne \mathbf{0}. \end{cases}$$

Due to Lemma 2.1.5, the linear system

$$\sum_{k=0}^{m-1} e^{2\pi i(r_k,q_l)} y_k = \sqrt{m}(2\pi i)^{[\alpha]}\lambda_\alpha \delta_{0l}, \quad l = 0, \dots, m-1,$$

has a unique solution $y_k = \frac{(2\pi i)^{[\alpha]}\lambda_\alpha}{\sqrt{m}}$, $k = 0, \dots, m-1$. It follows that

$$D^\alpha \mu_{0k}(0) - \frac{(2\pi i)^{[\alpha]}}{\sqrt{m}} \sum_{0\le\gamma<\alpha} \binom{\alpha}{\gamma} \lambda_\gamma (-M^{-1}s_k)^{\alpha-\gamma} = \frac{(2\pi i)^{[\alpha]}}{\sqrt{m}}\lambda_\alpha,$$

which was to be proved.◊

Condition (3.22) is usually referred as *sum rule* because of its equivalent form in terms of Fourier coefficients of m_0.

Note that in particular we established in Proposition 3.3.2 that given α, the set of parameters $\widetilde{\lambda}_\beta$, $\beta \in \mathbb{Z}_+^d$, $\beta \le \alpha$, in (3.20) is unique, and $\widetilde{\lambda}_\beta$ does not depend on α.

Theorem 3.3.3 *Let $\alpha \in \mathbb{Z}_+^d$, functions $\mu_{\nu,k}, \widetilde{\mu}_{\nu k} \in L_2(\mathbb{T}^d)$, $\nu, k = 0, \dots, r$, $r \ge m-1$, have continuous derivatives up to order α at the origin. Suppose square matrices $\mathcal{N} := \{\mu_{\nu k}\}_{\nu,k=0}^r$ and $\widetilde{\mathcal{N}} := \{\widetilde{\mu}_{\nu k}\}_{\nu,k=0}^r$ satisfy*

$$\mathcal{N}\widetilde{\mathcal{N}}^* = I_{r+1}; \tag{3.24}$$

$\widetilde{m}_\nu$, $\nu = 0, \dots, r$, is the function whose polyphase components are $\widetilde{\mu}_{\nu k}$, $k = 0, \dots, m-1$ and $\widetilde{m}_0(\mathbf{0}) = 1$. Then condition (3.15) holds if and only if

(a) there exist complex numbers λ_γ, $\gamma \in \mathbb{Z}_+^d$, $\gamma \le \alpha$, such that $\lambda_{\mathbf{0}} = 1$ and (3.20) is satisfied for all $\beta \in \mathbb{Z}_+^d$, $\beta \le \alpha$, and $k = 0, \dots, m-1$;
(b) $D^\gamma \mu_{0k}(\mathbf{0}) = 0$, $k = m, \dots, r$ for all $\gamma \in \mathbb{Z}_+^d$, $\gamma \le \alpha$.

Proof Suppose that (3.15) is valid and prove (a) and (b) by induction on $n := [\alpha]$. Check the initial step for $n = 0$. Let $\widetilde{m}_\nu(\mathbf{0}) = 0$, $\nu = 1, \dots, r$. It follows from (3.16) that

$$\sum_{k=0}^{m-1} \widetilde{\mu}_{\nu k}(\mathbf{0}) = 0, \quad \nu = 1, \dots, r. \tag{3.25}$$

On the other hand, by (3.24),

$$\sum_{k=0}^{r} \overline{\mu_{0k}(\mathbf{0})}\widetilde{\mu}_{\nu k}(\mathbf{0}) = 0, \quad \nu = 1, \ldots, m-1.$$

Because of linear independence of the vectors $(\widetilde{\mu}_{\nu 0}(\mathbf{0}), \ldots, \widetilde{\mu}_{\nu r}(\mathbf{0})) \in \mathbb{R}^{r+1}$, $\nu = 1, \ldots, r$, there exists $\lambda_{\mathbf{0}}$ so that

$$\mu_{00}(\mathbf{0}) = \cdots = \mu_{0,m-1}(\mathbf{0}) = \frac{\lambda_{\mathbf{0}}}{\sqrt{m}}, \quad \mu_{0m}(\mathbf{0}) = \cdots = \mu_{0r}(\mathbf{0}) = 0.$$

It follows from Proposition 3.3.1 that $m_0(\mathbf{0}) = 1$, where m_0 is the function whose polyphase components are μ_{0k}, $k = 0, \ldots, m-1$. Hence,

$$\frac{1}{\sqrt{m}}\left(\mu_{00}(\mathbf{0}) + \cdots + \mu_{0,m-1}(\mathbf{0})\right) = 1,$$

which yields $\lambda_{\mathbf{0}} = 1$.

For the inductive step $n \to n+1$, we assume that (3.15) is valid for some $\alpha \in \mathbb{Z}_+^d$, $[\alpha] = n+1$ and (a), (b) holds for all $\alpha' \in \mathbb{Z}_+^d$, $[\alpha'] \le n$, in particular, for every $\alpha' < \alpha$. So, due to Proposition 3.3.2, there exist constants $\lambda_\gamma \in \mathbb{C}$, $\gamma \in \mathbb{Z}_+^d$, $\gamma < \alpha$, such that (3.20) holds for all $\beta < \alpha$. If $\gamma \in \mathbb{Z}_+^d$, $\gamma < \alpha$, due to (3.16) and the Leibniz formula, we have

$$\frac{1}{\sqrt{m}} \sum_{\mathbf{0} \le \beta \le \alpha - \gamma} \binom{\alpha - \gamma}{\beta} \sum_{k=0}^{m-1} (2\pi i M^{-1} s_k)^{\alpha - \beta - \gamma} D^{\beta} \widetilde{\mu}_{\nu k}(\mathbf{0}) = D^{\alpha-\gamma}\widetilde{m}_\nu(M^{*-1}x)\Big|_{x=\mathbf{0}} = 0. \tag{3.26}$$

It follows from (3.24) that

$$\sum_{k=0}^{r} \overline{\mu_{0k}}\widetilde{\mu}_{\nu k} = 0, \quad \nu = 1, \ldots, m-1.$$

Differentiating this equality α times, we obtain

$$\sum_{\mathbf{0} \le \beta \le \alpha} \binom{\alpha}{\beta} \sum_{k=0}^{r} \overline{D^{\alpha-\beta}\mu_{0k}(\mathbf{0})} D^{\beta}\widetilde{\mu}_{\nu k}(\mathbf{0}) = 0.$$

Taking into account the inductive hypotheses, we have

$$\sum_{\mathbf{0}\le\beta\le\alpha}\binom{\alpha}{\beta}\sum_{k=0}^{m-1}\overline{D^{\alpha-\beta}\mu_{0k}(\mathbf{0})}D^{\beta}\widetilde{\mu}_{\nu k}(\mathbf{0})+\sum_{k=m}^{r}\overline{D^{\alpha}\mu_{0k}(\mathbf{0})}\widetilde{\mu}_{\nu k}(\mathbf{0})=0. \tag{3.27}$$

Multiply (3.26) by $(-2\pi i)^{[\gamma]}\binom{\alpha}{\alpha-\gamma}\overline{\lambda_\gamma}$ and subtract from (3.27). After the same manipulation with each $\gamma\in\mathbb{Z}_+^d$, $\gamma<\alpha$, we obtain

$$0=\sum_{k=m}^{r}\overline{D^{\alpha}\mu_{0k}(\mathbf{0})}\widetilde{\mu}_{\nu k}(\mathbf{0})+\sum_{\mathbf{0}\le\beta\le\alpha}\binom{\alpha}{\beta}\sum_{k=0}^{m-1}\overline{D^{\alpha-\beta}\mu_{0k}(\mathbf{0})}D^{\beta}\widetilde{\mu}_{\nu k}(\mathbf{0})$$

$$-\sum_{\mathbf{0}\le\gamma<\alpha}\frac{(-2\pi i)^{[\gamma]}}{\sqrt{m}}\binom{\alpha}{\alpha-\gamma}\overline{\lambda_\gamma}\sum_{\mathbf{0}\le\beta\le\alpha-\gamma}\binom{\alpha-\gamma}{\beta}\sum_{k=0}^{m-1}(2\pi i M^{-1}s_k)^{\alpha-\beta-\gamma}D^{\beta}\widetilde{\mu}_{\nu k}(\mathbf{0})$$

$$=\sum_{k=m}^{r}\overline{D^{\alpha}\mu_{0k}(\mathbf{0})}\widetilde{\mu}_{\nu k}(\mathbf{0})+\sum_{\mathbf{0}\le\beta\le\alpha}\binom{\alpha}{\beta}\sum_{k=0}^{m-1}\Bigg(\overline{D^{\alpha-\beta}\mu_{0k}(\mathbf{0})-}$$

$$\overline{\sum_{\substack{\gamma\ne\alpha\\ \mathbf{0}\le\gamma\le\alpha-\beta}}\frac{(2\pi i)^{[\gamma]}}{\sqrt{m}}\binom{\alpha-\gamma}{\beta}\binom{\alpha}{\alpha-\gamma}\binom{\alpha}{\beta}^{-1}\lambda_\gamma(-2\pi i M^{-1}s_k)^{\alpha-\beta-\gamma}}\Bigg)D^{\beta}\widetilde{\mu}_{\nu k}(\mathbf{0}).$$

From this, taking into account that

$$\binom{\alpha-\gamma}{\beta}\binom{\alpha}{\alpha-\gamma}\binom{\alpha}{\beta}^{-1}=\frac{(\alpha-\beta)!}{\gamma!(\alpha-\beta-\gamma)!}=\binom{\alpha-\beta}{\gamma}, \tag{3.28}$$

and using the inductive hypotheses, the sum over β is deduced to

$$\sum_{\mathbf{0}\le\beta\le\alpha}\binom{\alpha}{\beta}\sum_{k=0}^{m-1}\Bigg(\overline{D^{\alpha-\beta}\mu_{0k}(\mathbf{0})-}$$

$$\overline{\frac{(2\pi i)^{[\alpha-\beta]}}{\sqrt{m}}\sum_{\substack{\gamma\ne\alpha\\ \mathbf{0}\le\gamma\le\alpha-\beta}}\binom{\alpha-\beta}{\gamma}\lambda_\gamma(-M^{-1}s_k)^{\alpha-\beta-\gamma}}\Bigg)D^{\beta}\widetilde{\mu}_{\nu k}(\mathbf{0})=$$

$$\sum_{k=0}^{m-1}\left(\overline{D^{\alpha}\mu_{0k}(\mathbf{0})-\frac{(2\pi i)^{[\alpha]}}{\sqrt{m}}\sum_{\mathbf{0}\le\gamma<\alpha}\binom{\alpha}{\gamma}\lambda_\gamma(-M^{-1}s_k)^{\alpha-\gamma}}\right)\widetilde{\mu}_{\nu k}(\mathbf{0}).$$

So, we have

$$\sum_{k=0}^{m-1} \overline{\left(D^\alpha \mu_{0k}(\mathbf{0}) - \frac{(2\pi i)^{[\alpha]}}{\sqrt{m}} \sum_{0 \le \gamma < \alpha} \binom{\alpha}{\gamma} \lambda_\gamma (M^{-1} s_k)^{\alpha-\gamma} \right)} \widetilde{\mu}_{\nu k}(\mathbf{0}) + \sum_{k=m}^{r} \overline{D^\alpha \mu_{0k}(\mathbf{0})} \widetilde{\mu}_{\nu k}(\mathbf{0}) = 0.$$

Similarly to the arguments for the initial step, it follows from (3.25) that there exists $\widetilde{\lambda}_\alpha$ such that

$$D^\alpha \mu_{0k}(\mathbf{0}) - \frac{(2\pi i)^{[\alpha]}}{\sqrt{m}} \sum_{0 \le \gamma < \alpha} \binom{\alpha}{\gamma} \lambda_\gamma (-M^{-1} s_k)^{\alpha-\gamma} = \frac{(2\pi i)^{[\alpha]} \lambda_\alpha}{\sqrt{m}}, \quad k = 0, \dots, m-1,$$

$$D^\alpha \mu_{0k}(\mathbf{0}) = 0, \quad k = m, \dots, r.$$

Thus, (3.20) is valid for $\beta = \alpha$, as was to be proved.

Now we assume that (a), (b) are valid and prove (3.15) by induction on n. If (3.20) is valid for $\alpha = \mathbf{0}$, then $\mu_{0k}(\mathbf{0}) = 1/\sqrt{m}$, $k = 0, \dots, m-1$. It follows from (3.24) and (b) that

$$\widetilde{\mu}_{\nu 0}(\mathbf{0}) + \cdots + \widetilde{\mu}_{\nu, m-1}(\mathbf{0}) = 0, \quad \nu = 1, \dots, r.$$

Hence, on the basis of (3.16), $\widetilde{m}_\nu(\mathbf{0}) = 0$, $\nu = 1, \dots, r$, what proves the initial step for $n = 0$.

For the inductive step, we assume that (a), (b) is valid for some $\alpha \in \mathbb{Z}_+^d$, $[\alpha] = n+1$, and (3.15) holds for all $\alpha' \in \mathbb{Z}_+^d$, $[\alpha'] \le n$, in particular, for every $\alpha' < \alpha$, i.e.,

$$D^{\alpha-\gamma} \widetilde{m}_\nu (M^{*-1}\xi) \Big|_{\xi=\mathbf{0}} = 0, \quad \gamma \in \mathbb{Z}_+^d, \quad \gamma \ne \mathbf{0}, \quad \gamma \le \alpha.$$

This yields (3.26) for $\gamma \ne \mathbf{0}$. Multiply (3.26) by $(-2\pi i)^{[\gamma]} \binom{\alpha}{\alpha-\gamma} \overline{\lambda_\gamma}$ and add to (3.16) differentiated α times. After the same manipulation with each $\gamma \in \mathbb{Z}_+^d$, $\gamma < \alpha$, we obtain

$$D^\alpha \widetilde{m}_\nu (M^{*-1}\xi) \Big|_{\xi=\mathbf{0}} = \frac{1}{\sqrt{m}} \sum_{0 \le \beta \le \alpha} \binom{\alpha}{\beta} \sum_{k=0}^{m-1} (2\pi i M^{-1} s_k)^{\alpha-\beta} D^\beta \widetilde{\mu}_{\nu k}(\mathbf{0})$$

$$+ \sum_{0 < \gamma \le \alpha} \frac{(-2\pi i)^{[\gamma]}}{\sqrt{m}} \binom{\alpha}{\alpha-\gamma} \overline{\lambda_\gamma} \sum_{0 \le \beta \le \alpha-\gamma} \binom{\alpha-\gamma}{\beta} \sum_{k=0}^{m-1} (2\pi i M^{-1} s_k)^{\alpha-\beta-\gamma} D^\beta \widetilde{\mu}_{\nu k}(\mathbf{0})$$

$$= \sum_{0 \le \beta \le \alpha} \binom{\alpha}{\beta} \sum_{0 \le \gamma \le \alpha-\beta} \frac{(-2\pi i)^{[\gamma]}}{\sqrt{m}} \overline{\lambda_\gamma} \binom{\alpha}{\alpha-\gamma} \binom{\alpha-\gamma}{\beta} \binom{\alpha}{\beta}^{-1}$$

$$\times \sum_{k=0}^{m-1} (2\pi i M^{-1} s_k)^{\alpha-\beta-\gamma} D^\beta \widetilde{\mu}_{\nu k}(\mathbf{0}).$$

Due to (3.3), (3.20), and (*b*), this yields

$$D^{\alpha}\widetilde{m}_{\nu}(M^{*-1}\xi)\Big|_{\xi=\mathbf{0}} =$$

$$\overline{\sum_{\mathbf{0}\le\beta\le\alpha}\binom{\alpha}{\beta}\sum_{k=0}^{m-1}\frac{(2\pi i)^{[\alpha-\beta]}}{\sqrt{m}}\sum_{\mathbf{0}\le\gamma\le\alpha-\beta}\lambda_{\gamma}\binom{\alpha-\beta}{\gamma}(-M^{-1}s_k)^{\alpha-\beta-\gamma}}D^{\beta}\widetilde{\mu}_{\nu k}(\mathbf{0})$$

$$= \sum_{\mathbf{0}\le\beta\le\alpha}\binom{\alpha}{\beta}\sum_{k=0}^{m-1}\overline{D^{\alpha-\beta}\mu_{0k}(\mathbf{0})}D^{\beta}\widetilde{\mu}_{\nu k}(\mathbf{0}) = D^{\alpha}\left(\sum_{k=0}^{m-1}\overline{\mu_{0k}(\xi)}\widetilde{\mu}_{\nu k}(\xi)\right)\Bigg|_{\xi=\mathbf{0}}$$

$$= D^{\alpha}\left(\sum_{k=0}^{r}\overline{\mu_{0k}(\xi)}\widetilde{\mu}_{\nu k}(\xi)\right)\Bigg|_{\xi=\mathbf{0}}.$$

It follows from (3.24) that $D^{\alpha}\widetilde{m}_{\nu}(M^{*-1}\xi)\Big|_{\xi=\mathbf{0}} = 0$ as was to be proved.◊

We obtained a criterium for (3.15) which is equivalent to VM_{α} property, $\alpha \in \mathbb{Z}_+^d$, for the system $\{\widetilde{\psi}_{jk}^{(\nu)}\}_{j,k,\nu}$. But we are more interested in a similar result for the property VM^n, $n \in \mathbb{Z}_+$, which was required in Theorem 3.2.4 to provide a desirable approximation order. It is not difficult to see that $\{\widetilde{\psi}_{jk}^{(\nu)}\}_{j,k,\nu}$ has VM^n property if and only if (3.15) is valid for all $\alpha \in \mathbb{Z}^d$, $[\alpha] < n$. Taking this into account, we see that the following statement follows immediately from Theorem 3.3.3 and Proposition 3.3.2.

Theorem 3.3.4 *Let $n \in \mathbb{Z}_+$, functions $\mu_{\nu,k}, \widetilde{\mu}_{\nu k} \in L_2(\mathbb{T}^d)$, $\nu, k = 0, \dots, r$, $r \ge m-1$, have continuous derivatives up to order $n \in \mathbb{Z}_+$ at the origin. Suppose square matrices $\mathcal{N} := \{\mu_{\nu k}\}_{\nu,k=0}^{r}$ and $\widetilde{\mathcal{N}} := \{\widetilde{\mu}_{\nu k}\}_{\nu,k=0}^{r}$ satisfy*

$$\mathcal{N}\widetilde{\mathcal{N}}^{*} = I_{r+1}; \tag{3.29}$$

$\widetilde{m}_{\nu}$, $\nu = 0, \dots, r$, is the function whose polyphase components are $\widetilde{\mu}_{\nu k}$, $k = 0, \dots, m-1$, and $\widetilde{m}_0(\mathbf{0}) = 1$. Then condition (3.15) is valid for all $\alpha \in \mathbb{Z}^d$, $[\alpha] < n$, if and only if

(a) there exist complex numbers λ_{γ}, $\gamma \in \mathbb{Z}_+^d$, $[\gamma] < n$, such that $\lambda_{\mathbf{0}} = 1$ and (3.20) holds for all $\beta \in \mathbb{Z}^d$, $[\beta] < n$, $k = 0, \dots, m-1$;
(b) $D^{\gamma}\mu_{0k}(\mathbf{0}) = 0$, $k = m, \dots, r$ for all $\gamma \in \mathbb{Z}_+^d$, $[\gamma] < n$.

Theorem 3.3.4 gives a solution to our problem in the case $r = m-1$. Indeed, (3.15) is equivalent to condition (a), which depends only on m_0; in this case, i.e., only the first row of the matrix $\mathcal{M}$ (which coincides with $\mathcal{N}$ if $r = m-1$) is responsible for VM^n property for $\{\widetilde{\psi}_{jk}^{(\nu)}\}_{j,k,\nu}$. In the general case, the problem has not solved because a criterium for (3.15) is given in terms of $(r+1) \times (r+1)$ matrices $\mathcal{N}, \widetilde{\mathcal{N}}$, while Matrix Extension Principle is given in terms of $r \times (m-1)$ matrices $\mathcal{M}, \widetilde{\mathcal{M}}$. In the sequel, Theorem 3.3.4 will be used to find a sufficient condition for the existence of matrices $\mathcal{M}, \widetilde{\mathcal{M}}$ providing vanishing moments for $\{\widetilde{\psi}_{jk}^{(\nu)}\}_{j,k,\nu}$. Now we mention only that VM^n property for $\{\widetilde{\psi}_{jk}^{(\nu)}\}_{j,k,\nu}$ depends not only on m_0, but also on the matrices $\mathcal{M}$, $\widetilde{\mathcal{M}}$ in the case $r > m-1$. This may be illustrated by the following example.

Let $d = 1$, $M = m = 2$, $\mu_{00} = \mu_{01} = \widetilde{\mu}_{00} = \widetilde{\mu}_{01} \equiv \frac{1}{\sqrt{2}}$,

$$\mathcal{M} = \widetilde{\mathcal{M}} = \begin{pmatrix} \frac{1}{\sqrt{2}} & \frac{1}{\sqrt{2}} \\ \frac{1}{2} & -\frac{1}{2} \\ \frac{1}{2} & -\frac{1}{2} \end{pmatrix}, \quad \widetilde{\mathcal{M}}' = \begin{pmatrix} \frac{1}{\sqrt{2}} & \frac{1}{\sqrt{2}} \\ \frac{1}{\sqrt{2}} & 0 \\ 0 & \frac{1}{\sqrt{2}} \end{pmatrix}, \quad \mathcal{M}' = \begin{pmatrix} \frac{1}{\sqrt{2}} & \frac{1}{\sqrt{2}} \\ \frac{1}{\sqrt{2}} & -\frac{1}{\sqrt{2}} \\ -\frac{1}{\sqrt{2}} & \frac{1}{\sqrt{2}} \end{pmatrix}.$$

Either of pairs $\mathcal{M}, \widetilde{\mathcal{M}}$ and $\mathcal{M}', \widetilde{\mathcal{M}}'$ satisfies (3.19). The matrices $\mathcal{M}, \widetilde{\mathcal{M}}$ generate wavelet masks $m_1(\xi) = m_2(\xi) = \widetilde{m}_1(\xi) = \widetilde{m}_2(\xi) = \frac{1}{2\sqrt{2}} - \frac{1}{2\sqrt{2}}e^{2\pi i\xi}$. It is clear that $m_1(0) = m_2(0) = \widetilde{m}_1(0) = \widetilde{m}_2(0) = 0$; i.e., for the corresponding wavelet systems, VM^1 property is valid. The matrices $\mathcal{M}', \widetilde{\mathcal{M}}'$ generate wavelet masks

$$\widetilde{m}_1'(\xi) = \frac{1}{2}, \quad \widetilde{m}_2'(\xi) = \frac{1}{2}e^{2\pi i\zeta}, \quad m_1'(\xi) = \frac{1}{2} - \frac{1}{2}e^{2\pi i\xi}, m_2'(\xi) = -\frac{1}{2} + \frac{1}{2}e^{2\pi i\xi},$$

and we have $\widetilde{m}_1'(0) \neq 0$, $\widetilde{m}_2'(0) \neq 0$.

Let $n \in \mathbb{Z}_+$, we will denote by $L_\infty^{(n)}$ the class of complex-valued functions which are in $L_\infty(\mathbb{T}^d)$ and have continuous derivatives up to order n at the origin.

Lemma 3.3.5 *Let $\mu_{\nu 0}, \widetilde{\mu}_{\nu 0} \in L_\infty^{(n)}$, $\nu = 0, \ldots, r$, and*

$$\sum_{\nu=0}^{r} \mu_{\nu 0}\overline{\widetilde{\mu}_{\nu 0}} \equiv 1. \tag{3.30}$$

Then there exist functions $\mu_{\nu k}, \widetilde{\mu}_{\nu k} \in L_\infty^{(n)}$, $\nu = 0, \ldots, r$, $k = 1, \ldots, r$, such that

$$\sum_{\nu=0}^{r} \mu_{\nu l}\overline{\widetilde{\mu}_{\nu k}} \equiv \delta_{kl}, \quad k, l = 0, \ldots, r. \tag{3.31}$$

Proof Set

$$\mu_{\nu 0}' := \begin{cases} \mu_{\nu 0} \Big/ \sqrt{\sum\limits_{l=0}^{r} |\mu_{l0}|^2}, & \text{if } \sqrt{\sum\limits_{l=0}^{r} |\mu_{l0}|^2} \neq 0, \\ 1/\sqrt{m}, & \text{if } \sqrt{\sum\limits_{l=0}^{r} |\mu_{l0}|^2} = 0, \end{cases} \qquad \nu = 0, \ldots, r.$$

It is clear that the functions $\mu_{\nu 0}'$ are essentially bounded and

$$\sum_{\nu=0}^{r} |\mu'_{\nu 0}|^2 \equiv 1. \tag{3.32}$$

It follows from (3.30) that $\sum_{l=0}^{r} |\mu_{l0}(\mathbf{0})|^2 \neq 0$. So, $\mu'_{\nu 0} \in L_\infty^{(n)}$, $\nu = 0, \dots, r$. Let us extend the unit vector $\mu'_{00}, \dots, \mu'_{r0}$ to a unitary matrix. Due to (3.32), there exist ν_0 so that $\mu'_{\nu_0 0}(\mathbf{0}) \neq 1$. We may assume that $\nu_0 = 0$ (else we will interchange $\mu'_{\nu_0 0}$ and μ'_{00}, extend this new vector to a unitary matrix and interchange its 0-th and ν_0-th rows). Due to Householder transform, an extension to a unitary matrix may be found by:

$$\mu'_{0k} = \overline{\mu'_{k0}} \frac{1 - \mu'_{00}}{1 - \overline{\mu'_{00}}}, \quad \mu'_{\nu k} = \delta_{\nu k} - \frac{\mu'_{\nu 0} \overline{\mu'_{k0}}}{1 - \overline{\mu'_{00}}}, \quad \nu, k = 1, \dots, r.$$

Because of (3.32), we have $|\mu'_{\nu 0}| \leq \sqrt{1 - |\mu'_{00}|^2}$, $\nu = 1, \dots, r$. This yields essential boundedness of the functions $\mu'_{\nu k}$. Since $1 - \mu'_{00}(\mathbf{0}) \neq 0$, it follows that $\mu'_{\nu k} \in L_\infty^{(n)}$, $\nu, k = 0, \dots, r$. Set

$$\begin{aligned}
&\widetilde{\mu}_{\nu k} := \mu'_{\nu k}, \quad \nu = 0, \dots, r, \ k = 1, \dots, r, \\
&\widetilde{Q}_k := (\widetilde{\mu}_{0k}, \dots, \widetilde{\mu}_{rk}), \quad k = 0, \dots, r, \\
&Q_0 := (\mu_{00}, \dots, \mu_{r0}), \quad Q_k := \widetilde{Q}_k - \widetilde{Q}_k \widetilde{Q}_0^* Q_0, \quad k = 1, \dots, r.
\end{aligned}$$

It is not difficult to see that the entries of Q_k are in $L_\infty^{(n)}$ and $Q_k \widetilde{Q}_l^* = \delta_{kl}$, $k, l = 0, \dots, r$. It remains to denote by $\mu_{\nu k}$ the ν-th component of Q_k.◊

Lemma 3.3.6 *Let A be a class of complex-valued functions such that*

(i) if $f, g \in A$, $a, b \in \mathbb{C}$, then $af + bg \in A$,
(ii) if $f, g \in A$, then $fg \in A$,
and let $\mathcal{A}$ be a class of matrices whose entries are in A. If any two $n \times 1$ matrices Q, $\widetilde{Q} \in \mathcal{A}$ satisfying $Q^* \widetilde{Q} = 1$ can be extended to $n \times n$ matrices $\mathcal{N}, \widetilde{\mathcal{N}} \in \mathcal{A}$ satisfying $\mathcal{N}^* \widetilde{\mathcal{N}} = I_n$, then any two $n \times j$ matrices $\mathcal{M}$, $\widetilde{\mathcal{M}} \in \mathcal{A}$, $1 < j < n$, satisfying $\mathcal{M}^* \widetilde{\mathcal{M}} = I_j$, can be extended to $n \times n$ matrices $\mathcal{N}, \widetilde{\mathcal{N}} \in \mathcal{A}$ satisfying $\mathcal{N}^* \widetilde{\mathcal{N}} = I_n$.

Proof We will prove by induction on j. The base for $j = 1$ is given. Let us check the inductive step $j-1 \to j$. Let $j \times n$ matrices $\mathcal{M}$, $\widetilde{\mathcal{M}} \in \mathcal{A}$ satisfy $\mathcal{M}^* \widetilde{\mathcal{M}} = I_j$. Denote by Q_k, $\widetilde{Q}_k$ the k-th columns, respectively, of $\mathcal{M}, \widetilde{\mathcal{M}}$. Due to the statement of the base, the matrices Q_1, $\widetilde{Q}_1 \in \mathcal{A}$ can be extended to $n \times n$ matrices $\mathcal{N}', \widetilde{\mathcal{N}'} \in \mathcal{A}$ satisfying $\mathcal{N}'^* \widetilde{\mathcal{N}'} = I_n$. Let Q'_k, $\widetilde{Q}'_k$, $k = 2, \dots, n$, denote the k-th columns, respectively, of $\mathcal{N}', \widetilde{\mathcal{N}'} \in \mathcal{A}$. Fix a point x for which $Q'_l(x)^* \widetilde{Q}'_k(x) = \delta_{kl}$. Since the vectors $Q'_2(x), \dots, Q'_n(x)$ form a basis for the orthogonal complement to $\widetilde{Q}'_1(x)$ in $\mathbb{R}^n$, we have

$$Q_k(x) = \sum_{l=2}^{n} \alpha_{lk}(x) Q'_l(x), \quad k = 2, \dots, j.$$

Similarly,

$$\widetilde{Q}_k(x) = \sum_{l=2}^{n} \widetilde{\alpha}_{lk}(x)\widetilde{Q}'_l(x), \quad k = 2, \ldots, j.$$

It is clear that $\alpha_{lk}, \widetilde{\alpha}_{lk} \in A$ and $\sum_{l=2}^{n} \alpha_{lk}\overline{\widetilde{\alpha}}_{lk'} = \delta_{kk'}$, $k, k' = 2, \ldots, j$. Due to the inductive hypotheses, there exist functions $\alpha_{lk}, \widetilde{\alpha}_{lk} \in A$, $l = 2, \ldots, n$, $k = j + 1, \ldots, n$, such that

$$\sum_{l=2}^{n} \alpha_{lk}\overline{\widetilde{\alpha}}_{lk'} = \delta_{kk'}, \quad k, k' = 2, \ldots, n. \tag{3.33}$$

Set

$$Q_k := \sum_{l=2}^{n} \alpha_{lk} Q'_l, \quad \widetilde{Q}_k := \sum_{l=2}^{n} \widetilde{\alpha}_{lk} \widetilde{Q}'_l, \quad k = j+1, \ldots, n.$$

Because of (3.33) and biorthogonality of the systems $Q_1, Q'_2, \ldots, Q'_n$ and $\widetilde{Q}_1, \widetilde{Q}'_2, \ldots, \widetilde{Q}'_n$, we obtain

$$Q_l(x)^* \widetilde{Q}_k(x) = \delta_{kl}, \quad k, l = 1, \ldots, n.$$

To complete the proof, it remains to introduce matrices $\mathcal{N}$ and $\widetilde{\mathcal{N}}$ whose columns are, respectively, $Q_1, \ldots, Q_n$ and $\widetilde{Q}_1, \ldots, \widetilde{Q}_n$.$\Diamond$

Now we are ready to give a necessary condition for VM^n property.

Theorem 3.3.7 *Let $\varphi, \widetilde{\varphi} \in L_2(\mathbb{R}^d)$ be refinable functions whose Fourier transforms have continuous derivatives up to order n at the origin, and $\widehat{\varphi}(\mathbf{0}) = \widehat{\widetilde{\varphi}}(\mathbf{0}) = 1$. Suppose $\{\psi_{jk}^{(\nu)}\}_{j,k,\nu}$, $\{\widetilde{\psi}_{jk}^{(\nu)}\}_{j,k,\nu}$ are generated from $\varphi, \widetilde{\varphi}$ by MEP, and let the entries $\mu_{\nu k}, \widetilde{\mu}_{\nu k}$ of the corresponding polyphase matrices $\mathcal{M}, \widetilde{\mathcal{M}}$ be in $L_\infty^{(n)}$. If $\{\widetilde{\psi}_{jk}^{(\nu)}\}_{j,k,\nu}$ has VM^n property, then condition (3.20) is satisfied for all $\beta \in \mathbb{Z}_+^d$, $[\beta] < n$, and $k = 0, \ldots, m-1$.*

Proof Due to Lemmas 3.3.5, 3.3.6, the $(r+1) \times m$ matrices $\mathcal{M}, \widetilde{\mathcal{M}}$ can be extended to $(r+1) \times (r+1)$ matrices $\mathcal{N}, \widetilde{\mathcal{N}}$ such that their entries are in $L_\infty^{(n)}$ and (3.24) is satisfied. It remains to apply conditions (a) of Theorem 3.3.4.$\Diamond$

Next let us discuss how to construct a dual wavelet system with VM^n property generated from refinable functions $\varphi, \widetilde{\varphi}$ by MEP.

We know from Theorem 3.3.7 that the polyphase components of m_0 should satisfy (3.20) to provide VM^n property for the system $\{\widetilde{\psi}_{jk}^{(\nu)}\}_{j,k,\nu}$. Due to Theorem 3.3.4, this condition is also sufficient for the existence of proper polyphase matrices $\mathcal{M}, \widetilde{\mathcal{M}}$ in the case $r = m - 1$. The situation is different if $r > m - 1$. Now we are interested in sufficient conditions for the general case to develop a method for the construction of dual wavelet systems with vanishing moments.

Firstly, let us show that it is not sufficient to have an appropriate mask m_0 for the construction in the general case. Consider the following example. Let $d = 1$, $M = m = 2$, $s_0 = 0$, $s_1 = 1$, $r > 1$,

$$\mu_{00} = \widetilde{\mu}_{00} = \widetilde{\mu}_{01} \equiv \frac{1}{\sqrt{2}}, \quad \mu_{01} = \frac{1}{\sqrt{2}}\left(1 - \frac{i}{2}\sin 2\pi\xi\right).$$

It is clear that (3.20) holds for μ_{00} and μ_{01} with $n = 2$ and $\lambda_0 = 1$, $\lambda_1 = 0$. Assume that there exist dual wavelet systems $\{\psi_{jk}^{(\nu)}\}_{j,k,\nu}$, $\{\widetilde{\psi}_{jk}^{(\nu)}\}_{j,k,\nu}$ generated by MEP with VM^2 property for $\{\widetilde{\psi}_{jk}^{(\nu)}\}_{j,k,\nu}$. This means that the polyphase matrices $\mathcal{M}, \widetilde{\mathcal{M}}$, whose first rows are, respectively (μ_{00}, μ_{01}), $(\widetilde{\mu}_{00}, \widetilde{\mu}_{01})$, satisfy (3.19) and such that (3.15) holds for the corresponding wavelet masks. Due to Lemmas 3.3.5, 3.3.6, the matrices $\mathcal{M}, \widetilde{\mathcal{M}}$ can be extended to matrices $\mathcal{N}, \widetilde{\mathcal{N}}$ such that their entries $\mu_{\nu k}, \widetilde{\mu}_{\nu k}$, $\nu, k = 0, \dots, r$, are in $L_\infty^{(2)}$ and $\mathcal{N}^*, \widetilde{\mathcal{N}}$ are mutually inverse. It follows from Theorem 3.3.4 that

$$\frac{d}{d\xi}\left(\sum_{k=2}^{r} \mu_{0k}(\xi)\overline{\widetilde{\mu}_{0k}(\xi)}\right)\Bigg|_{\xi=0} = 0. \tag{3.34}$$

But

$$\frac{d}{d\xi}\left(\sum_{k=2}^{r} \mu_{0k}(\xi)\overline{\widetilde{\mu}_{0k}(\xi)}\right)\Bigg|_{\xi=0} = \frac{d}{d\xi}\left(1 - \sum_{k=0}^{1} \mu_{0k}(\xi)\overline{\widetilde{\mu}_{0k}(\xi)}\right)\Bigg|_{\xi=0} =$$
$$\frac{d}{d\xi}\left(1 - \frac{1}{2} - \frac{1}{2}\left(1 - \frac{i}{2}\sin 2\pi\xi\right)\right)\Bigg|_{\xi=0} = \frac{d}{d\xi}\left(\frac{i}{4}\sin 2\pi\xi\right)\Bigg|_{\xi=0} \neq 0.$$

So, we see that a generating refinable function $\widetilde{\varphi}$ (or its mask $\widetilde{m}_0$) should be also chosen properly to provide VM^n property for $\{\widetilde{\psi}_{jk}^{(\nu)}\}_{j,k,\nu}$.

The following statement gives a sufficient condition for a pair of refinable functions to generate dual wavelet systems with VM^n property.

Theorem 3.3.8 *Let $\varphi, \widetilde{\varphi} \in L_2(\mathbb{R}^d)$ be refinable functions, their Fourier transforms $\widehat{\varphi}, \widehat{\widetilde{\varphi}}$ have continuous derivatives up to order n at the origin, $\widehat{\varphi}(\mathbf{0}) = \widehat{\widetilde{\varphi}}(\mathbf{0}) = 1$, and let $\mu_{00}, \dots, \mu_{0,m-1}, \widetilde{\mu}_{00}, \dots, \widetilde{\mu}_{0,m-1} \in L_\infty^{(n)}$ be the polyphase components of their masks. If there exist complex numbers λ_γ, $\gamma \in \mathbb{Z}_+^d$, $[\gamma] < n$, such that $\lambda_{\mathbf{0}} = 1$ and (3.20) holds for all $\beta \in \mathbb{Z}_+^d$, $[\beta] < n$ and $k = 0, \dots, m-1$, and there exist functions $\mu_{0k}, \widetilde{\mu}_{0k} \in L_\infty^{(n)}$, $k = m, \dots, r$, such that*

$$\sum_{k=0}^{r} \mu_{0k}\overline{\widetilde{\mu}_{0k}} = 1,$$
$$D^\beta \mu_{0k}(\mathbf{0}) = 0, \ k = m, \dots, r, \quad \forall \beta \in \mathbb{Z}_+^d, [\beta] < n.$$

then the functions $\varphi, \widetilde{\varphi}$ *generate dual wavelet systems* $\{\psi_{jk}^{(\nu)}\}_{j,k,\nu}$, $\{\widetilde{\psi}_{jk}^{(\nu)}\}_{j,k,\nu}$ *with* VM^n *property for* $\{\widetilde{\psi}_{jk}^{(\nu)}\}_{j,k,\nu}$.

Proof Set $Q = (\mu_{00}, \dots, \mu_{0r})$, $\widetilde{Q} = (\widetilde{\mu}_{00}, \dots, \widetilde{\mu}_{0r})$. Due to Lemma 3.3.5, the $1 \times (r+1)$ matrices $Q, \widetilde{Q}$ can be extended to $(r+1) \times (r+1)$ matrices $\mathcal{N} = \{\mu_{\nu k}\}_{\nu,k=0}^{r}, \widetilde{\mathcal{N}} = \{\widetilde{\mu}_{\nu k}\}_{\nu,k=0}^{r}$ satisfying (3.24), and such that their entries are in $L_\infty^{(n)}$. So, the polyphase matrices $\mathcal{M}, \widetilde{\mathcal{M}}$ satisfy (3.19). It follows from Theorem 3.3.4 that the corresponding wavelet masks $m_1, \dots, m_{m-1}$ satisfy (3.15) for all $\beta \in \mathbb{Z}^d$, $[\beta] < n$, what was to be proved.◊

Let us return to the example before Theorem 3.3.8. We could not succeed with VM^2 property because the derivative of $\sum_{k=0}^{1} \mu_{0k}(\xi)\overline{\widetilde{\mu}_{0k}(\xi)}$ did not vanish at the origin. Try to change $\widetilde{\mu}_{01} \equiv \frac{1}{\sqrt{2}}$ for $\widetilde{\mu}_{01} = \mu_{01}$. Now we have

$$\frac{d}{d\xi}\left(\sum_{k=0}^{1} \mu_{0k}(\xi)\overline{\widetilde{\mu}_{0k}(\xi)}\right)\Bigg|_{\xi=0} = \frac{d}{d\xi}\left(\frac{1}{8}\sin^2 2\pi\xi\right)\Bigg|_{\xi=0} = 0.$$

So, providing condition (3.20) for both the masks $m_0, \widetilde{m}_0$ improved the situation. Taking into account this observation, let us consider generating masks $m_0, \widetilde{m}_0$ whose polyphase components satisfy

$$D^\beta \mu_{0k}(\mathbf{0}) = \frac{(2\pi i)^{[\beta]}}{\sqrt{m}} \sum_{\mathbf{0} \le \gamma \le \beta} \lambda_\gamma \binom{\beta}{\gamma} (-M^{-1}s_k)^{\beta-\gamma} \quad \forall \beta \in \mathbb{Z}_+^d, [\beta] < n, \tag{3.35}$$

$$D^\beta \widetilde{\mu}_{0k}(\mathbf{0}) = \frac{(2\pi i)^{[\beta]}}{\sqrt{m}} \sum_{\mathbf{0} \le \gamma \le \beta} \widetilde{\lambda}_\gamma \binom{\beta}{\gamma} (-M^{-1}s_k)^{\beta-\gamma} \quad \forall \beta \in \mathbb{Z}_+^d, [\beta] < n. \tag{3.36}$$

Theorem 3.3.9 *Let* $\varphi, \widetilde{\varphi} \in L_2(\mathbb{R}^d)$ *be refinable functions whose Fourier transforms have continuous derivatives up to order n at the origin, and* $\widehat{\varphi}(\mathbf{0}) = \widehat{\widetilde{\varphi}}(\mathbf{0}) = 1$. *Suppose* $\{\psi_{jk}^{(\nu)}\}_{j,k,\nu}$, $\{\widetilde{\psi}_{jk}^{(\nu)}\}_{j,k,\nu}$ *are generated from* $\varphi, \widetilde{\varphi}$ *by MEP, and let the entries* $\mu_{\nu k}, \widetilde{\mu}_{\nu k}$ *of the corresponding polyphase matrices* $\mathcal{M}, \widetilde{\mathcal{M}}$ *be in* $L_\infty^{(n)}$. *If there exist complex numbers* $\lambda_\gamma, \widetilde{\lambda}_\gamma$, $\gamma \in \mathbb{Z}_+^d$, $[\gamma] < n$, $\lambda_\mathbf{0} = \widetilde{\lambda}_\mathbf{0} = 1$, *such that (3.35), (3.36) are satisfied for* $k = 0, \dots, m-1$, *and at least one of the systems* $\{\psi_{jk}^{(\nu)}\}_{j,k,\nu}$, $\{\widetilde{\psi}_{jk}^{(\nu)}\}_{j,k,\nu}$ *has* VM^n *property, then*

$$\sum_{\mathbf{0} \le \gamma \le \alpha} (-1)^{[\alpha-\gamma]} \binom{\alpha}{\gamma} \lambda_\gamma \overline{\widetilde{\lambda}_{\alpha-\gamma}} = 0 \quad \forall \alpha \in \mathbb{Z}_+^d, \quad 0 < [\alpha] < n. \tag{3.37}$$

Proof Let $\alpha \in \mathbb{Z}_+^d$, $0 < [\alpha] < n$, $k = 0, \dots, m-1$, $\rho = \rho_k := -M^{-1}s_k$. Due to (3.35), (3.36), we have

$$\frac{m}{(2\pi i)^{[\alpha]}} D^\alpha \left(\mu_{0k}(\xi)\overline{\widetilde{\mu}_{0k}(\xi)}\right)\Big|_{\xi=\mathbf{0}} = \frac{m}{(2\pi i)^{[\alpha]}} \sum_{\mathbf{0}\le\beta\le\alpha} \binom{\alpha}{\beta} D^\beta \mu_{0k}(\mathbf{0})\overline{D^{\alpha-\beta}\widetilde{\mu}_{0k}(\mathbf{0})}$$

$$= \sum_{\mathbf{0}\le\beta\le\alpha} \binom{\alpha}{\beta} \sum_{\mathbf{0}\le\gamma\le\beta} \binom{\beta}{\gamma} \lambda_{\beta-\gamma}\rho^\gamma(-1)^{[\alpha-\beta]} \sum_{\mathbf{0}\le\delta\le\alpha-\beta} \binom{\alpha-\beta}{\delta} \overline{\widetilde{\lambda}_{\alpha-\beta-\delta}\rho^\delta}$$

$$= \sum_{\mathbf{0}\le\gamma\le\alpha}\sum_{\gamma\le\beta\le\alpha}\sum_{\mathbf{0}\le\delta\le\alpha-\beta} \binom{\alpha}{\beta}\binom{\beta}{\gamma}\binom{\alpha-\beta}{\delta}(-1)^{[\alpha-\beta]}\rho^{\gamma+\delta}\lambda_{\beta-\gamma}\overline{\widetilde{\lambda}_{\alpha-\beta-\delta}}$$

$$= \sum_{\mathbf{0}\le\gamma\le\alpha}\sum_{\mathbf{0}\le\delta\le\alpha-\gamma}\sum_{\gamma\le\beta\le\alpha-\delta} \binom{\alpha}{\beta}\binom{\beta}{\gamma}\binom{\alpha-\beta}{\delta}(-1)^{[\alpha-\beta]}\rho^{\gamma+\delta}\lambda_{\beta-\gamma}\overline{\widetilde{\lambda}_{\alpha-\beta-\delta}}$$

$$= \sum_{\mathbf{0}\le\gamma\le\alpha}\sum_{\gamma\le\varepsilon\le\alpha}\sum_{\gamma\le\beta\le\alpha-\varepsilon+\gamma} \binom{\alpha}{\beta}\binom{\beta}{\gamma}\binom{\alpha-\beta}{\varepsilon-\gamma}(-1)^{[\alpha-\beta]}\rho^{\varepsilon}\lambda_{\beta-\gamma}\overline{\widetilde{\lambda}_{\alpha-\beta-\varepsilon+\gamma}}$$

$$= \sum_{\mathbf{0}\le\varepsilon\le\alpha}\rho^\varepsilon\sum_{\mathbf{0}\le\gamma\le\varepsilon}\sum_{\gamma\le\beta\le\alpha-\varepsilon+\gamma} \binom{\alpha}{\beta}\binom{\beta}{\gamma}\binom{\alpha-\beta}{\varepsilon-\gamma}(-1)^{[\alpha-\beta]}\lambda_{\beta-\gamma}\overline{\widetilde{\lambda}_{\alpha-\beta-\varepsilon+\gamma}}$$

$$= \sum_{\mathbf{0}\le\varepsilon\le\alpha}\rho^\varepsilon\sum_{\mathbf{0}\le\gamma\le\varepsilon}\sum_{\mathbf{0}\le\kappa\le\alpha-\varepsilon} \binom{\alpha}{\kappa+\gamma}\binom{\kappa+\gamma}{\gamma}\binom{\alpha-\kappa-\gamma}{\varepsilon-\gamma}(-1)^{[\alpha-\kappa-\gamma]}\lambda_{\kappa}\overline{\widetilde{\lambda}_{\alpha-\kappa-\varepsilon}}$$

$$= \sum_{\mathbf{0}\le\varepsilon\le\alpha}\rho^\varepsilon\sum_{\mathbf{0}\le\kappa\le\alpha-\varepsilon}\lambda_\kappa\overline{\widetilde{\lambda}_{\alpha-\kappa-\varepsilon}}\sum_{\mathbf{0}\le\gamma\le\varepsilon}\frac{\alpha!}{\kappa!\gamma!(\varepsilon-\gamma)!(\alpha-\kappa-\varepsilon)!}(-1)^{[\alpha-\kappa-\gamma]}$$

$$= \sum_{\mathbf{0}\le\varepsilon\le\alpha}\rho^\varepsilon\sum_{\mathbf{0}\le\kappa\le\alpha-\varepsilon}\lambda_\kappa\overline{\widetilde{\lambda}_{\alpha-\kappa-\varepsilon}}\binom{\kappa+\varepsilon}{\varepsilon}\binom{\alpha}{\kappa+\varepsilon}\sum_{\mathbf{0}\le\gamma\le\varepsilon}\binom{\varepsilon}{\gamma}(-1)^{[\alpha-\kappa-\gamma]} =$$

$$\sum_{\mathbf{0}\le\varepsilon\le\alpha}\rho^\varepsilon\sum_{\mathbf{0}\le\kappa\le\alpha-\varepsilon}(-1)^{[\alpha-\kappa]}\lambda_\kappa\overline{\widetilde{\lambda}_{\alpha-\kappa-\varepsilon}}\binom{\kappa+\varepsilon}{\varepsilon}\binom{\alpha}{\kappa+\varepsilon}\prod_{j=1}^{d}(1-1)^{\varepsilon_i} =$$

$$\sum_{\mathbf{0}\le\kappa\le\alpha}(-1)^{[\alpha-\kappa]}\binom{\alpha}{\kappa}\lambda_\kappa\overline{\widetilde{\lambda}_{\alpha-\kappa}}. \tag{3.38}$$

So, $D^\alpha\left(\mu_{0k}(\xi)\overline{\widetilde{\mu}_{0k}(\xi)}\right)\Big|_{\xi=\mathbf{0}}$ does not depend on k and

$$(2\pi i)^{[\alpha]}\sum_{\mathbf{0}\le\kappa\le\alpha}(-1)^{[\alpha-\kappa]}\binom{\alpha}{\kappa}\lambda_\kappa\overline{\widetilde{\lambda}_{\alpha-\kappa}} = D^\alpha\left(\sum_{k=0}^{m-1}\mu_{0k}(\xi)\overline{\widetilde{\mu}_{0k}(\xi)}\right)\Bigg|_{\xi=\mathbf{0}}. \tag{3.39}$$

Due to Lemmas 3.3.5, 3.3.6, the $(r+1)\times m$ matrices $\mathcal{M}, \widetilde{\mathcal{M}}$ can be extended to $(r+1)\times(r+1)$ matrices $\mathcal{N}, \widetilde{\mathcal{N}}$ such that their entries are in $L_\infty^{(n)}$ and $\mathcal{N}^*\widetilde{\mathcal{N}} = I_{r+1}$.

Due to conditions (b) of Theorem 3.3.4,

$$D^{\alpha}\left(\mu_{0k}(\xi)\overline{\widetilde{\mu}_{0k}(\xi)}\right)\Big|_{\xi=\mathbf{0}} = 0,\ k = m, \dots, r.$$

From this, taking into account that $\sum\limits_{k=0}^{r} \mu_{0k}(\xi)\overline{\widetilde{\mu}_{0k}(\xi)} = 1$, we obtain

$$D^{\alpha}\left(\sum_{k=0}^{m-1} \mu_{0k}(\xi)\overline{\widetilde{\mu}_{0k}(\xi)}\right)\Bigg|_{\xi=\mathbf{0}} = D^{\alpha}\left(\sum_{k=0}^{r} \mu_{0k}(\xi)\overline{\widetilde{\mu}_{0k}(\xi)}\right)\Bigg|_{\xi=\mathbf{0}} = 0.$$

It remains to combine this with (3.39).$\diamondsuit$

Corollary 3.3.10 *Let $\mu_{0k}, \widetilde{\mu}_{0k} \in L_\infty^{(n)}$ for some $k = 0, \dots, m-1$. If there exist complex numbers $\lambda_\gamma, \widetilde{\lambda}_\gamma$, $\gamma \in \mathbb{Z}_+^d$, $[\gamma] < n$, such that (3.35), (3.36), (3.37) are satisfied, then*

$$D^{\alpha}\left(\overline{\mu_{0k}(\xi)}\widetilde{\mu}_{0k}(\xi)\right)\Big|_{\xi=\mathbf{0}} = 0,$$

for all $\alpha \in \mathbb{Z}_+^d$, $0 < [\alpha] < n$.

The proof follows from (3.38).

3.4 Construction of Compactly Supported Frames

Applied mathematicians and engineers are especially interested in construction of compactly supported wavelet systems. To provide this property, generating refinable functions should be compactly supported and all wavelet masks should be trigonometric polynomials. Next we discuss whether it is possible to construct compactly supported dual wavelet frames generated from $\varphi, \widetilde{\varphi}$, provided these functions are chosen properly. The following statement gives a positive answer to this question.

Theorem 3.4.1 *Let $\varphi, \widetilde{\varphi} \in L_2(\mathbb{R}^d)$ be compactly supported refinable functions with polynomial masks, $\widehat{\varphi}(\mathbf{0}) = \widehat{\widetilde{\varphi}}(\mathbf{0}) = 1$, and let $\mu_{00}, \dots, \mu_{0,m-1}, \widetilde{\mu}_{00}, \dots, \widetilde{\mu}_{0,m-1}$ be the polyphase components of their masks. If there exist complex numbers λ_γ, $\gamma \in \mathbb{Z}_+^d$, $[\gamma] < n$, such that (3.20) holds for all $\gamma \in \mathbb{Z}_+^d$, $[\gamma] < n$, and $k = 0, \dots, m-1$, and there exist trigonometric polynomials $\mu_{0k}, \widetilde{\mu}_{0k}$, $k = m, \dots, r$, such that*

$$\sum_{k=0}^{r} \mu_{0k}\overline{\widetilde{\mu}_{0k}} = 1,$$

$$D^{\beta}\mu_{0k}(\mathbf{0}) = 0,\ k = m, \dots, r,\quad \forall \beta \in \mathbb{Z}_+^d, [\beta] < n.$$

then the functions $\varphi, \widetilde{\varphi}$ *generate dual compactly supported wavelet systems* $\{\psi_{jk}^{(\nu)}\}_{j,k,\nu}$, $\{\widetilde{\psi}_{jk}^{(\nu)}\}_{j,k,\nu}$ *with* VM^n *property for* $\{\widetilde{\psi}_{jk}^{(\nu)}\}_{j,k,\nu}$. *If, moreover,*

$$\widetilde{\mu}_{0k}(\mathbf{0}) = 0, \ k = m, \dots, r, \tag{3.40}$$

then the systems $\{\psi_{jk}^{(\nu)}\}_{j,k,\nu}$, $\{\widetilde{\psi}_{jk}^{(\nu)}\}_{j,k,\nu}$ *are dual frames.*

Proof Set $Q = (\mu_{00}, \dots, \mu_{0r})$, $\widetilde{Q} = (\widetilde{\mu}_{00}, \dots, \widetilde{\mu}_{0r})$. Using Suslin's solution of a generalized Serre conjecture, it is possible to construct $(r+1) \times (r+1)$ matrices $\mathcal{N}, \widetilde{\mathcal{N}}$ extending Q, $\widetilde{Q}$, satisfying (3.24), and such that their entries are trigonometric polynomials (see Sect. 2.4). So, the polyphase matrices $\mathcal{M}, \widetilde{\mathcal{M}}$ satisfy (3.19). It follows from Theorem 3.3.4 that the corresponding wavelet masks $\widetilde{m}_1, \dots, \widetilde{m}_r$ ($\widetilde{m}_\nu$ is the function whose polyphase components are $\widetilde{\mu}_{\nu 0}, \dots, \widetilde{\mu}_{\nu, m-1}$) satisfy (3.15) for all $\beta \in \mathbb{Z}^d, [\beta] < n$. Hence, VM^n property for $\{\widetilde{\psi}_{jk}^{(\nu)}\}_{j,k,\nu}$ is satisfied. If, moreover, (3.40) holds, then, again by Theorem 3.3.4, the wavelet masks $m_1, \dots, m_r$ satisfy (3.15) for $\beta = \mathbf{0}$. Thus VM^1 property for $\{\psi_{jk}^{(\nu)}\}_{j,k,\nu}$ is also satisfied. It follows from Theorem 3.1.8 that the systems $\{\psi_{jk}^{(\nu)}\}_{j,k,\nu}$, $\{\widetilde{\psi}_{jk}^{(\nu)}\}_{j,k,\nu}$ are dual frames.◊

Due to Theorem 3.4.1 and Corollary 3.3.10, refinable functions $\varphi, \widetilde{\varphi}$ whose masks are trigonometric polynomials satisfying (3.35), (3.36), (3.37) generate dual compactly supported dual wavelet systems $\{\psi_{jk}^{(\nu)}\}_{j,k,\nu}$, $\{\widetilde{\psi}_{jk}^{(\nu)}\}_{j,k,\nu}$ with VM^n property for $\{\widetilde{\psi}_{jk}^{(\nu)}\}_{j,k,\nu}$. Indeed, we can set $r = m$,

$$\mu_{0m} \equiv 1, \quad \widetilde{\mu}_{0m} = 1 - \sum_{l=0}^{m-1} \overline{\mu_{0l}}\, \widetilde{\mu}_{0l}.$$

But, this is a bad construction. The system $\{\psi_{jk}^{(\nu)}\}_{j,k,\nu}$ does not have VM^1 property because $\mu_{0m}(\mathbf{0}) \neq 0$. The system $\{\psi_{jk}^{(\nu)}\}_{j,k,\nu}$ cannot be a frame in this case, due to Theorem 3.2.1.

The following method allows to provide VM^n property for each of the systems $\{\psi_{jk}^{(\nu)}\}_{j,k,\nu}$, $\{\widetilde{\psi}_{jk}^{(\nu)}\}_{j,k,\nu}$.

Let functions $\mu_{00}, \dots, \mu_{0\,m-1}$ and $\widetilde{\mu}'_{00}, \dots, \widetilde{\mu}'_{0\,m-1}$ be defined by

$$\mu_{0k}(\xi) = \frac{1}{\sqrt{m}} \sum_{[\alpha]<n} g_\alpha(\xi)(2\pi i)^{[\alpha]} \sum_{\mathbf{0} \le \beta \le \alpha} \binom{\alpha}{\beta} \left(-M^{-1} s_k\right)^\beta \lambda_{\alpha-\beta} + T_k(\xi), \tag{3.41}$$

$$\widetilde{\mu}'_{0k}(\xi) = \frac{1}{\sqrt{m}} \sum_{[\alpha]<n} g_\alpha(\xi)(2\pi i)^{[\alpha]} \sum_{\mathbf{0} \le \beta \le \alpha} \binom{\alpha}{\beta} \left(-M^{-1} s_k\right)^\beta \widetilde{\lambda}_{\alpha-\beta} + \widetilde{T}_k(\xi), \tag{3.42}$$

where $T_k, \widetilde{T}_k$ and g_α are trigonometric polynomials such that $D^\beta T_k(\mathbf{0}) = 0$, $D^\beta \widetilde{T}_k(\mathbf{0}) = 0$, $D^\beta g_\alpha(\mathbf{0}) = \delta_{\alpha\beta}$ for all $\beta \in \mathbb{Z}_+^d$, $[\beta] < n$. It is clear that condition (3.35) is satisfied.

Set

$$\sigma := \sum_{l=0}^{m-1} \overline{\mu_{0l}}\,\widetilde{\mu}'_{0l}, \quad \widetilde{\mu}_{0k} := (2-\sigma)\widetilde{\mu}'_{0k}, k = 0, \ldots, m-1, \tag{3.43}$$

$$\mu_{0m} := \overline{(1-\sigma)}, \; \widetilde{\mu}_{0m} := 1-\sigma. \tag{3.44}$$

Due to Corollary 3.3.10, we have $D^{\beta}\sigma(\mathbf{0}) = 0$ for all $\beta \in \mathbb{Z}_+^d$, $0 < [\beta] < n$. Therefore, $D^{\beta}\,\overline{(1-\sigma)}\,(\mathbf{0}) = 0$ for all $\beta \in \mathbb{Z}_+^d$, $0 \le [\beta] < n$. It follows that $D^{\beta}\widetilde{\mu}_{0k}(\mathbf{0}) = D^{\beta}\widetilde{\mu}'_{0k}(\mathbf{0})$ for all $\beta \in \mathbb{Z}_+^d$, $[\beta] < n$. It is not difficult to see that (3.36) holds and

$$1 - \sum_{k=0}^{m-1} \overline{\mu_{0k}}\widetilde{\mu}_{0k} = 1 - (2-\sigma)\sum_{k=0}^{m-1} \overline{\mu_{0k}}\widetilde{\mu}'_{0k} = (1-\sigma)^2,$$

which yields $\sum_{k=0}^{m} \overline{\mu_{0k}}\widetilde{\mu}_{0k} = 1$.

Now one can find matrices

$$\mathcal{N} := \begin{pmatrix} \mu_{00} & \cdots & \mu_{0,m-1} & \mu_{0m} \\ \mu_{1,0} & \cdots & \mu_{1,m-1} & * \\ \vdots & \ddots & \vdots & \vdots \\ \mu_{m,0} & \cdots & \mu_{m,m-1} & * \end{pmatrix}, \quad \widetilde{\mathcal{N}} := \begin{pmatrix} \widetilde{\mu}_{00} & \cdots & \widetilde{\mu}_{0,m-1} & \widetilde{\mu}_{0m} \\ \widetilde{\mu}_{1,0} & \cdots & \widetilde{\mu}_{1,m-1} & * \\ \vdots & \ddots & \vdots & \vdots \\ \widetilde{\mu}_{m,0} & \cdots & \widetilde{\mu}_{m,m-1} & * \end{pmatrix},$$

such that their entries are trigonometric polynomials and $\mathcal{N}\widetilde{\mathcal{N}}^* = I_{m+1}$. Though the matrices $\mathcal{N}, \widetilde{\mathcal{N}}$ can be constructed theoretically (see Sect. 2.4), it is very complicate to implement the algorithm in practice. Instead, we suggest the following explicit way (the payment of simplicity of this way is increasing the redundancy).

Set $r = m$ if $\sigma \equiv 1$, otherwise we set $r = m+1$, $\mu_{0,m+1} \equiv 0$, $\widetilde{\mu}_{0,m+1} \equiv 0$. For each $\nu = 1, \ldots, r$, define

$$\widetilde{\mu}_{\nu r} := \overline{\mu_{0,r-\nu}}, \quad \mu_{\nu,m+1} := \overline{\widetilde{\mu}_{0,r-\nu}},$$

$$\mu_{\nu k} := \delta_{r-\nu,k} - \mu_{0k}\overline{\widetilde{\mu}_{0,r-\nu}}, \quad k = 0, \ldots, r-1, \tag{3.45}$$

$$\widetilde{\mu}_{\nu k} := \delta_{r-\nu,k} - \widetilde{\mu}_{0k}\overline{\mu_{0,r-\nu}}, \quad k = 0, \ldots, r-1. \tag{3.46}$$

Or, alternatively, the extended matrices $\mathcal{N} := \{\mu_{\nu k}\}_{\nu,k=0}^{r}$, $\widetilde{\mathcal{N}} := \{\widetilde{\mu}_{\nu k}\}_{\nu,k=0}^{r}$ can be written as

$$\mathcal{N} := \begin{pmatrix} P & 0 \\ I_r - \widetilde{P}^* P & \widetilde{P}^* \end{pmatrix}, \quad \widetilde{\mathcal{N}} := \begin{pmatrix} \widetilde{P} & 0 \\ I_r - P^* \widetilde{P} & P^* \end{pmatrix},$$

where $P = (\mu_{00}, \ldots, \mu_{0r-1})$, $\widetilde{P} = (\widetilde{\mu}_{00}, \ldots, \widetilde{\mu}_{0r-1})$. It is not difficult to see that these matrices satisfy $\mathcal{N}\widetilde{\mathcal{N}}^* = I_{r+1}$. This yields that the columns of the polyphase matrices $\mathcal{M}, \widetilde{\mathcal{M}}$ are biorthonormal.

Now we can give a constructive description of our method.

Algorithm 1
Step 1. Given dilation matrix M and $n \in \mathbb{N}$, choose an arbitrary set of digits $D(M) = \{s_0, \dots, s_{m-1}\}$, $s_0 = \mathbf{0}$, and an arbitrary set of numbers $\lambda_\beta \in \mathbb{C}$, $\beta \in \mathbb{Z}_+^d$, $[\beta] < n$, $\lambda_\mathbf{0} = 1$; find a dual set of numbers $\widetilde{\lambda}_\beta \in \mathbb{C}$ satisfying (3.37) by the following recursive formulas

$$\widetilde{\lambda}_\mathbf{0} = 1, \quad \widetilde{\lambda}_\alpha = -(-1)^{[\alpha]}\overline{\lambda_\alpha} - \sum_{\mathbf{0}<\beta\le\alpha} (-1)^{[\beta]} \binom{\alpha}{\beta} \overline{\lambda_\beta} \widetilde{\lambda}_{\alpha-\beta}.$$

Step 2. For every $\alpha \in \mathbb{Z}_+^d$, $[\alpha] < n$, choose a trigonometric polynomial g_α such that $D^\beta g_\alpha(\mathbf{0}) = \delta_{\alpha\beta}$ for all $\beta \in \mathbb{Z}_+^d$, $[\beta] < n$ (explicit formulas for g_α are given below). Choose arbitrary trigonometric polynomials T_k, $\widetilde{T}_k$ such that $D^\beta T_k(\mathbf{0}) = D^\beta \widetilde{T}_k(\mathbf{0}) = 0$ for all $\beta \in \mathbb{Z}_+^d$, $[\beta] < n$. Define functions $\mu_{00}, \dots, \mu_{0\,m-1}$ and $\widetilde{\mu}'_{00}, \dots, \widetilde{\mu}'_{0\,m-1}$, respectively, by (3.41), (3.42).
Step 3. Define $\widetilde{\mu}_{00}, \dots, \widetilde{\mu}_{0m}$ and μ_{0m} by (3.43), (3.44).
Step 4. Set $r = m$ if $\sigma \equiv 1$, otherwise set $r = m + 1$. Find $\mu_{\nu k}$ $\widetilde{\mu}_{\nu k}$, $\nu = 1, \dots, r$, $k = 0, \dots, m-1$, by (3.45), (3.46), respectively. Find refinable mask m_0, $\widetilde{m}_0$ and wavelet masks m_ν, $\widetilde{m}_\nu$, $\nu = 1, \dots, r$, by (3.16) and a similar formula for $\widetilde{m}_\nu$.$\Diamond$
Note that Step 4 of Algorithm 1 may be improved if $\lambda_\beta = 0$ for all $\beta \in \mathbb{Z}_+^d$, $0 < [\beta] < n$, and $\sigma = 1$. In this case, using a method described in Sect. 2.5, one can reduce r to $m - 1$, see Example 1 below.

Suppose polyphase matrices $\mathcal{M}$, $\widetilde{\mathcal{M}}$ are constructed according to the above algorithm. Let m_ν, $\widetilde{m}_\nu$ be the functions whose polyphase components are $\mu_{\nu 0}, \dots, \mu_{\nu,m-1}$, $\widetilde{\mu}_{\nu 0}, \dots, \widetilde{\mu}_{\nu,m-1}$, respectively, and let functions φ, $\widetilde{\varphi}$ be defined by

$$\widehat{\varphi}(\xi) = \prod_{j=1}^{\infty} m_0(M^{*-j}\xi), \quad \widehat{\widetilde{\varphi}}(\xi) = \prod_{j=1}^{\infty} \widetilde{m}_0(M^{*-j}\xi).$$

It remains to check whether the functions φ, $\widetilde{\varphi}$ are in $L_2(\mathbb{R}^d)$. Unfortunately, most often, it is not easy to do (see Sect. 2.6). But if we succeeded, i.e., $\varphi, \widetilde{\varphi} \in L_2(\mathbb{R}^d)$, then setting

$$\widehat{\psi^{(\nu)}}(\xi) = m_\nu(M^{*-1}\xi)\widehat{\varphi}(M^{*-1}\xi), \quad \widehat{\widetilde{\psi}^{(\nu)}}(\xi) = \widetilde{m}_\nu(M^{*-1}\xi)\widehat{\widetilde{\varphi}}(M^{*-1}\xi), \tag{3.47}$$

we have dual compactly supported wavelet frames $\{\psi_{jk}^{(\nu)}\}_{j,k,\nu}$, $\{\widetilde{\psi}_{jk}^{(\nu)}\}_{j,k,\nu}$ with VM^n property for each of them.

Thus, we have an algorithmic method for the constructions of polyphase matrices $\mathcal{M}$, $\widetilde{\mathcal{M}}$ providing vanishing moments. The method leads to wavelet frames whenever the corresponding refinable functions φ, $\widetilde{\varphi}$ are in $L_2(\mathbb{R}^d)$. The latter property can be checked for each individual function using hard numerical calculations. It is impossible to guess in advance how to choose the parameters for the construction of polyphase matrix to succeed. In this aspect, the situation with tight frames is better. To construct compactly supported tight wavelet frame, one needs a polyphase matrix

$$\mathcal{M} := \begin{pmatrix} \mu_{00} & \cdots & \mu_{0,m-1} \\ \mu_{1,0} & \cdots & \mu_{1,m-1} \\ \vdots & \ddots & \vdots \\ \mu_{r,0} & \cdots & \mu_{r,m-1} \end{pmatrix},$$

such that its entries are trigonometric polynomials and its columns form an orthonormal system. Obviously, this yields that

$$\sum_{k=0}^{m-1} |\mu_{0k}|^2 \le 1.$$

It follows from Lemma 2.5.1 and Proposition 2.6.6 that the corresponding refinable function φ is in $L_2(\mathbb{R}^d)$. If, moreover, $\mu_{0k}(\mathbf{0}) = 1/\sqrt{m}$, $k = 0, \dots, m-1$, then $\hat{\varphi}(\mathbf{0}) = 1$ (because the refinable mask m_0, whose polyphase components are $\mu_{00}, \dots, \mu_{0,m-1}$, satisfies $m_0(\mathbf{0}) = 1$). This is sufficient for the wavelet system $\{\psi_{jk}^{(\nu)}\}_{j,k,\nu}$ generated from φ (or from m_0) by MEP to be a tight frame in $L_2(\mathbb{R}^d)$.

Now we are interested on how to provide vanishing moments to a tight wavelet frame $\{\psi_{jk}^{(\nu)}\}_{j,k,\nu}$ generated by MEP. Due to Theorems 3.3.7 and 3.3.8, we have the following necessary condition for a tight wavelet frame to have vanishing moments.

Theorem 3.4.2 *Let $\varphi \in L_2(\mathbb{R}^d)$ be a compactly supported refinable function with a polynomial mask m_0 whose polyphase components are $\mu_{00}, \dots, \mu_{0m-1}$, and $\widehat{\varphi}(\mathbf{0}) = 1$. Suppose $\{\psi_{jk}^{(\nu)}\}_{j,k,\nu}$ is a tight frame generated from φ by MEP. If $\{\psi_{jk}^{(\nu)}\}_{j,k,\nu}$ has VM^n property, $n \in \mathbb{Z}_+$, then there exists complex numbers λ_γ, $\gamma \in \mathbb{Z}_+^d$, $[\gamma] < n$, $\lambda_0 = 1$, such that the relations*

$$D^\beta \mu_{0k}(\mathbf{0}) = \frac{(2\pi i)^{[\beta]}}{\sqrt{m}} \sum_{0 \le \gamma \le \beta} \lambda_\gamma \binom{\beta}{\gamma} (-M^{-1}s_k)^{\beta-\gamma} \quad \forall \beta \in \mathbb{Z}_+^d, [\beta] < n \tag{3.48}$$

are satisfied for all $k = 0, \dots, m-1$, and the equalities

$$\sum_{0 \le \gamma \le \alpha} (-1)^{[\alpha-\gamma]} \binom{\alpha}{\gamma} \lambda_\gamma \overline{\lambda_{\alpha-\gamma}} = 0 \quad \forall \alpha \in \mathbb{Z}_+^d, 0 < [\alpha] < n. \tag{3.49}$$

hold for the numbers λ_γ.

A sufficient condition for a compactly supported tight wavelet frame generated by MEP to have vanishing moments is given as follows.

Theorem 3.4.3 *Let m_0 be a trigonometric polynomial whose polyphase components are $\mu_{00}, \dots, \mu_{0m-1}$. If there exist complex numbers λ_γ, $\gamma \in \mathbb{Z}_+^d$, $[\gamma] < n$, $n \in \mathbb{Z}_+$, $\lambda_0 = 1$, satisfying (3.48) for all $k = 0, \dots, m-1$, and there exist trigonometric polynomials μ_{0k}, $k = m, \dots, r$, $\mu_{\nu k}$, $\nu = 1, \dots, r$, $k = 0, \dots, r$, such that the matrix $\mathcal{N} := \{\mu_{\nu k}\}_{\nu,k=0}^r$ is unitary and*

$$D^{\beta}\mu_{0k}(\mathbf{0}) = 0, \; k = m, \ldots, r, \quad \forall \beta \in \mathbb{Z}_+^d, [\beta] < n, \tag{3.50}$$

then the wavelet system $\{\psi_{jk}^{(\nu)}\}$ generated by MEP from m_0 is a compactly supported tight wavelet frame with VM^n property.

Proof It follows from Lemma 2.5.1 and Proposition 2.6.6 that the refinable function φ given by (3.47) is in $L_2(\mathbb{R}^d)$. By Theorem 3.3.4, the wavelet masks $m_\nu, \nu = 1, \ldots, r$, ($m_\nu$ is the function whose polyphase components are $\mu_{\nu,0}, \ldots, \mu_{\nu,m-1}$) satisfy (3.15) for all $\beta \in \mathbb{Z}^d$, $[\beta] < n$. Hence, VM^n property is satisfied for $\{\psi_{jk}^{(\nu)}\}_{j,k,\nu}$. It follows from Theorem 2.4.5 and Proposition 1.1.8 that the system $\{\psi_{jk}^{(\nu)}\}_{j,k,\nu}$ is a tight wavelet frame in $L_2(\mathbb{R}^d)$.$\Diamond$

The following well-known statement will be actively used for our constructions of tight frames.

Theorem 3.4.4 (Riesz Lemma) *For any nonnegative trigonometric polynomial of one variable $h(t) = \sum_{k=-N}^{N} h_k e^{2\pi i k t}$ there exists a trigonometric polynomial $m_0(t) = \sum\limits_{k=0}^{N} c_k e^{-2\pi i k t}$ such that $|m_0(t)|^2 = h(t)$. In addition, if the coefficients of the polynomial h are real, it takes the form*

$$h(t) = \sum_{k=0}^{N} a_k \cos 2\pi k t, \qquad a_k \in \mathbb{R},$$

and the polynomial m_0 can be chosen in such a way that its coefficients c_k are also real.

Proof Setting

$$\mathbf{h}(z) = \sum_{k=-N}^{N} h_k z^k \quad z \in \mathbb{C},$$

we have $h(\xi) = \mathbf{h}(e^{-2\pi i \xi})$. Since h is a nonnegative polynomial, it takes only real values, which yields $h_{-k} = \overline{h_k}$. Consequently, $\mathbf{h}(1/\bar{z}) = \overline{\mathbf{h}(z)}$. Hence, the set of roots of the equation $\mathbf{h}(z) = 0$ is symmetric with respect to inversion (the change of variables $z \to 1/\bar{z}$). If we single out the roots on the unit circle and divide the other roots in pairs, then $\mathbf{h}$ can be represented in the form

$$\mathbf{h}(z) = a z^n \prod_{l=1}^{q} (z - e^{-2\pi i \alpha_l}) \prod_{k=1}^{p} \big(z - z_k\big)\big(z - (\overline{z_k})^{-1}\big), \tag{3.51}$$

where $n + q + 2p = 2N$, α_k being roots of the polynomial h on the fundamental period. As $h \geq 0$, every such root has even multiplicity. Hence, $k = 2r$ and $\prod\limits_{l=1}^{q}(z -$

$e^{-2\pi i\alpha_l}) = \prod_{l=1}^{r}(z - e^{-2\pi i\alpha_l})^2$. Equality (3.51) takes the form

$$\mathbf{h}(z) = b\prod_{l=1}^{r}\left(z - e^{-2\pi i\alpha_l}\right)\left(z^{-1} - e^{2\pi i\alpha_l}\right)\prod_{k=1}^{p}\left(z - z_k\right)\left(z^{-1} - \overline{z_k}\right),$$

where $r + p = N$. Since both products are positive for all $z = e^{-2\pi i\xi}$, it follows that b is a real positive number. It remains to set

$$\mathbf{m}_0(z) = \sqrt{b}\prod_{l=1}^{r}\left(z - e^{-2\pi i\alpha_l}\right)\prod_{k=1}^{p}\left(z - z_k\right).$$

Thus, the set of roots of $\mathbf{h}$ can be divided into two mutually inverse sets A and $1/\bar{A}$. For every such partition, there is a corresponding polynomial $\mathbf{m}_0(z) = \sqrt{b}\prod\limits_{z_k\in A}\left(z - z_k\right)$, and $m_0(\xi) = \mathbf{m}_0(e^{-2\pi i\xi})$.

If all coefficients of h_k are real, the set of roots of $\mathbf{h}(\mathbf{z})$ is symmetric with respect to the real axis. Hence for A, we can also take a set symmetric with respect to the real axis. In this case, the coefficients of the polynomial m_0 will be real.$\Diamond$

Theorem 3.4.3 gives a theoretical method for the construction of compactly supported tight wavelet frames. First of all we need a refinable mask whose polyphase components satisfy necessary conditions of Theorem 3.4.2 and the row $\mu_{00}, \ldots, \mu_{0m-1}$ may be extended by trigonometric polynomials $\mu_{0m}, \ldots, \mu_{0r}$ such that (3.50) holds and $\sum\limits_{k=0}^{r} |\mu_{0k}|^2 = 1$. After that the row $\mu_{00}, \ldots, \mu_{0r}$ should be extended to a unitary m5atrix. In contrast to the case of dual frames, it is not known whether such an extension always exists. A positive answer to this question is known only for the case $d = 1$ (see, e.g., [2, Sect. 2.6]). But this is an open problem if $d > 1$. However, we will be able to avoid this difficulty by increasing the redundancy. Namely, if one of the entries of the first row is identical zero, then the problem has a simple solution. Since (3.50) is satisfied for such an element, it can be added. So the main problem is in finding an appropriate mask m_0 and in possibility to extend properly its polyphase row. Clearly, it is not enough to provide (3.48), (3.49). Matrix extension is possible only if $\sum\limits_{k=0}^{m-1} |\mu_{0k}|^2 \le 1$. If this condition is fulfilled, then $T := 1 - \sum_{k=0}^{m-1} |\mu_{0k}|^2$ is a nonnegative function, and T should be represented in the form $T = \sum_{k=m}^{r} |\mu_{0k}|^2$, where $\mu_{0m}, \ldots, \mu_{0r}$ are trigonometric polynomials. If $d = 1$, then such a representation exists with $r = m$ due to the Riesz Theorem 3.4.4. In the case $d = 2$, the existence of representation is known because it was proved in [3] that any nonnegative trigonometric polynomial of two variables can be represented as a finite sum of squared magnitudes of trigonometric polynomials. However, an analogue of this statement does not hold for $d > 2$, see [4–7].

Theorem 3.4.5 *For any dilation matrix M and any $n \in \mathbb{N}$, there exist compactly supported wavelet functions $\psi^{(\nu)}$, $\nu = 1, \dots, r$, $r \le d(m-1)+m$, such that $\{\psi_{jk}^{(\nu)}\}_{j,k,\nu}$ is a tight wavelet frame with* $\mathbf{VM^n}$ *property.*

Proof For every $k = 0, \dots, m-1$ and every $j = 1, \dots, d$, we set

$$T_{kj}(t) := \sum_{l=0}^{n-1} g_l(t)\big(-2\pi i (M^{-1}s_k)_j\big)^l, \quad t \in \mathbb{R}, \tag{3.52}$$

where g_l is a trigonometric polynomial such that

$$\frac{d^i g_l}{dx^i}(0) = \delta_{il}, \quad i, l = 0, \dots, n-1, \quad g_0 \equiv 1.$$

It is easy to see that such polynomials exist for every n, explicit recursive formulas for g_l are given below. Obviously, we have

$$\frac{d^l T_{kj}}{dt^l}(0) = \big(-2\pi i (M^{-1}s_k)_j\big)^l, \quad l = 0, \dots, n-1. \tag{3.53}$$

Note that $T_{0j} \equiv 1$ and set

$$T'_{0j}(t) := T_{0j}(t) = 1, \quad \forall t \in \mathbb{R}, j = 1, \dots, n-1.$$

Setting $A := \frac{4m-1}{m-1}$, choose $\varepsilon \in (0, \sqrt{A^{1/d}-1})$ and define trigonometric polynomials T'_{kj}, $k = 1, \dots, m-1$, $j = 1, \dots, d$, by

$$T'_{kj}(t) := \big(1 - \sin^{2L} \pi t\big)^N T_{kj}(t), \quad t \in \mathbb{R}, \tag{3.54}$$

choosing nonnegative integers L and N so that

$$\frac{d^l T'_{kj}}{dt^l}(0) = \frac{d^l T_{kj}}{dt^l}(0), \quad l = 0, \dots, n. \tag{3.55}$$

$$|T'_{kj}(t)|^2 \le 1 + \varepsilon^2, \quad \forall t \in \mathbb{R}, \tag{3.56}$$

Such L and N exist. Indeed, obviously, if L is big enough, then (3.55) holds for any N. Since

$$Im T_{kj}(0) = 0, \quad Re T_{kj}(0) = 1, \quad \frac{d(Re T_{kj})}{dt}(0) = 0, \quad \frac{d^2(Re T_{kj})}{dt^2}(0) < 0,$$

inequality (3.56) is satisfied on some neighborhood of 0 for all L and N. But if N is big enough, then (3.56) remains to be true for all $t \in [-1/2, 1/2]$.

Next we set

$$\mu'_{0k}(\xi) := \frac{1}{\sqrt{m}} \prod_{j=1}^{d} T'_{kj}(\xi_j), \quad x \in \mathbb{Z}^d, k = 0, \dots, m-1, \tag{3.57}$$

$$\sigma := \sum_{k=0}^{m-1} |\mu'_{0k}|^2, \quad \mu_{0k} := \frac{3-\sigma}{2} \mu'_{0k}, \quad k = 0, \dots, m-1, \tag{3.58}$$

and note that

$$1 - \sum_{k=0}^{m-1} |\mu_{0k}|^2 = 1 - \frac{(3-\sigma)^2 \sigma}{4} = \left(\frac{1-\sigma}{2}\right)^2 (4-\sigma), \tag{3.59}$$

Using (3.52) and (3.55), it is easy to see that

$$D^{\beta} \mu'_{0k}(\mathbf{0}) = \frac{(2\pi i)^{[\beta]}}{\sqrt{m}} (-M^{-1} s_k)^{\beta}, \ k = 0, \dots, m-1, \ \forall \beta \in \mathbb{Z}_+^d, [\beta] < n.$$

Since, due to Corollary 3.3.10,

$$D^{\alpha} \sigma(\mathbf{0}) = 0 \quad \forall \alpha \in \mathbb{Z}_+^d, \ 0 < [\alpha] < n, \tag{3.60}$$

and $\sigma(\mathbf{0}) = 1$, we have

$$D^{\beta} \mu_{0k}(\mathbf{0}) = \frac{(2\pi i)^{[\beta]}}{\sqrt{m}} (-M^{-1} s_k)^{\beta}, \ k = 0, \dots, m-1, \ \forall \beta \in \mathbb{Z}_+^d, [\beta] < n. \tag{3.61}$$

It follows that condition (3.48) is satisfied with $\lambda_{\mathbf{0}} = 1$, $\lambda_{\gamma} = 0$ for all $\gamma \in \mathbb{Z}_+^d$, $0 < [\gamma] < n$.

Let $k = 1, \dots, m-1$. Since $|T'_{kj}|^2 \le 1 + \varepsilon^2$, $(1+\varepsilon^2)^d \le A$ and T'_{kj} is a trigonometric polynomial of one variable, due to the Riesz Lemma, there exists a trigonometric polynomial Q_{kj} such that

$$1 - \frac{1}{1+\varepsilon^2} |T'_{kj}(t)|^2 = |Q_{kj}(t)|^2, \quad j = 1, \dots, d-1, \tag{3.62}$$

$$1 - \frac{(1+\varepsilon^2)^{d-1}}{A} |T'_{kd}(t)|^2 = |Q_{kd}(t)|^2. \tag{3.63}$$

Taking into account that $\mu'_{00} \equiv 1/\sqrt{m}$, we have

$$4 - \sigma(\xi) = 4 - \frac{1}{m} - \sum_{k=1}^{m-1} |\mu'_{0k}(\xi)|^2 = \frac{1}{m} \sum_{k=1}^{m-1} \left(A - \prod_{j=1}^{d} |T'_{kj}(\xi_j)|^2 \right)$$

Using that, by (3.62),

$$\frac{A}{(1+\varepsilon^2)^{l-1}} - \prod_{j=l}^{d} |T'_{kj}(\xi_j)|^2 = \frac{A}{(1+\varepsilon^2)^{l-1}} |Q_{kl}(\xi_l)|^2 + \tag{3.64}$$

$$|T'_{kl}(\xi_l)|^2 \left(\frac{A}{(1+\varepsilon^2)^{l}} - \prod_{j=l+1}^{d} T'_{kj}(\xi_j) \right) \tag{3.65}$$

for all $k = 1, \ldots, m-1$, $l = 1, \ldots, d-1$, and, by (3.63),

$$\frac{A}{(1+\varepsilon^2)^{d-1}} - |T'_{kd}(\xi_d)|^2 = \frac{A}{(1+\varepsilon^2)^{d-1}} |Q_{kd}(\xi_d)|^2,$$

for all $k = 1, \ldots, m-1$, we obtain

$$4 - \sigma(\xi) = \sum_{k=1}^{m-1} \sum_{l=1}^{d} \frac{A}{m(1+\varepsilon^2)^{l-1}} \left| Q_{kl}(\xi_l) \prod_{j=1}^{l-1} T'_{kj}(\xi_j) \right|^2 . \tag{3.66}$$

Numerating the summands of this double sum in an arbitrary order, rewrite them as $|P_i(\xi)|^2$, $i = 1, \ldots, d(m-1)$, and set $r = d(m-1) + m$,

$$\mu_{0k} := \frac{1-\sigma}{2} P_{k-m+1}, \quad k = m, \ldots, r-1, \quad \mu_{0r} \equiv 0. \tag{3.67}$$

It follows from (3.59) and (3.66) that

$$\sum_{k=0}^{r} |\mu_{0k}|^2 \equiv 1.$$

Since $\sigma(\mathbf{0}) = 1$, equalities (3.50) follow from 3.60.

Extend the row $(\mu_{00}, \ldots, \mu_{0r})$ up to a unitary matrix by

$$\mu_{\nu r} := \overline{\mu_{0,r-\nu}}, \; \mu_{\nu k} := \delta_{r-\nu,k} - \mu_{0k}\overline{\mu_{0,r-\nu}}, \; k = 0, \ldots, r-1, \nu = 1, \ldots, r. \tag{3.68}$$

Thus all conditions of Theorem 3.4.3 are satisfied, and hence, the matrix $\mathcal{M}$ consisting of the first m columns of the constructed unitary matrix $\{\mu_{\nu k}\}_{\nu,k=0}^{r}$ generates a compactly supported tight wavelet frame with $\mathbf{VM^n}$ property◊

Analyzing the proof of Theorem 3.4.5, we can give a constructive description of our method.

Algorithm 2

Step 1. Given dilation matrix M and $n \in \mathbb{N}$, choose an arbitrary set of digits $D(M) = \{s_0, \dots, s_{m-1}\}$, $s_0 = \mathbf{0}$, and trigonometric polynomials of one variable g_l, $l = 0, \dots, n-1$, such that

$$\frac{d^i g_l}{dx^i}(0) = \delta_{il}$$

for all $i = 0, \dots, n$, $g_0 \equiv 1$, (see formulas for g_l below). Find trigonometric polynomials T_{kj}, $k = 0, \dots, m-1$, $j = 1, \dots, d$, by (3.52).
Step 2. Set $A = \frac{4m-1}{m-1}$, choose $\varepsilon > 0$ satisfying $1+\varepsilon^2 < A^{1/d}$ and find trigonometric polynomials T'_{kj}, $k = 0, \dots, m-1$, $j = 1, \dots, d$, satisfying (3.55) and (3.56) as follows. If $|T_{kj}|^2 \le 1+\varepsilon^2$, then $T'_{kj} = T_{kj}$, otherwise, by choosing L and N properly, define T'_{kj} by (3.54).
Step 3. Find μ'_{0k} and μ_{0k}, $k = 0, \dots, m-1$, by (3.57) and (3.58), respectively.
Step 4. Find trigonometric polynomials $|Q_{kj}|^2$, $k = 1, \dots, m-1$, $j = 1, \dots, d$, satisfying (3.62) or (3.63), and trigonometric polynomials $P_1, \dots, P_{d(m-1)}$ whose squared magnitudes are the summands of the right-hand side of (3.66). Set $r = d(m-1) + m$ and find μ_{0k}, $k = m, \dots, r$, by (3.67).
Step 5. Find $\mu_{\nu k}$, $\nu = 1, \dots, r$, $k = 0, \dots, m-1$, by (3.68). Find refinable mask m_0 and wavelet masks m_ν, $\nu = 1, \dots, r$, by (3.16).$\Diamond$

The above algorithm is applicable for any dilation matrix M, but it may be essentially simplified for some class of matrices. Suppose that all entries of some column, say j_0-th, of M are divisible by m. There exist many such dilation matrices, for example:

$$m=2:\ M = \begin{pmatrix} 2 & -2 \\ 1 & 0 \end{pmatrix}, \quad M = \begin{pmatrix} 5 & 2 \\ -11 & -4 \end{pmatrix}, \quad M = \begin{pmatrix} -2 & 3 & 1 \\ -2 & 2 & 1 \\ 2 & 0 & 0 \end{pmatrix};$$

$$m=3:\ M = \begin{pmatrix} 3 & -1 \\ 3 & -2 \end{pmatrix}, \quad M = \begin{pmatrix} -3 & 5 & 1 \\ -3 & 4 & 1 \\ 3 & 0 & 0 \end{pmatrix}; \quad m=4: M = \begin{pmatrix} -4 & -2 \\ 8 & 3 \end{pmatrix}.$$

For such matrices, digits $s_0, , \dots, s_{m-1}$ can be chosen so that $(M^{-1}s_k)_j = 0$ for all $j \ne j_0$. It follows that $T_{kj} \equiv 1$ whenever $j \ne j_0$ (see the proof of Theorem 3.4.5). Hence, the functions $\mu'_{0k}, \mu_{0k}, k = 0, \dots, m-1$, depend on only one variable x_{j_0}, and extension of the row $(\mu_{00}, \dots, \mu_{0m-1})$ can be improved as follows. Instead of (3.62) or (3.63), using the Riesz Lemma, we find only one trigonometric polynomial Q such that

$$1 - \frac{1}{4m-1}\sum_{k=1}^{m-1} |T'_{kj_0}(t)|^2 = |Q(t)|^2, \tag{3.69}$$

replace (3.66) by

$$4 - \sigma(\xi) = \frac{4m-1}{m} \left|Q(\xi_{j_0})\right|^2 =: |P_1(\xi)|^2 \tag{3.70}$$

and find μ_{0m} by (3.67). Adding $\mu_{0m+1} \equiv 0$ we find polyphase matrix by (3.68) with $r = m+1$ and the corresponding wavelet masks m_ν, $\nu = 1, \dots, r$, by (3.16). Since all entries of the row $(\mu_{00}, \dots, \mu_{0m})$ are trigonometric polynomials of one variable, the row can be extended to a unitary matrix without adding μ_{0m+1}. A constructive algorithm for such extension is known (see [8] or [2, Sect. 2.6]). Thus, the number of wavelet functions may be reduced to m in this way. Note that the restriction for choosing ε is now much weaker than in the general situation. It is clear from (3.69) that it suffices for it to satisfy $1 + \varepsilon^2 \le \frac{4m-1}{m-1}$.

Now we give a constructive description of our method.

Algorithm 3

Step 1. Given $n \in \mathbb{N}$ and a dilation matrix M such that all entries of j_0-th column are divisible by m, choose a set of digits $D(M) = \{s_0, \dots, s_{m-1}\}$, $s_k = \frac{l_k}{m} M e_{j_0}$, where $l_0 = 0$, $l_k \not\equiv l_i (\mod m)$ for $k \ne i$, e_{j_0} is the j_0-th unit vector (column) in $\mathbb{R}^d$, and trigonometric polynomials of one variable g_l, $l = 0, \dots, n$, such that

$$\frac{d^i g_l}{dx^i}(0) = \delta_{il}$$

for all $i = 0, \dots, n$, $g_0 \equiv 1$, (see formulas for g_l below). Find trigonometric polynomials T_{kj_0}, $k = 0, \dots, m-1$, by (3.52).
Step 2. Choose $\varepsilon > 0$ satisfying

$$\varepsilon \le \sqrt{\frac{3m}{m-1}}$$

and find a trigonometric polynomial T'_{kj_0}, $k = 0, \dots, m-1$, $j = 1, \dots, d$, satisfying (3.55) and (3.56) as follows. If $|T_{kj_0}|^2 \le 1 + \varepsilon^2$, then $T'_{kj_0} = T_{kj}$, otherwise, by choosing L and N properly, define T'_{kj_0} by (3.54). Set $T'_{kj} \equiv 1$, $j = 1, \dots, d$, $j \ne j_0$.
Step 3. Find μ'_{0k} and μ_{0k}, $k = 0, \dots, m-1$, by (3.57) and (3.58), respectively.
Step 4. Define a trigonometric polynomial $|Q|^2$ by (3.69), and find a trigonometric polynomials P_1 satisfying (3.70). Set $r = m+1$ and find μ_{0m}, μ_{km+1} by (3.67).
Step 5. Find $\mu_{\nu k}$, $\nu = 1, \dots, r$, $k = 0, \dots, m-1$, by (3.68). Find refinable mask m_0 and wavelet masks m_ν, $\nu = 1, \dots, r$, by (3.16).$\Diamond$

Suppose m_0^* is a trigonometric polynomial of one variable, and it is an orthogonal mask with respect to a scalar dilation m, that is

$$\sum_{k=0}^{m-1} \left| m_0^* \left(t + \frac{k}{m} \right) \right|^2 = 1 \quad \forall t \in \mathbb{R}.$$

Now we show how such masks can be used for the construction of multivariate tight frames with minimal possible wavelet functions for some dilation matrices.

Let M be a dilation matrix such that there is a column, say the first one, whose each entry is divisible by m, and let e denote the 1-st unit vector (column) in $\mathbb{R}^d$. It is easy to see that the vectors $s_k := \frac{k}{m}Me, k = 0, \dots, m-1$, can be chosen as digits of M. Assume also that m_0^* satisfies the sum rule of order n, i.e.,

$$\frac{d^l m_0^*}{dt^l}\left(\frac{k}{m}\right) = 0 \quad \forall k = 1, \dots, m-1, \ l = 0, \dots, n-1. \tag{3.71}$$

Choose $s_k^* := k, k = 0, \dots, m-1$, as digits for the dilation m, and let $\mu_{00}^*, \dots, \mu_{0m-1}^*$ be the corresponding polyphase components of m_0^*.

Due to Proposition 3.3.2 and Theorems 3.4.2 and 3.4.3, there exist complex numbers $\lambda_l^*, l = 0, \dots, n$, such that

$$\lambda_0^* = 1, \quad \sum_{\gamma=0}^{\alpha}(-1)^{[\alpha-\gamma]}\binom{\alpha}{\gamma}\lambda_\gamma^*\overline{\lambda_{\alpha-\gamma}^*} = 0 \ \ \forall \alpha - 1, \dots, n-1,$$

and

$$\frac{d^\beta \mu_{0k}^*}{dt^\beta}(0) = \frac{(2\pi i)^\beta}{\sqrt{m}}\sum_{\gamma=0}^{\beta}\lambda_\gamma^*\binom{\beta}{\gamma}\left(-\frac{k}{m}\right)^{\beta-\gamma} \quad \forall \beta = 0, \dots, n-1, \ k = 0, \dots, m-1.$$

Let $\gamma \in \mathbb{Z}^d$, $[\gamma] < n$. Define λ_γ by

$$\lambda_\gamma := \begin{cases} \lambda_{\gamma_1}^*, & \text{if } \gamma = \gamma_1 e, \\ 0, & \text{if } \gamma \neq \gamma_1 e, \end{cases}$$

and set

$$\mu_{0k}(\xi) := \mu_{0k}^*(\xi_1), \quad \forall x \in \mathbb{R}^d, \ k = 0, \dots, m-1,$$

and check that relations (3.21) hold.

If $\beta \in \mathbb{Z}_+^d$, $[\beta] < n$, then

$$D^\beta \mu_{0k}(\mathbf{0}) = \begin{cases} \frac{(2\pi i)^{\beta_1}}{\sqrt{m}}\sum\limits_{j=0}^{\beta_1}\binom{\beta_1}{j}\lambda_j^*\left(-\frac{k}{m}\right)^{\beta_1-j}, & \text{if } \beta = \beta_1 e, \\ 0, & \text{if } \beta \neq \beta_1 e. \end{cases}$$

On the other hand,

$$\frac{(2\pi i)^{[\beta]}}{\sqrt{m}}\sum_{\mathbf{0}\leq\gamma\leq\beta}\binom{\beta}{\gamma}\lambda_\gamma(-M^{-1}s_k)^{\beta-\gamma} = \frac{(2\pi i)^{[\beta]}}{\sqrt{m}}\sum_{j=0}^{\beta_1}\binom{\beta_1}{j}\lambda_j^*\left(-\frac{k}{m}e\right)^{\beta-je}$$

$$= \begin{cases} \frac{(2\pi i)^{\beta_1}}{\sqrt{m}} \sum_{j=0}^{\beta_1} \binom{\beta_1}{j} \lambda_j^* \left(-\frac{k}{m}\right)^{\beta_1 - j}, & \text{if } \beta = \beta_1 e, \\ 0, & \text{if } \beta \neq \beta_1 e, \end{cases}$$

which yields (3.20).

Since $(M^*\xi)_1 = (e, M^*\xi) = m(s_1, \xi)$ and $(s_k, \xi) = \frac{k}{m}(Me, \xi) = \frac{k}{m}(e, M^*\xi)$, we have

$$m_0(\xi) = \sum_{k=0}^{m-1} e^{2\pi i (s_k, \xi)} \mu_{0k}(M^*\xi) = \sum_{k=0}^{m-1} e^{2\pi i k (s_1, \xi)} \mu_{0k}^*(m(s_1, \xi)),$$

which yields

$$m_0(\xi) = m_0^*((s_1, \xi)). \tag{3.72}$$

It follows from (3.71) and Proposition 2.6.6 that

$$\sum_{k=0}^{m-1} |\mu_{0k}(\xi)|^2 = \sum_{k=0}^{m-1} |\mu_{0k}^*(\xi_1)|^2 \equiv 1.$$

All entries of the row $(\mu_{00}, \ldots, \mu_{0m-1})$ are trigonometric polynomials of one variable; hence, the row can be extended to a unitary matrix $\{\mu_{\nu k}\}_{\nu,k=0}^{m-1}$ (see [8] or [2, Sect. 2.6]). In the case $m = 2$, this extension can be given by

$$\mu_{10} = \overline{-\mu_{01}}, \quad \mu_{11} = \overline{\mu_{00}}, \tag{3.73}$$

and the wavelet mask m_1 is given by

$$m_1(\xi) = \mu_{00}(M^*\xi) + e^{2\pi i (s_1, \xi)} \mu_{01}(M^*\xi) = -\overline{\mu_{00}^*(2(s_1, \xi))} + e^{2\pi i (s_1, \xi)} \overline{\mu_{0k}^*(2(s_1, \xi))}.$$

Taking into account that

$$-\overline{\mu_{00}^*(2t)} + e^{2\pi i t} \overline{\mu_{0k}^*(2t)} = e^{2\pi i t} \overline{\left(\mu_{00}^*(2t) + e^{2\pi i (t+1/2)} \mu_{01}^*(2t)\right)} = e^{2\pi i t} \overline{m_0^*(t + 1/2)},$$

we obtain

$$m_1(\xi) = e^{2\pi i (s_1, \xi)} \overline{m_0^*((s_1, \xi) + 1/2)}. \tag{3.74}$$

For $m = 2$, orthogonal masks m_0^* satisfying (3.71) are well known; they are called Daubechies masks and actively used in signal processing. The Daubechies masks of minimal degree can be found as follows.

Let

$$q_n(t) := 1 - c_n \int_0^t \sin^{2n-1} 2\pi u \, du, \quad t \in \mathbb{R}, \tag{3.75}$$

where $c_n = \left(\int_0^{1/2} \sin^{2n-1} 2\pi u\, du\right)^{-1}$. Since q_n is an even nonnegative trigonometric polynomial, by the Riesz Lemma, there exists a trigonometric polynomial

$$m_0^*(t) = \sum_{k=0}^{2n-1} h_k e^{2\pi i k t}, \quad h_k \in \mathbb{R},$$

such that $|m_0^*(t)|^2 = q_n(t)$. It is easy to check that $q_n(t) + q_n(t+1/2) = 1$. Hence, m_0^* is an orthogonal mask. Since

$$\frac{d^l q_n}{dt^l}\left(\frac{k}{m}\right) = 0, \quad l = 0, \dots, 2n-1,$$

we have

$$\frac{d^l m_0^*}{dt^l}\left(\frac{1}{2}\right) = 0, \quad l = 0, \dots, n-1;$$

i.e., (3.71) is satisfied for m_0^*.

Now we give formulas for functions g_β, $\beta \in \mathbb{Z}_+^d$, which are used in all our algorithms. Recall that given $n \in \mathbb{Z}_+$, g_β is a trigonometric polynomial such that $D^\beta g_\beta(\mathbf{0}) = 1$ and

$$D^\gamma g_\beta(\mathbf{0}) = 0 \quad \forall\, \gamma \in \mathbb{Z}_+^d,\ \gamma \neq \beta,\ [\gamma] < n. \tag{3.76}$$

We replace (3.76) by

$$D^\gamma g_\beta(\mathbf{0}) = 0 \quad \forall\, \gamma \in \mathbb{Z}_+^d,\ \gamma \neq \beta,\ \gamma \leq \alpha,$$

and construct functions g_β, $\beta \leq \alpha$, for every $\alpha \in \mathbb{Z}_+^d$. If $\alpha_j = n$, $j = 1, \dots, d$, then (3.76) is satisfied.

Given $\alpha \in \mathbb{Z}_+^d$, functions g_β can be defined, for example, by

$$g_\beta(x) = \prod_{j=1}^d g_{\beta_j}(x_j),$$

$$g_{\beta_j}(u) = \frac{1}{\beta_j!(-2\pi i)^{\beta_j}} \left(\left(1 - e^{2\pi i u}\right)^{\beta_j} - \sum_{l=1}^{\alpha_j - \beta_j} a_l \left(1 - e^{2\pi i u}\right)^{\beta_j + l} \right),$$

$$a_l = \frac{(-2\pi i)^{-\beta_j - l}}{(\beta_j + l)!} \frac{d^{\beta_j + l}}{du^{\beta_j + l}} \left(\left(1 - e^{2\pi i u}\right)^{\beta_j} - \sum_{r=1}^{l-1} a_r \left(1 - e^{2\pi i u}\right)^{\beta_j + r} \right)\Bigg|_{u=0}$$

In particular, by this formula, we have

$$\alpha_j = 1: \quad g_1(u) = -\frac{1}{2\pi i}\left(1 - e^{2\pi i u}\right)$$

$$\alpha_j = 2: \quad g_1(u) = -\frac{1}{2\pi i}\left(\left(1 - e^{2\pi i u}\right) + \frac{1}{2}\left(1 - e^{2\pi i u}\right)^2\right)$$

$$g_2(u) = -\frac{1}{8\pi^2}\left(\left(1 - e^{2\pi i u}\right)^2 + \left(1 - e^{2\pi i u}\right)^3\right)$$

$$\alpha_j = 3: \quad g_1(u) = -\frac{1}{2\pi i}\left(\left(1 - e^{2\pi i u}\right) + \frac{1}{2}\left(1 - e^{2\pi i u}\right)^2 + \frac{1}{3}\left(1 - e^{2\pi i u}\right)^3\right)$$

$$g_2(u) = -\frac{1}{8\pi^2}\left(\left(1 - e^{2\pi i u}\right)^2 + \left(1 - e^{2\pi i u}\right)^3 + \frac{11}{12}\left(1 - e^{2\pi i u}\right)^4\right)$$

$$g_3(u) = \frac{1}{48\pi^3 i}\left(\left(1 - e^{2\pi i u}\right)^3 + \frac{3}{2}\left(1 - e^{2\pi i u}\right)^4 + \frac{7}{4}\left(1 - e^{2\pi i u}\right)^5\right)$$

Similarly, real functions g_{β_j} can be defined by

$$g_{\beta_j}(u) = \frac{1}{\beta_j!(2\pi)^{\beta_j}}\left(\sin^{\beta_j} 2\pi u - \sum_{l=1}^{\alpha_j - \beta_j} a_l \sin^{\beta_j + l} 2\pi u\right),$$

$$a_l = \frac{(2\pi)^{-\beta_j - l}}{(\beta_j + l)!}\frac{d^{\beta_j + l}}{du^{\beta_j + l}}\left(\sin^{\beta_j} 2\pi u - \sum_{r=1}^{l-1} a_r \sin^{\beta_j + r} 2\pi u\right)\Bigg|_{u=0}.$$

Finally we give a generalization of Matrix Extension Principle, which also can be used for the constructions of compactly supported tight wavelet frames.

Theorem 3.4.6 *Let $\varphi \in L_2(\mathbb{R}^d)$ be a compactly supported refinable function with mask m_0, $\widehat{\varphi}(\mathbf{0}) = 1$, and let S be a continuous nonnegative 1-periodic (with respect to each variable) function. Suppose there exist trigonometric polynomials m_ν, $\nu = 1, \ldots, r$, such that*

$$S(M^*\xi) m_0(\xi + M^{*-1}q))\overline{m_0(\xi + M^{*-1}q'))} + \sum_{\nu=1}^{r} m_\nu(\xi + M^{*-1}q))\overline{m_\nu(\xi + M^{*-1}q'))} = S(\xi + M^{*-1}q)\delta_{qq'} \tag{3.77}$$

for all $q, q' \in D(M^*)$, and define wavelet functions $\psi^{(\nu)}$, $\nu = 1, \ldots, r$, by

$$\widehat{\psi^{(\nu)}}(\xi) = m_\nu(M^{*-1}\xi)\widehat{\varphi}(M^{*-1}\xi). \tag{3.78}$$

Then the wavelet system $\{\psi_{jk}^{(\nu)}\}_{j,k,\nu}$ is a tight frame consisting of compactly supported functions.

Proof Set

$$\widetilde{m}_0(\xi) = \sqrt{S(M^*\xi))} m_0(\xi)$$

and define a function $\psi^{(0)}$ by

$$\widehat{\psi^{(0)}}(\xi) = \widetilde{m}_0(M^{*-1}\xi)\widehat{\varphi}(M^{*-1}\xi). \tag{3.79}$$

Obviously, $\psi^{(0)} \in L_2(\mathbb{R}^d)$. Since φ is a compactly supported function, the system $\{\varphi_{0k}\}_k$ is Bessel, due to by Proposition 1.1.8. It follows from Remark 1.1.7 that the function $[\widehat{\varphi}, \widehat{\varphi}]$ is essentially bounded, which yields that $[\widehat{\psi^{(0)}}, \widehat{\psi^{(0)}}]$ is also essentially bounded.

Let us show that for any function $f \in L_2(\mathbb{R}^d)$ and any $j, j' \in \mathbb{Z}$, $j < j'$, the following equality holds:

$$\sum_{k\in\mathbb{Z}^d} |\langle f, \psi^{(0)}_{jk}\rangle|^2 + \sum_{i=j}^{j'-1}\sum_{\nu=1}^{r}\sum_{k\in\mathbb{Z}^d} |\langle f, \psi^{(\nu)}_{ik}\rangle|^2 = \sum_{k\in\mathbb{Z}^d} |\langle f, \psi^{(0)}_{j'k}\rangle|^2. \tag{3.80}$$

Clearly, it suffices for it to check (3.80) for $j' = j + 1$. For convenience, we also set

$$\widetilde{m}_\nu = m_\nu, \quad \nu = 1, \dots, r.$$

Using (3.78), (3.79), and Lemma 2.2.4, taking into account periodicity of the functions $\widetilde{m}_\nu$, we have

$$\sum_{k\in\mathbb{Z}^d} |\langle f, \psi^{(0)}_{jk}\rangle|^2 + \sum_{\nu=1}^{r}\sum_{k\in\mathbb{Z}^d} |\langle f, \psi^{(\nu)}_{jk}\rangle|^2 = m^j \sum_{\nu=0}^{r} \int_{\mathbb{T}^d} \left|[\widehat{f}(M^{*j}\cdot), \widehat{\psi^{(\nu)}}](\xi)\right|^2 d\xi$$

$$m^j \sum_{\nu=0}^{r} \int_{\mathbb{T}^d} \sum_{k\in\mathbb{Z}^d} \widehat{f}\left(M^{*j}(\xi+k)\right) \overline{\widetilde{m}_\nu\left(M^{*-1}(\xi+k)\right)\widehat{\varphi}\left(M^{*-1}(\xi+k)\right)} \times$$

$$\overline{\sum_{l\in\mathbb{Z}^d} \widehat{f}\left(M^{*j}(\xi+l)\right) \overline{\widetilde{m}_\nu\left(M^{*-1}(\xi+l)\right)\widehat{\varphi}\left(M^{*-1}(\xi+l)\right)}}\, d\xi =$$

$$m^j \sum_{\nu=0}^{r} \int_{\mathbb{T}^d} \left(\sum_{q\in D(M^*)} [\widehat{f}(M^{*(j+1)}\cdot), \widehat{\varphi}](M^{*-1}(\xi+q))\overline{\widetilde{m}_\nu\left(M^{*-1}(\xi+q)\right)}\right)\cdot$$

$$\overline{\left(\sum_{q'\in D(M^*)} [\widehat{f}(M^{*(j+1)}\cdot), \widehat{\varphi}](M^{*-1}(\xi+q'))\overline{\widetilde{m}_\nu\left(M^{*-1}(\xi+q')\right)}\right)}\, d\xi. \tag{3.81}$$

By (3.77), we have

$$\sum_{\nu=0}^{r} \widetilde{m}_\nu\left(M^{*-1}(\xi+q)\right)\overline{\widetilde{m}_\nu\left(M^{*-1}(\xi+q')\right)} = S(\left(M^{*-1}(\xi+q)\right))\delta_{qq'}, \quad q, q' \in D(M^*).$$

Hence, the right-hand side of (3.81) can be reduced to

$$m^j \int\limits_{[0,1)^d} \sum_{q\in D(M^*)} \left|[\widehat{f}(M^{*(j+1)}\cdot), \widehat{\varphi}](M^{*-1}(\xi+q))\right|^2 S\left(M^{*-1}(\xi+q)\right) d\xi.$$

Using Lemma 2.1.4, we obtain

$$\sum_{k\in\mathbb{Z}^d} |\langle f, \psi_{jk}^{(0)}\rangle|^2 + \sum_{\nu=1}^{r}\sum_{k\in\mathbb{Z}^d} |\langle f, \psi_{jk}^{(\nu)}\rangle|^2 =$$

$$m^{j+1} \sum_{s\in D(M^*)} \int\limits_{M^{*-1}(\mathbb{T}^d+s)} \left|[\widehat{f}(M^{*(j+1)}\cdot), \widehat{\varphi}](\xi)\right|^2 S(\xi)\, d\xi =$$

$$m^{j+1} \int\limits_{\mathbb{T}^d} \left|[\widehat{f}(M^{*(j+1)}\cdot), \widehat{\varphi}](\xi)\right|^2 S(\xi)\, d\xi = m^{j+1} \int\limits_{\mathbb{T}^d} \left|[\widehat{f}(M^{*(j+1)}\cdot), \widehat{\psi}^{(0)}](\xi)\right|^2 d\xi.$$

To prove (3.80) for $j' = j+1$, it remains to use Lemma 2.2.4 and Remark 1.1.7 once more.

It follows from Lemma 2.2.7 that

$$\lim_{j\to-\infty} \sum_{k\in\mathbb{Z}^d} \left|\langle f, \psi_{jk}^{(0)}\rangle\right|^2 = 0.$$

Passing to the limit as $j \to -\infty$ in (3.80), for any $j' \in \mathbb{Z}$, we get the equality

$$\sum_{i=-\infty}^{j'} \sum_{\nu=1}^{r} \sum_{k\in\mathbb{Z}^d} \left|\langle f, \psi_{ik}^{(\nu)}\rangle\right|^2 = \sum_{k\in\mathbb{Z}^d} \left|\langle f, \psi_{j',k}^{(0)}\rangle\right|^2.$$

It follows from Theorem 2.2.6 that

$$\lim_{j'\to+\infty} \sum_{i=-\infty}^{j'} \sum_{\nu=1}^{r} \sum_{k\in\mathbb{Z}^d} \left|\langle f, \psi_{ik}^{(\nu)}\rangle\right|^2 = S(0)\|f\|^2,$$

as was to be proved.◊

3.5 Examples

In this section, we shall give several examples of compactly supported dual and tight wavelet frames to illustrate the results of this chapter. All examples are based on the constructions of refinable and wavelet masks described above.

1. Let $n = 3$, $M = \begin{pmatrix} 2 & 1 \\ -1 & 1 \end{pmatrix}$. For this matrix $M^{-1} = \begin{pmatrix} 1/3 & -1/3 \\ 1/3 & 2/3 \end{pmatrix}$, $m = 3$.

Following Algorithm 1, choose digits $s_0 = \begin{pmatrix} 0 \\ 0 \end{pmatrix}$, $s_1 = \begin{pmatrix} 1 \\ 0 \end{pmatrix}$, $s_2 = \begin{pmatrix} -1 \\ 0 \end{pmatrix}$, numbers $\lambda_{00} = 1$, $\lambda_\gamma = \widetilde{\lambda}_\gamma = 0$ for all $\gamma \in \mathbb{Z}_+^2$, $[\gamma] \le 2, \gamma \ne (0, 0)$ and functions

$$g_{00} \equiv 1, \quad g_{10}(\xi) = \frac{\sin 2\pi\xi_1}{2\pi}, \quad g_{01}(\xi) = \frac{\sin 2\pi\xi_2}{2\pi},$$
$$g_{20}(\xi) = \frac{\sin^2 2\pi\xi_1}{8\pi^2}, \quad g_{02}(\xi) = \frac{\sin^2 2\pi\xi_2}{8\pi^2}, \quad g_{11}(\xi) = \frac{\sin 2\pi\xi_1 \sin 2\pi\xi_2}{4\pi^2},$$

$T_0 = T_1 = \widetilde{T}_1 = T_2 = \widetilde{T}_2 \equiv 0$, $\widetilde{T}_0(\xi) = -\frac{1}{162\sqrt{3}}(\sin 2\pi\xi_1 + \sin 2\pi\xi_2)^4$. Using (3.41), (3.42), we obtain

$$\mu_{00}(\xi) = \frac{1}{\sqrt{3}}, \quad \widetilde{\mu}'_{00}(\xi) = \frac{1}{\sqrt{3}}\left(1 - \frac{1}{162}(\sin 2\pi\xi_1 + \sin 2\pi\xi_2)^4\right),$$

$$\mu_{01}(\xi) = \widetilde{\mu}'_{01}(\xi) = \overline{\mu_{02}(\xi)} = \overline{\widetilde{\mu}'_{02}(\xi)} =$$

$$\frac{1}{\sqrt{3}}\left(1 - \frac{1}{18}(\sin 2\pi\xi_1 + \sin 2\pi\xi_2)^2 - \frac{i}{3}(\sin 2\pi\xi_1 + \sin 2\pi\xi_2)\right).$$

Since $\sigma \equiv 1$, following Step 3 of Algorithm 1, we set $\widetilde{\mu}_{0k} = \widetilde{\mu}'_{0k}$, $k = 0, 1, 2$. The refinable masks m_0, $\widetilde{m}_0$ are trigonometric polynomials whose Fourier coefficients are given by the following tables:

$$m_0 : \quad \begin{pmatrix} \frac{1}{216} & 0 & \frac{1}{216} & -\frac{1}{108} & 0 & -\frac{1}{108} & \frac{1}{216} & 0 & \frac{1}{216} & 0 & 0 \\ 0 & 0 & -\frac{1}{18} & 0 & \frac{1}{18} & \frac{1}{18} & 0 & -\frac{1}{18} & 0 & 0 & 0 \\ 0 & \frac{1}{108} & 0 & \frac{1}{108} & \frac{17}{54} & \frac{1}{3} & \frac{17}{54} & \frac{1}{108} & 0 & \frac{1}{108} & 0 \\ 0 & 0 & 0 & -\frac{1}{18} & 0 & \frac{1}{18} & \frac{1}{18} & 0 & -\frac{1}{18} & 0 & 0 \\ 0 & 0 & \frac{1}{216} & 0 & \frac{1}{216} & -\frac{1}{108} & 0 & -\frac{1}{108} & \frac{1}{216} & 0 & \frac{1}{216} \end{pmatrix}$$

$$\widetilde{m}_0 : \left(\begin{matrix} -\frac{1}{7776} & 0 & 0 & \frac{1}{1944} & 0 & 0 & -\frac{1}{1296} & 0 \\ 0 & 0 & 0 & 0 & 0 & 0 & 0 & 0 \\ 0 & -\frac{1}{1944} & 0 & \frac{1}{216} & \frac{1}{486} & \frac{1}{216} & -\frac{1}{108} & -\frac{1}{324} \\ 0 & 0 & 0 & 0 & 0 & -\frac{1}{18} & 0 & \frac{1}{18} \\ 0 & 0 & -\frac{1}{1296} & 0 & \frac{1}{108} & \frac{1}{324} & \frac{1}{108} & \frac{17}{54} \\ 0 & 0 & 0 & 0 & 0 & 0 & -\frac{1}{18} & 0 \\ 0 & 0 & 0 & -\frac{1}{1944} & 0 & \frac{1}{216} & \frac{1}{486} & \frac{1}{216} \\ 0 & 0 & 0 & 0 & 0 & 0 & 0 & 0 \\ 0 & 0 & 0 & 0 & -\frac{1}{7776} & 0 & 0 & \frac{1}{1944} \end{matrix}\right. \cdots$$

$$\cdots \left.\begin{matrix} 0 & \frac{1}{1944} & 0 & 0 & -\frac{1}{7776} & 0 & 0 & 0 & 0 \\ 0 & 0 & 0 & 0 & 0 & 0 & 0 & 0 & 0 \\ -\frac{1}{108} & \frac{1}{216} & \frac{1}{486} & \frac{1}{216} & 0 & -\frac{1}{1944} & 0 & 0 & 0 \\ \frac{1}{18} & 0 & -\frac{1}{18} & 0 & 0 & 0 & 0 & 0 & 0 \\ \mathbf{\frac{71}{216}} & \frac{17}{54} & \frac{1}{108} & \frac{1}{324} & \frac{1}{108} & 0 & -\frac{1}{1296} & 0 & 0 \\ \frac{1}{18} & \frac{1}{18} & 0 & -\frac{1}{18} & 0 & 0 & 0 & 0 & 0 \\ -\frac{1}{108} & -\frac{1}{324} & -\frac{1}{108} & \frac{1}{216} & \frac{1}{486} & \frac{1}{216} & 0 & -\frac{1}{1944} & 0 \\ 0 & 0 & 0 & 0 & 0 & 0 & 0 & 0 & 0 \\ 0 & 0 & -\frac{1}{1296} & 0 & 0 & \frac{1}{1944} & 0 & 0 & -\frac{1}{7776} \end{matrix}\right).$$

Checking shows (see Sect. 6.8) that the corresponding refinable functions $\varphi, \widetilde{\varphi}$ are in $L_2(\mathbb{R}^d)$.

Next, following Algorithm 1, we can set $r = 3$, $\mu_{03} = \widetilde{\mu}_{03} \equiv 0$ and find functions $\mu_{\nu k}$, $\nu = 1, 2, 3$, $k = 0, 1, 2$ by (3.45), (3.46). However, r may be reduced to 2, what is possible because the function μ_{00} identically equals $\frac{1}{\sqrt{3}}$. Due to this feature, the row $(\sqrt{3}\mu_{00}, \sqrt{3}\mu_{01}, \sqrt{3}\mu_{02})$ can be easily extended to a unimodular matrix. Using a method described in Sect. 2.5, one can extend this row and the row $(\frac{1}{\sqrt{3}}\widetilde{\mu}_{00}, \frac{1}{\sqrt{3}}\widetilde{\mu}_{00}, \frac{1}{\sqrt{3}}\widetilde{\mu}_{00})$ to square matrices $\mathcal{M}, \widetilde{\mathcal{M}}$ satisfying $\mathcal{M}^*\widetilde{\mathcal{M}} = I_3$. In this way, we obtain

$$\begin{aligned} \mu_{\nu 0} &= -\frac{1}{3}\overline{\widetilde{\mu}_{0\nu}}, \quad \widetilde{\mu}_{\nu 0} = -3\overline{\mu_{0\nu}}, \\ \mu_{\nu k} &= \frac{1}{\sqrt{3}}(\delta_{\nu k} - \mu_{0k}\overline{\widetilde{\mu}_{0\nu}}), \quad \nu, k = 1, 2, \\ \widetilde{\mu}_{\nu k} &= \sqrt{3}\delta_{\nu k}, \quad \nu, k = 1, 2. \end{aligned}$$

To get wavelet masks $m_\nu, \widetilde{m}_\nu, \nu = 1, 2, 3$, it remains to substitute their polyphase components $\mu_{\nu k}, \widetilde{\mu}_{\nu k}$ into (3.16) (and a similar formula for $\widetilde{m}_\nu$).

2. Let $n = 2$, $M = \begin{pmatrix} 1 & 1 \\ 1 & -1 \end{pmatrix}$ be the quincunx dilation matrix. For this matrix $m = 2$, we can choose digits $s_0 = \begin{pmatrix} 0 \\ 0 \end{pmatrix}$, $s_1 = \begin{pmatrix} 1 \\ 0 \end{pmatrix}$.

Let φ be a box spline whose Fourier transform is defined by

$$\widehat{\varphi}(\xi) = \left(\frac{1-e^{-2\pi i\xi_1}}{2\pi i\xi_1}\right)\left(\frac{1-e^{-2\pi i\xi_2}}{2\pi i\xi_2}\right)^2\left(\frac{1-e^{-2\pi i(\xi_1+\xi_2)}}{2\pi i(\xi_1+\xi_2)}\right)\left(\frac{1-e^{-2\pi i(\xi_1-\xi_2)}}{2\pi i(\xi_1-\xi_2)}\right)^2.$$

It is not difficult to check that this box spline is in C^2 (see [9]) and satisfies the refinement equation with the mask

$$m_0(\xi) = \left(\frac{1+e^{-2\pi i\xi_1}}{2}\right)\left(\frac{1+e^{-2\pi i\xi_2}}{2}\right)^2,$$

whose polyphase components are

$$\mu_{00}(\xi) = \frac{1}{\sqrt{2}}\left(\frac{1}{4}+\frac{1}{4}e^{-2\pi i(\xi_1-\xi_2)}+\frac{1}{2}e^{-2\pi i\xi_1}\right),$$
$$\mu_{01}(\xi) = \frac{1}{\sqrt{2}}\left(\frac{1}{2}e^{-2\pi i\xi_1}+\frac{1}{4}e^{-2\pi i(\xi+\xi_2)}+\frac{1}{4}e^{-4\pi i\xi_1}\right).$$

Since (3.22) holds for $\alpha \in \mathbb{Z}^2$, $[\alpha] < 2$, the polyphase components satisfy condition (3.20) and $\lambda_{00} = 1$, $\lambda_{10} = -\frac{3}{4}$, $\lambda_{01} = \frac{1}{4}$, due to Proposition 3.3.2. Following Algorithm 1, find $\widetilde{\lambda}_{00} = 1$, $\widetilde{\lambda}_{10} = -\frac{3}{4}$, $\widetilde{\lambda}_{01} = \frac{1}{4}$, and choosing

$$g_{00} \equiv 1, \quad g_{10}(\xi) = \frac{1-e^{-2\pi i\xi_1}}{2\pi i}, \quad g_{01}(\xi) = \frac{1-e^{-2\pi i\xi_2}}{2\pi i},$$

$$\widetilde{T}_0 = \frac{3}{4}\cos(2\pi\xi_2) - \frac{3}{4}, \quad \widetilde{T}_1 = -\frac{1}{2}\cos(2\pi\xi_1) + \frac{1}{2},$$

we obtain

$$\widetilde{\mu}'_{00}(\xi) = \frac{1}{\sqrt{2}}\left(-\frac{1}{4}+\frac{3}{4}e^{-2\pi i\xi_1}+\frac{1}{8}e^{-2\pi i\xi_2}+\frac{3}{8}e^{2\pi i\xi_2}\right),$$
$$\widetilde{\mu}'_{01}(\xi) = \frac{1}{\sqrt{2}}\left(e^{-2\pi i\xi_1}-\frac{1}{4}e^{2\pi i\xi_1}+\frac{1}{4}e^{-2\pi i\xi_2}\right).$$

Next we find

$$\sigma(\xi) = \frac{1}{64}e^{-2i\pi(\xi_1+2\xi_2)}(7e^{2i\pi(\xi_1+\xi_2)}+9e^{4i\pi(\xi_1+\xi_2)}-2e^{6i\pi(\xi_1+\xi_2)}+4e^{2i\pi(2\xi_1+\xi_2)}-$$
$$2e^{4i\pi(2\xi_1+\xi_2)}+2e^{2i\pi(3\xi_1+\xi_2)}+26e^{2i\pi(\xi_1+2\xi_2)}-4e^{2i\pi(3\xi_1+2\xi_2)}+$$
$$11e^{2i\pi(\xi_1+3\xi_2)}+6e^{2i\pi(2\xi_1+3\xi_2)}+e^{4i\pi\xi_1}+6e^{4i\pi\xi_2}),$$

and, following Step 3 of Algorithm 1, find μ_{02}, μ_{03} and $\widetilde{\mu}_{0k}$, $k = 0, 1, 2, 3$. The refinable mask $\widetilde{m}_0$ is a trigonometric polynomial whose Fourier coefficients are given by the following table:

$$\widetilde{m}_0 : \begin{pmatrix} -\frac{1}{1024} & -\frac{1}{512} & -\frac{1}{512} & -\frac{1}{512} & \frac{1}{512} & \frac{1}{128} & 0 & -\frac{1}{256} \\ 0 & -\frac{1}{512} & -\frac{1}{128} & \frac{1}{128} & \frac{1}{64} & -\frac{1}{256} & -\frac{1}{128} & 0 \\ -\frac{13}{1024} & -\frac{11}{512} & -\frac{1}{64} & -\frac{5}{256} & 0 & \frac{19}{512} & \frac{3}{512} & -\frac{1}{256} \\ 0 & \frac{23}{256} & \frac{43}{256} & \frac{3}{128} & -\frac{23}{128} & \frac{1}{128} & \frac{3}{256} & 0 \\ -\frac{3}{64} & -\frac{17}{256} & -\frac{145}{512} & -\frac{41}{512} & -\frac{3}{1024} & \frac{19}{512} & \frac{3}{512} & 0 \\ 0 & \frac{39}{64} & \frac{51}{64} & \frac{73}{256} & -\frac{3}{64} & -\frac{9}{512} & 0 & 0 \\ -\frac{9}{256} & -\frac{3}{64} & -\frac{21}{256} & -\frac{11}{128} & -\frac{33}{1024} & 0 & 0 & 0 \end{pmatrix},$$

Checking shows (see Sect. 6.8) that the corresponding refinable function $\widetilde{\varphi}$ is in $L_2(\mathbb{R}^d)$.

To get wavelet masks $m_\nu, \widetilde{m}_\nu$, $\nu = 1, 2, 3$, it remains to find polyphase components using Step 4 of Algorithm 1 and substitute them into (3.16) (and a similar formula for $\widetilde{m}_\nu$).

3. Let $n = 3$, $M = \begin{pmatrix} 1 & 3 \\ 1 & 0 \end{pmatrix}$. For this matrix $M^{-1} = \begin{pmatrix} 0 & 1 \\ 1/3 & -1/3 \end{pmatrix}$, $m = 3$. Following Algorithm 2, set $A = \frac{11}{2}$, choose digits $s_0 = \begin{pmatrix} 0 \\ 0 \end{pmatrix}$, $s_1 = \begin{pmatrix} 1 \\ 0 \end{pmatrix}$, $s_2 = \begin{pmatrix} -1 \\ 0 \end{pmatrix}$, and functions

$$g_0 \equiv 1, \quad g_1(t) = \frac{\sin 2\pi t}{2\pi}, \quad g_2(t) = \frac{\sin^2 2\pi t}{8\pi^2}.$$

Then $M^{-1}s_1 = \begin{pmatrix} 0 \\ 1/3 \end{pmatrix}$, $M^{-1}s_2 = \begin{pmatrix} 0 \\ -1/3 \end{pmatrix}$,

$$T_{11} = T_{21} \equiv 1, \quad T_{12}(t) = \overline{T_{22}(t)} = 1 - \frac{\sin^2 2\pi t}{18} - i\frac{\sin 2\pi t}{3}.$$

Since $|T_{11}(t)|^2 = |T_{12}(t)| \le 1$, $|T_{12}(t)|^2 = |T_{22}(t)|^2 = 1 + \frac{1}{18^2}\sin^4 2\pi t < 5/4$, we can choose ε so that $1 + \varepsilon^2 = 9/4$, and set $T'_{kj}(t) = T_{kj}(t)$, $k, j = 1, 2$.

Following Step 3 of Algorithm 2, we get

$$\mu'_{00}(\xi) \equiv \frac{1}{\sqrt{3}}, \quad \mu'_{01}(\xi) = \overline{\mu'_{02}(\xi)} = \frac{1}{\sqrt{3}} T_{12}(\xi_2),$$

$$\sigma(\xi) = \frac{1}{3}\left(1 + 2|T_{12}(\xi_2)|^2\right) = 1 + \frac{1}{486}\sin^4 2\pi\xi_2.$$

and define $\mu_{0\nu}$, $\nu = 0, 1, 2$, by 3.58. Next we have

$$|Q_{11}(t)|^2 = |Q_{21}(t)|^2 = 1 - \frac{4}{9} = \left(\frac{\sqrt{5}}{3}\right)^2$$

$$|Q_{12}(t)|^2 = |Q_{22}(t)|^2 = 1 - \frac{81}{88}\left(1 + \frac{1}{18^2}\sin^4 2\pi t\right) = \frac{1}{88}\left(7 - \frac{1}{4}\sin^4 2\pi t\right) =$$
$$\frac{1}{88}\left(\sqrt{7} + \frac{1}{2}\sin^2 2\pi t\right)\left(\sqrt{7} - \frac{1}{2} + \frac{1}{2}\cos^2 2\pi t\right) =$$
$$\left|\frac{1}{2\sqrt{22}}\left(\sqrt[4]{7} + \frac{i}{\sqrt{2}}\sin 2\pi t\right)\left(\sqrt{\sqrt{7} - 1/2} + \frac{i}{\sqrt{2}}\cos 2\pi t\right)\right|^2$$

Following Step 4 of Algorithm 2, we should set $r = 7$ and find $\mu_{0\nu}$, $\nu = 3\ldots 6$ by 3.67, where

$$P_1(\xi) = P_2(\xi) \equiv \frac{\sqrt{110}}{9\sqrt{3}},$$
$$P_3(\xi) = P_4(\xi) = \frac{1}{6\sqrt{3}}\left(\sqrt[4]{7} + \frac{i}{\sqrt{2}}\sin 2\pi\xi_2\right)\left(\sqrt{\sqrt{7} - 1/2} + \frac{i}{\sqrt{2}}\cos 2\pi\xi_2\right)$$

However, since $P_1 = P_2$ and $P_3 = P_4$, one can reduce r to 5 and set

$$\mu_{03} = \frac{1-\sigma}{\sqrt{2}}P_1, \quad \mu_{04} = \frac{1-\sigma}{\sqrt{2}}P_3 \quad \mu_{05} \equiv 0.$$

It remains to find $\mu_{\nu k}$, $\nu = 1, \ldots, 5$, $k = 0, 1, 2$, by (3.68) and get masks m_ν, $\nu = 0, \ldots, 5$, by (3.16).

4. Let $n = 3$, $M = \begin{pmatrix} -3 & 5 & 1 \\ -3 & 4 & 1 \\ 3 & 0 & 0 \end{pmatrix}$. For this matrix $m = 3$, all entries of the first column of M are divisible by m and $M^{-1} = \begin{pmatrix} 0 & 0 & \frac{1}{3} \\ 1 & -1 & 0 \\ -4 & 5 & 1 \end{pmatrix}$. Following Algorithm 3, choose digits $s_0 = \begin{pmatrix} 0 \\ 0 \\ 0 \end{pmatrix}$, $s_1 = \begin{pmatrix} -1 \\ -1 \\ 1 \end{pmatrix}$, $s_2 = \begin{pmatrix} 1 \\ 1 \\ -1 \end{pmatrix}$, and functions

$$g_0 \equiv 1, \quad g_1(t) = \frac{\sin 2\pi t}{2\pi}, \quad g_2(t) = \frac{\sin^2 2\pi t}{8\pi^2}.$$

Then $M^{-1}s_1 = \begin{pmatrix} 1/3 \\ 0 \\ 0 \end{pmatrix}$, $M^{-1}s_2 = \begin{pmatrix} -1/3 \\ 0 \\ 0 \end{pmatrix}$,

$$T_{11}(t) = \overline{T_{21}(t)} = 1 - \frac{\sin^2 2\pi t}{18} - i\frac{\sin 2\pi t}{3}; \quad T_{k2}(t) = T_{k3}(t) \equiv 1, \quad k = 1, 2.$$

Since

$$|T_{11}(t)|^2 = |T_{21}(t)|^2 = \frac{1}{3} + \frac{2}{3}|T(t)|^2 = 1 + \frac{1}{18^2}\sin^4 2\pi t < 2,$$

we can take $\varepsilon = 1$, and set $T'_{kj}(t) = T_{kj}(t)$, $k = 1, 2$, $j = 1, 2, 3$.

Following Step 3 of Algorithm 3, we get

$$\mu'_{00}(\xi) = \frac{1}{\sqrt{3}}, \quad \mu'_{01}(\xi) = \overline{\mu'_{02}(\xi)} = \frac{1}{\sqrt{3}} T_{11}(\xi_1),$$

$$\sigma(\xi) = \frac{1}{3}\left(1 + 2|T_{11}(\xi_1)|^2\right) = 1 + \frac{1}{486}\sin^4 2\pi\xi_1.$$

Next, using (3.69), we have

$$|Q(t)|^2 = 1 - \frac{2}{11}|T_{11}(t)|^2 = \frac{2}{11\cdot(18)^2}\left(27\sqrt{2} - 1 + \cos^2 2\pi t\right)\left(27\sqrt{2} + \sin^2 2\pi t\right) =$$
$$\left|\frac{\sqrt{2}}{18\sqrt{11}}\left(\sqrt{27\sqrt{2}-1} + i\cos 2\pi t\right)\left(\sqrt{27\sqrt{2}} + i\sin 2\pi t\right)\right|^2,$$

Following Step 4 of Algorithm 3, we find

$$P_1(\xi) = \sqrt{\frac{11}{3}}\,Q(\xi_1) = \frac{\sqrt{2}}{18\sqrt{3}}\left(\sqrt{27\sqrt{2}-1} + i\cos 2\pi\xi_1\right)\left(\sqrt{27\sqrt{2}} + i\sin 2\pi\xi_1\right),$$

set $r = 4$ and find $\mu_{0\nu}$, $\nu = 3, 4$ by 3.67.

It remains to find $\mu_{\nu k}$, $\nu = 1, \dots, 4$, $k = 0, 1, 2$, by (3.68) and get masks m_ν, $\nu = 0, \dots, 4$, by (3.16).

5. Let $n = 2$, $M = \begin{pmatrix} 0 & 1 \\ 6 & -1 \end{pmatrix}$. For this matrix $m = 6$, all entries of the first column of M are divisible by m and $M^{-1} = \begin{pmatrix} \frac{1}{6} & \frac{1}{6} \\ 1 & 0 \end{pmatrix}$. Following Algorithm 3, choose digits $s_0 = \begin{pmatrix} 0 \\ 0 \end{pmatrix}$, $s_1 = -s_2 = \begin{pmatrix} 0 \\ 1 \end{pmatrix}$, $s_3 = -s_4 = \begin{pmatrix} 0 \\ 2 \end{pmatrix}$, $s_5 = \begin{pmatrix} 0 \\ 3 \end{pmatrix}$, and functions $g_0 \equiv 1$, $\quad g_1(t) = \frac{\sin 2\pi t}{2\pi}$. Next we have

$$M^{-1}s_1 = -M^{-1}s_2 = \begin{pmatrix} 1/6 \\ 0 \end{pmatrix}, \quad M^{-1}s_3 = -M^{-1}s_4 = \begin{pmatrix} 1/3 \\ 0 \end{pmatrix}, \quad M^{-1}s_5 = \begin{pmatrix} 1/2 \\ 0 \end{pmatrix},$$

$$T_{11}(t) = \overline{T_{21}(t)} = 1 - i\frac{\sin 2\pi t}{6}; \quad T_{31}(t) = \overline{T_{41}(t)} = 1 - i\frac{\sin 2\pi t}{3};$$
$$T_{51}(t) = 1 - i\frac{\sin 2\pi t}{2}; \quad T_{k2}(t) \equiv 1, \;\; k = 1, \dots, 5.$$

Since $|T_{k1}(t)|^2 < 2$, $k = 1, \ldots, 5$, we can take $\varepsilon = 1$, and set $T'_{kj} = T_{kj}$, $k = 1, \ldots, 5$, $j = 1, 2$. Following Step 3 of Algorithm 3, we get

$$\mu'_{00}(\xi) = \frac{1}{\sqrt{6}}, \quad \mu'_{01}(\xi) = \overline{\mu'_{02}(\xi)} = \frac{1}{\sqrt{6}}\left(1 - i\frac{\sin 2\pi\xi_1}{6}\right),$$

$$\mu'_{03}(\xi) = \overline{\mu'_{04}(\xi)} = \frac{1}{\sqrt{6}}\left(1 - i\frac{\sin 2\pi\xi_1}{3}\right), \quad \mu'_{05}(\xi) = \frac{1}{\sqrt{6}}\left(1 - i\frac{\sin 2\pi\xi_1}{2}\right),$$

$$\sigma(\xi) = \frac{1}{6^3}(216 + 19\sin^2 2\pi\xi_1),$$

Next, using (3.69), we have

$$|Q(t)|^2 = \frac{1}{6^3}\left(648 - 19\sin^2 2\pi t\right) = \frac{1}{6^3}\left(629 + 19\cos^2 2\pi t\right).$$

Following Step 4 of Algorithm 3, we find

$$P_1(\xi) = \frac{1}{6\sqrt{6}}\left(\sqrt{629} + i\sqrt{19}\cos 2\pi\xi_1\right)$$

set $r = 7$ and find $\mu_{0\nu}$, $\nu = 6, 7$ by 3.67.

It remains to find $\mu_{\nu k}$, $\nu = 1, \ldots, 7$, $k = 0 \ldots 5$, by (3.68) and get masks m_ν, $\nu = 0, \ldots, 7$, by (3.16).

6. Let $n = 2$, $M = \begin{pmatrix} -2 & 3 & 1 \\ -2 & 2 & 1 \\ 2 & 0 & 0 \end{pmatrix}$. For this matrix $m = 2$, all entries of the first column of M are divisible by m. Thus a tight wavelet frame can be constructed on the basis of the Daubechies mask m_0^* satisfying $|m_0^*(t)|^2 = q_2(t)$. Choose digits $s_0 = \begin{pmatrix} 0 \\ 0 \\ 0 \end{pmatrix}$, $s_1 = \begin{pmatrix} -1 \\ -1 \\ 1 \end{pmatrix}$. Due to (3.75), we have

$$q_2(t) = \frac{1}{4}(2 + 3\cos 2\pi t - \cos^3 2\pi t).$$

It is not difficult to check that m_0^* can be taken as $m_0^*(t) = \sum_{k=0}^{3} c_k e^{2\pi i t}$, where

$$c_0 = \frac{1+\sqrt{3}}{8}, \quad c_1 = \frac{3+\sqrt{3}}{8}, \quad c_2 = \frac{3-\sqrt{3}}{8}, \quad c_3 = \frac{1-\sqrt{3}}{8}.$$

Using (3.74) and (3.75), we get refinable and wavelet masks:

$$m_0(\xi) = \sum_{k=0}^{3} c_k e^{-2\pi i(\xi_1+\xi_2-\xi_3)}, \quad m_1(\xi) = \sum_{k=-1}^{2} (-1)^{k+1} c_k e^{2\pi i(\xi_1+\xi_2-\xi_3)}.$$

7. Let $n \in \mathbb{N}$, M be a dilation matrix in $\mathbb{R}^d$ such that $m = 2$ and $(M^{*-1}q)_{j_0} = 1/2$, where q is the nonzero digit of M^*, $j_0 = 1, \dots, d$, and let

$$g(t) = e^{\pi int} \cos^n \pi t.$$

Since $|g(t)|^2 + |g(t+1/2)|^2 \le 1$, by the Riesz Lemma, there exists a trigonometric polynomial h such that $|h(t)|^2 = 1 - |g(t)|^2 - |g(t+1/2)|^2$. Set

$$m_0(\xi) = g(\xi_{j_0}), \quad m_1(\xi) = e^{2\pi i \xi_{j0}} \overline{g(\xi_{j_0} + 1/2)}, \quad m_2(\xi) = g(\xi_{j_0}) h((M^*\xi)_{j_0}),$$

$$S(\xi) = |g(\xi_{j_0})|^2 + |g(\xi_{j_0} + 1/2)|^2.$$

We have

$$S(M^*\xi)|m_0(\xi)|^2 + |m_1(\xi)|^2 + |m_2(\xi)|^2 =$$

$$S(M^*(\xi)|g(\xi_{j_0})|^2 + |g(\xi_{j_0} + 1/2)|^2 + |g(\xi_{j_0})|^2(1 - S(M^*\xi)) = S(\xi);$$

$$S(M^*\xi)m_0(\xi)\overline{m_0(\xi + M^{*-1}q)} + m_1(\xi)\overline{m_1(\xi + M^{*-1}q)} + m_2(\xi)\overline{m_2(\xi + M^{*-1}q)} =$$

$$S(M^*\xi)g(\xi_{j0})\overline{g(\xi_{j0} + 1/2)} - g(\xi_{j0})\overline{g(\xi_{j0} + 1/2)} + g(\xi_{j0})\overline{g(\xi_{j0} + 1/2)}(1 - S(M^*\xi)) = 0.$$

Thus (3.77) is satisfied for the masks m_0, m_1, m_2. The function φ defined by

$$\widehat{\varphi}(\xi) = \prod_{j=1}^{\infty} m_0(M^{*-j}\xi)$$

is in $L_2(\mathbb{R}^d)$ due to Proposition 2.6.6. Hence, by Theorem 3.4.6, the functions ψ_1, ψ_2 defined by (3.78) generate a tight wavelet frame. It is easy to see that

$$D^\alpha m_\nu(M^{*-1}\xi)\big|_{\xi=\mathbf{0}} = 0 \forall \alpha \in \mathbb{Z}^d, \ [\alpha] < n, \quad \nu = 1, 2,$$

which yields that the frame has $\mathbf{VM^n}$ property.

References

1. Sadovnichii, V.A.: Theory of Operators. Roger Cooke Publisher, New York, Consultants Bureau (1991)
2. Novikov, I.Y., Protasov, V.Y., Skopina, M.A.: Wavelet Theory, vol. 239. AMS, Providence, RI, Translations Mathematical Monographs (2011)

3. Scheiderer, C.: Sums of squares on real algebraic surfaces. Manuscripta Math. **119**(4), 395–410 (2006)
4. Charina, M., Putinar, M., Scheiderer, C., Stöckler, J.: An algebraic perspective on multivariate tight wavelet frames. Constr. Approx. **38**, 253–276 (2013)
5. Charina, M., Putinar, M., Scheiderer, C., Stöckler, J.: An algebraic perspective on multivariate tight wavelet frames II. Appl. Comput. Harmon. Anal. **11**, 185–213 (2006)
6. Dritschel, M.A.: On factorization of trigonometric polynomials. Integr. Eqn. Oper. Theory **49**, 11–42 (2004)
7. Geronimo, J., Lai, M.J.: Factorization of multivariate positive laurent polynomials. J. Approx. Theory **139**(1–2), 327–345 (2006)
8. Lawton, W., Lee, S.L., Shen, Z.: An algorithm for matrix extension and wavelet construction. Math. Comp. **37**, 271–300 (1996)
9. Ron, A., Shen, Z.: Compactly supported tight affine spline frames in $L_2(R^d)$. Math. Comp. **67**, 191–207 (1998)

Chapter 4
Frame-Like Wavelet Expansions

Abstract A notion of frame-like wavelet systems is introduced and analyzed. Those systems are not dual wavelet frames although preserve important properties of frames. The construction of such systems is based on the matrix extension principle (MEP), but it is simpler than the construction of wavelet frames.

4.1 Dual Frame-Like Systems

An important property of a frame $\{f_n\}_n$ in a Hilbert space H is the following: Every $f \in H$ can be decomposed as $f = \sum_n \langle f, \widetilde{f_n}\rangle f_n$, where $\{\widetilde{f_n}\}_n$ is a dual frame to $\{f_n\}_n$ (see Corollary 1.2.7). In particular, if $\{\psi_{jk}^{(\nu)}\}_{j,k,\nu}$, $\{\widetilde{\psi}_{jk}^{(\nu)}\}_{j,k,\nu}$ are dual wavelet frames in $L_2(\mathbb{R}^d)$ generated by wavelet functions $\psi^{(\nu)}$, $\widetilde{\psi}^{(\nu)}$, $\nu = 1, \dots, r$, then every $f \in L_2(\mathbb{R}^d)$ can be decomposed as

$$f = \sum_{j=-\infty}^{\infty} \sum_{\nu=1}^{r} \sum_{k \in \mathbb{Z}^d} \langle f, \widetilde{\psi}_{jk}^{(\nu)}\rangle \psi_{jk}^{(\nu)}.$$

In Chap. 3, we discussed how to construct compactly supported dual wavelet frames with a given approximation order. Analyzing the methods of construction, one can see that we have to overcome substantial difficulty to provide vanishing moments for all wavelet functions $\psi^{(\nu)}$, $\widetilde{\psi}^{(\nu)}$, which is a necessary condition for the systems $\{\psi_{jk}^{(\nu)}\}_{j,k,\nu}$, $\{\widetilde{\psi}_{jk}^{(\nu)}\}_{j,k,\nu}$ to be frames in $L_2(\mathbb{R}^d)$ (see Theorem 3.2.1). However, engineers often do not take care of this. Providing vanishing moments only for the functions $\widetilde{\psi}^{(\nu)}$, they successfully apply such "frames" (which are really not frames) for signal processing (see, e.g., [1]). Thus, it makes sense to study a wider class of dual wavelet systems which are not dual frames but preserve important properties of frames. Han and Shen [2] and Ehler [3] considered a class of dual wavelet systems $\{\psi_{jk}^{(\nu)}\}_{j,k,\nu}$, $\{\widetilde{\psi}_{jk}^{(\nu)}\}_{j,k,\nu}$ which are, respectively, frames in the Sobolev spaces $W_2^s(\mathbb{R}^d)$, $W_2^{-s}(\mathbb{R}^d)$ but not a pair of dual frames in $L_2(\mathbb{R}^d)$. In this chapter, we study frame-type decomposition for a wide class of compactly supported dual wavelet systems $\{\psi_{jk}^{(\nu)}\}_{j,k,\nu}$, $\{\widetilde{\psi}_{jk}^{(\nu)}\}_{j,k,\nu}$, where $\psi^{(\nu)}$ and $\widetilde{\psi}^{(\nu)}$ are functions or distributions.

A. Krivoshein et al., *Multivariate Wavelet Frames*,
Industrial and Applied Mathematics, DOI 10.1007/978-981-10-3205-9_4

First of all, we have to extend some notions and notations to tempered distributions. Recall that the space of tempered distributions is the dual space of the Schwartz class S of functions defined on $\mathbb{R}^d$. If $f \in S, g \in S'$, then $\langle f, g\rangle := \overline{\langle g, f\rangle} := \overline{g(f)}$. If $f \in L_p(\mathbb{R}^d), g \in L_q(\mathbb{R}^d), \frac{1}{p} + \frac{1}{q} = 1$, then $\langle f, g\rangle := \int_{\mathbb{R}^d} f\overline{g}$. If $f \in S'$, then $\widehat{f}$ denotes its Fourier transform defined by $\langle \widehat{f}, \widehat{g}\rangle = \langle f, g\rangle, g \in S$. Note that if $f \in S'$ is compactly supported, then, due to the Paley–Wiener theorem for tempered distributions (see, e.g., [4, 5]), $\widehat{f}$ is infinitely differentiable on $\mathbb{R}^d$. If $f \in S', j \in \mathbb{Z}, k \in \mathbb{R}^d$, we define f_{jk} by

$$\langle f_{jk}, g\rangle = \langle f, g_{-j,-M^{-j}k}\rangle \quad \forall g \in S.$$

A compactly supported tempered distribution φ is called refinable, if there exists a trigonometric polynomial m_0 (refinable mask) such that

$$\widehat{\varphi}(\xi) = m_0(M^{*-1}\xi)\widehat{\varphi}(M^{*-1}\xi). \tag{4.1}$$

This relation is a refinement equation. By Theorem 2.6.4, for any trigonometric polynomial m_0 satisfying $m_0(\mathbf{0}) = 1$, there exists a unique solution (up to a factor) of the refinement Eq. (4.1) in S'. The solution is compactly supported and given by

$$\widehat{\varphi}(\xi) := \prod_{j=1}^{\infty} m_0(M^{*-j}\xi). \tag{4.2}$$

Now, we define dual wavelet systems generated from two arbitrary compactly supported refinable tempered distributions by MEP. Actually, we just repeat the method of construction by MEP in $L_2(\mathbb{R}^d)$. Let $\varphi, \widetilde{\varphi} \in S'$ be refinable and compactly supported, polynomials $m_0, \widetilde{m}_0$ be their masks, respectively. Find trigonometric polynomials (wavelet masks) $m_\nu, \widetilde{m}_\nu, \nu = 1, \dots, r, r \geq m - 1$, such that the corresponding polyphase matrices (see Sect. 3.3)

$$\mathcal{M} := \begin{pmatrix} \mu_{00} & \cdots & \mu_{0,m-1} \\ \vdots & \ddots & \vdots \\ \mu_{r,0} & \cdots & \mu_{r,m-1} \end{pmatrix}, \quad \widetilde{\mathcal{M}} := \begin{pmatrix} \widetilde{\mu}_{00} & \cdots & \widetilde{\mu}_{0,m-1} \\ \vdots & \ddots & \vdots \\ \widetilde{\mu}_{r,0} & \cdots & \widetilde{\mu}_{r,m-1} \end{pmatrix}$$

satisfy

$$\mathcal{M}^T \overline{\widetilde{\mathcal{M}}} = I_m, \tag{4.3}$$

and define wavelet functions by

$$\widehat{\psi^{(\nu)}}(\xi) = m_\nu(M^{*-1}\xi)\widehat{\varphi}(M^{*-1}\xi), \quad \widehat{\widetilde{\psi}^{(\nu)}}(\xi) = \widetilde{m}_\nu(M^{*-1}\xi)\widehat{\widetilde{\varphi}}(M^{*-1}\xi). \tag{4.4}$$

If wavelet functions $\psi^{(\nu)}, \widetilde{\psi}^{(\nu)}, \ \nu = 1, \dots, r$, are constructed as above, then $\{\psi_{jk}^{(\nu)}\}_{j,k,\nu}, \{\widetilde{\psi}_{jk}^{(\nu)}\}_{j,k,\nu}$ are said to be dual wavelet systems generated from $\varphi, \widetilde{\varphi}$ by MEP. We shall also say that wavelet functions $\psi^{(\nu)}, \widetilde{\psi}^{(\nu)}$ *are associated* with $\varphi, \widetilde{\varphi}$.

Generally speaking, these systems are in S', but of course both the systems or one of them can be in any subspace of S', e.g., in $L_p(\mathbb{R}^d)$, $1 \le p \le \infty$.

Let $\{\psi_{jk}^{(\nu)}\}_{j,k,\nu}$, $\{\widetilde{\psi}_{jk}^{(\nu)}\}_{j,k,\nu}$ be dual wavelet systems generated from φ, $\widetilde{\varphi}$ by MEP, and let $\mathcal{A}$ be a class of functions f for which $\langle f, \widetilde{\varphi}_{0k}\rangle$, $\langle f, \widetilde{\psi}_{jk}^{(\nu)}\rangle$ have meaning. We say that $\{\psi_{jk}^{(\nu)}\}_{j,k,\nu}$ is *frame-like*, if

$$f = \sum_{i=-\infty}^{\infty} \sum_{\nu=1}^{r} \sum_{k \in \mathbb{Z}^d} \langle f, \widetilde{\psi}_{jk}^{(\nu)}\rangle \psi_{jk}^{(\nu)} \quad \forall f \in \mathcal{A}, \tag{4.5}$$

and $\{\psi_{jk}^{(\nu)}\}_{j,k,\nu}$ is *almost frame-like*, if

$$f = \sum_{k \in \mathbb{Z}^d} \langle f, \widetilde{\varphi}_{0k}\rangle \varphi_{0k} + \sum_{i=0}^{\infty} \sum_{\nu=1}^{r} \sum_{k \in \mathbb{Z}^d} \langle f, \widetilde{\psi}_{jk}^{(\nu)}\rangle \psi_{jk}^{(\nu)} \quad \forall f \in \mathcal{A}, \tag{4.6}$$

where the series in (4.5) and (4.6) converge in some natural sense.

4.2 Scaling Expansions

If $\langle f, \widetilde{\varphi}_{jk}\rangle$ has meaning and the series $\sum_{k \in \mathbb{Z}^d} \langle f, \widetilde{\varphi}_{jk}\rangle \varphi_{jk}$ converges in some sense, we set

$$Q_j(\varphi, \widetilde{\varphi}, f) = Q_j(f) := \sum_{k \in \mathbb{Z}^d} \langle f, \widetilde{\varphi}_{jk}\rangle \varphi_{jk}$$

and call Q_j the scaling operator, $j \in \mathbb{Z}^d$.

In this section, we study the scaling operators Q_j defined on the spaces S' and $L_p(\mathbb{R}^d)$, $1 \le p \le \infty$.

Lemma 4.2.1 *Let $\varphi \in S'$ be compactly supported, $f \in S$, $j \in \mathbb{Z}$,*

$$G_j(\xi) = G_j(\varphi, f, \xi) := \sum_{l \in \mathbb{Z}^d} \widehat{f}(M^{*j}(\xi + l))\overline{\widehat{\varphi}(\xi + l)}.$$

Then G_j is a bounded 1-periodic function, in particular, $G_j \in L_2(\mathbb{T}^d)$, and

$$\langle f, \varphi_{jk}\rangle = m^{j/2}\widehat{G}_j(k).$$

Moreover, if $j \ge 0$, then

$$|G_j(\xi)| \le C_{\varphi,f,M}, \forall \xi \in \mathbb{R}^d, \tag{4.7}$$

and for any $N \in \mathbb{N}$ there exists a constant $C_{N,\varphi,f,M} > 0$ such that

$$\left|G_j(\xi) - \widehat{f}(M^{*j}\xi)\overline{\widehat{\varphi}(\xi)}\right| \le C_{N,\varphi,f,M}\|M^{*-j}\|^N \tag{4.8}$$

for all $\xi \in [-\frac{1}{2}; \frac{1}{2}]^d$.

Proof By the Paley–Wiener theorem for tempered distributions (see, e.g., [4, 5]), there exist $N_0 \in \mathbb{N}$ and $C_\varphi > 0$ so that $|\widehat{\varphi}(\xi)| \le C_\varphi\, |\xi|^{N_0}$ for all $\xi \notin [-\frac{1}{2}; \frac{1}{2}]^d$.

Clearly, it suffices to check (4.8) for big enough N. Let $N > N_0 + d$, $N \in \mathbb{N}$, $C_{N,f} = \sup\limits_{\xi\in\mathbb{R}^d} |\xi|^N |\widehat{f}(\xi)|$. If $\xi \in [-\frac{1}{2}; \frac{1}{2}]^d$, then

$$\left|\sum_{l\in\mathbb{Z}^d,\, l\ne\mathbf{0}} \widehat{f}(M^{*j}(\xi+l))\overline{\widehat{\varphi}(\xi+l)}\right| \le C_\varphi \sum_{l\in\mathbb{Z}^d,\, l\ne\mathbf{0}} \frac{|\xi+l|^N |\widehat{f}(M^{*j}(\xi+l))|}{|\xi+l|^{N-N_0}} \le$$

$$C_\varphi C_{N,f}\|M^{*-j}\|^N \sum_{l\in\mathbb{Z}^d,\, l\ne\mathbf{0}} \frac{1}{|\xi+l|^{N-N_0}} = C_{N,\varphi,f}\|M^{*-j}\|^N,$$

which yields the boundedness of G_j for all $j \in \mathbb{N}$ and (4.8) for all $N \in \mathbb{N}$ and $j \ge 0$. Since

$$|\widehat{f}(M^{*j}\xi)\overline{\widehat{\varphi}(\xi)}| \le C_\varphi\|M^{*-j}\|^{N_0}|\widehat{f}(M^{*j}\xi)||M^{*j}\xi|^{N_0} \le C_{N_0,f}C_\varphi\|M^{*-j}\|^{N_0},$$

taking into account (2.1) and (4.8), we obtain (4.7). Besides, we proved that the series

$$\sum_{l\in\mathbb{Z}^d} |\widehat{f}(M^{*j}(\xi+l))\overline{\widehat{\varphi}(\xi+l)}|$$

uniformly converges on $[-\frac{1}{2}; \frac{1}{2}]^d$ to a bounded function, which implies

$$\langle f, \varphi_{jk}\rangle = \langle \widehat{f}, \widehat{\varphi_{jk}}\rangle = m^{-j/2}\int\limits_{\mathbb{R}^d} \widehat{f}(\xi)\overline{\widehat{\varphi}(M^{*-j}\xi)}e^{-2\pi i\,(k,\, M^{*-j}\xi)}d\xi =$$

$$m^{j/2}\int\limits_{[-\frac{1}{2};\frac{1}{2}]^d} G_j(\xi)e^{-2\pi i\,(k,\xi)}d\xi = m^{j/2}\widehat{G}_j(k),$$

where $k \in \mathbb{Z}^d$.◊

Theorem 4.2.2 *Let* $\varphi, \widetilde{\varphi} \in S'$ *be compactly supported,* $f \in S$. *Then*

(a) for every $j \in \mathbb{Z}$, *the series* $\sum_{k\in\mathbb{Z}^d}\langle f, \widetilde{\varphi}_{jk}\rangle\varphi_{jk}$ *converges unconditionally in* S', *in particular,* $Q_j(\varphi, \widetilde{\varphi}, f) \in S'$.

(b) $Q_j(\varphi, \widetilde{\varphi}, f)$ *tends to* $\widehat{\varphi}(\mathbf{0})\overline{\widehat{\widetilde{\varphi}}(\mathbf{0})}f$ *in* S' *as* $j \to +\infty$.

Proof Let $\widetilde{f} \in S$. To prove (a), it suffices to check that the series $\sum_{k\in\mathbb{Z}^d}\langle f, \widetilde{\varphi}_{jk}\rangle \langle\varphi_{jk}, \widetilde{f}\rangle$ is absolutely convergent, but Lemma 4.2.1 yields that

$$\sum_{k\in\mathbb{Z}^d} \left|\langle f, \widetilde{\varphi}_{jk}\rangle\langle\varphi_{jk}, \widetilde{f}\rangle\right| \le m^j \|G_j(\widetilde{\varphi}, f, \cdot)\|_2 \|G_j(\varphi, \widetilde{f}, \cdot)\|_2 < \infty.$$

To prove (b), we should check that

$$\lim_{j\to+\infty} \sum_{k\in\mathbb{Z}^d} \langle f, \widetilde{\varphi}_{jk}\rangle\langle\varphi_{jk}, \widetilde{f}\rangle = \widehat{\varphi}(\mathbf{0})\overline{\widehat{\widetilde{\varphi}}(\mathbf{0})}\langle f, \widetilde{f}\rangle = \widehat{\varphi}(\mathbf{0})\overline{\widehat{\widetilde{\varphi}}(\mathbf{0})} \int_{\mathbb{R}^d} \widehat{f}(\xi)\overline{\widehat{\widetilde{f}}}(\xi)d\xi.$$

Due to Lemma 4.2.1,

$$\sum_{k\in\mathbb{Z}^d} \langle f, \widetilde{\varphi}_{jk}\rangle\langle\varphi_{jk}, \widetilde{f}\rangle = m^j \int_{[-\frac{1}{2};\frac{1}{2}]^d} G_j(\widetilde{\varphi}, f, \xi)\overline{G_j(\varphi, \widetilde{f}, \xi)}d\xi =$$

$$m^j \int_{[-\frac{1}{2};\frac{1}{2}]^d} \widehat{f}(M^{*j}\xi)\overline{\widehat{\widetilde{\varphi}}(\xi)}\overline{\widehat{\widetilde{f}}(M^{*j}\xi)\widehat{\varphi}(\xi)}d\xi + m^j \int_{[-\frac{1}{2};\frac{1}{2}]^d} \widehat{f}(M^{*j}\xi)\overline{\widehat{\widetilde{\varphi}}}(\xi)\overline{H_j(\varphi, \widetilde{f}, \xi)}d\xi +$$

$$m^j \int_{[-\frac{1}{2};\frac{1}{2}]^d} H_j(\widetilde{\varphi}, f, \xi)\overline{\widehat{\widetilde{f}}(M^{*j}\xi)\widehat{\varphi}(\xi)}d\xi + m^j \int_{[-\frac{1}{2};\frac{1}{2}]^d} H_j(\widetilde{\varphi}, f, \xi)\overline{H_j(\varphi, \widetilde{f}, \xi)}d\xi, \tag{4.9}$$

where

$$H_j(\widetilde{\varphi}, f, \xi) = G_j(\widetilde{\varphi}, f, \xi) - \widehat{f}(M^{*j}\xi)\overline{\widehat{\widetilde{\varphi}}(\xi)}, \quad H_j(\varphi, \widetilde{f}, \xi) = G_j(\varphi, \widetilde{f}, \xi) - \widehat{\widetilde{f}}(M^{*j}\xi)\overline{\widehat{\varphi}(\xi)}.$$

By changing of variable in the first term of the right-hand side of (4.9), we have

$$m^j \int_{[-\frac{1}{2};\frac{1}{2}]^d} \widehat{f}(M^{*j}\xi)\overline{\widehat{\widetilde{\varphi}}(\xi)}\overline{\widehat{\widetilde{f}}(M^{*j}\xi)\widehat{\varphi}(\xi)}d\xi$$

$$= \int_{\mathbb{R}^d} \chi_{M^{*j}[-\frac{1}{2};\frac{1}{2}]^d}(\xi)\widehat{f}(\xi)\overline{\widehat{\widetilde{\varphi}}(M^{*-j}\xi)}\overline{\widehat{\widetilde{f}}(\xi)\widehat{\varphi}(M^{*-j}\xi)}d\xi.$$

Due to the Paley–Wiener theorem for tempered distributions and (2.1), there exist positive numbers N, $\widetilde{N}$ and C, $\widetilde{C}$ such that

$$|\widehat{\varphi}(M^{*-j}\xi)| \le C(1+|\xi|)^N, \quad |\widehat{\widetilde{\varphi}}(M^{*-j}\xi)| \le \widetilde{C}(1+|\xi|)^{\widetilde{N}} \tag{4.10}$$

for all $\xi \in \mathbb{R}^d$ and all $j > 0$. Hence,

$$|\chi_{M^{*j}[-\frac{1}{2};\frac{1}{2}]^d}(\xi)\widehat{f}(\xi)\overline{\widehat{\widetilde{\varphi}}(M^{*-j}\xi)}\overline{\widehat{\widetilde{f}}(\xi)\widehat{\varphi}(M^{*-j}\xi)}| \le C\widetilde{C}(1+|\xi|)^{N+\widetilde{N}}|\widehat{f}(\xi)||\overline{\widehat{\widetilde{f}}(\xi)}|.$$

Since for any $\xi \in \mathbb{R}^d$

$$\lim_{j\to+\infty} \chi_{M^{*j}[-\frac{1}{2};\frac{1}{2}]^d}(\xi)\overline{\widehat{\widetilde{\varphi}}(M^{*-j}\xi)}\widehat{\varphi}(M^{*-j}\xi) = \overline{\widehat{\widetilde{\varphi}}(\mathbf{0})}\widehat{\varphi}(\mathbf{0})$$

due to (2.1) and since $(1+|\xi|)^{N+\widetilde{N}}|\widehat{f}(\xi)||\overline{\widehat{\widetilde{f}}(\xi)}|$ is summable on $\mathbb{R}^d$, using Lebesgue's dominated convergence theorem, we obtain

$$\lim_{j\to+\infty}\left(m^j \int\limits_{[-\frac{1}{2};\frac{1}{2}]^d} \widehat{f}(M^{*j}\xi)\overline{\widehat{\widetilde{\varphi}}(\xi)}\overline{\widehat{\widetilde{f}}(M^{*j}\xi)}\widehat{\varphi}(\xi)d\xi - \overline{\widehat{\widetilde{\varphi}}(\mathbf{0})}\widehat{\varphi}(\mathbf{0}) \int\limits_{\mathbb{R}^d} \widehat{f}(\xi)\overline{\widehat{\widetilde{f}}(\xi)}d\xi \right) = 0. \tag{4.11}$$

It follows from Lemma 4.2.1 that

$$\sup_{\xi\in[-\frac{1}{2};\frac{1}{2}]^d} |H_j(\varphi, \widetilde{f}, \xi)| \le C_{1,\varphi,\widetilde{f},M}\|M^{*-j}\|.$$

Hence, by changing of variable and using (4.10), we have

$$\left| m^j \int\limits_{[-\frac{1}{2};\frac{1}{2}]^d} \widehat{f}(M^{*j}\xi)\overline{\widehat{\widetilde{\varphi}}(\xi)H_j(\varphi, \widetilde{f}, \xi)}d\xi \right| \le C_{1,\varphi,\widetilde{f},M}\|M^{*-j}\|m^j \int\limits_{[-\frac{1}{2};\frac{1}{2}]^d} |\widehat{f}(M^{*j}\xi)\widehat{\widetilde{\varphi}}(\xi)|\, d\xi =$$

$$C_{1,\varphi,\widetilde{f},M}\|M^{*-j}\| \int\limits_{M^{*j}[-\frac{1}{2};\frac{1}{2}]^d} |\widehat{f}(\xi)\widehat{\widetilde{\varphi}}(M^{*-j}\xi)|\, d\xi \le C_{1,\varphi,\widetilde{f},M}\widetilde{C}\|M^{*-j}\| \int\limits_{\mathbb{R}^d} |\widehat{f}(\xi)|(1+|\xi|)^{\widetilde{N}}\, d\xi.$$

Thus, due to (2.1), the second term of the right-hand side of (4.9) tends to 0 as $j \to +\infty$. Similarly, the third and the forth terms also tend to 0.$\Diamond$

Corollary 4.2.3 *Let* $\varphi, \widetilde{\varphi} \in S'$ *be compactly supported,* $f \in S$, $j \in \mathbb{Z}$, $Q_j := Q_j(\varphi, \widetilde{\varphi}, f)$. *Then* $\widehat{Q}_j$ *is a function on* $\mathbb{R}^d$ *and*

$$\widehat{Q}_j(\xi) = \widehat{\varphi}(M^{*-j}\xi) \sum_{l\in\mathbb{Z}^d} \widehat{f}(\xi + M^{*j}l)\overline{\widehat{\widetilde{\varphi}}(M^{*-j}\xi + l)} \quad a.e. \tag{4.12}$$

If, moreover, $\widehat{\varphi} \in L_p(\mathbb{R}^d)$, $1 \le p \le \infty$, *then* $\widehat{Q}_j \in L_p(\mathbb{R}^d)$.

Proof Due to condition (a) of Theorem 4.2.2, we have

$$\widehat{Q}_j(\xi) = \sum_{k\in\mathbb{Z}^d} \langle f, \widetilde{\varphi}_{jk}\rangle \widehat{\varphi_{jk}} = \sum_{k\in\mathbb{Z}^d} \widehat{G_j}(k) e^{2\pi i\,(k,\, M^{*-j}\xi)}\widehat{\varphi}(M^{*-j}\xi),$$

where

$$G_j(\xi) = G_j(\widetilde{\varphi}, f, \xi)$$

is a function from Lemma 4.2.1. Since $G_j \in L_2(\mathbb{T}^d)$, by the Carleson theorem, $G_j(M^{*-j}\xi) = \sum_{k\in\mathbb{Z}^d} \widehat{G_j}(k)e^{2\pi i\,(k,\,M^{*-j}\xi)}$ for almost all $\xi \in \mathbb{R}^d$, which proves (4.12). The boundedness of G_j yields the second statement.◊

Next, we are going to consider the scaling operator Q_j in the spaces $L_p(\mathbb{R}^d)$. First, we give several simple lemmas.

Lemma 4.2.4 *Let* $1 \le p < \infty$, $\frac{1}{p} + \frac{1}{q} = 1$, $f \in L_p(\mathbb{R}^d)$, $\varphi \in L_q(\mathbb{R}^d)$, φ *be compactly supported. Then*

$$\left(\sum_{k\in\mathbb{Z}^d} |\langle f, \varphi_{jk}\rangle|^p\right)^{1/p} \le C_{p,\varphi}\, m^{\frac{j}{p}-\frac{j}{2}} \|f\|_p. \tag{4.13}$$

Proof Assume that supp $\varphi \subset [-N,\ N]^d$. Then,

$$\left(\sum_{k\in\mathbb{Z}^d} |\langle f, \varphi_{jk}\rangle|^p\right)^{1/p} = \left(\sum_{k\in\mathbb{Z}^d} \left| m^{-j/2} \int\limits_{[-N,N]^d-k} f(M^{-j}x)\varphi(x+k)\,dx \right|^p\right)^{1/p} \le$$

$$m^{-j/2}\|\varphi\|_q \left(\sum_{k\in\mathbb{Z}^d} \int\limits_{[-N,N]^d-k} |f(M^{-j}x)|^p\,dx\right)^{1/p} \le$$

$$m^{-\frac{j}{2}}\|\varphi\|_q (2N+1)^{\frac{d}{p}} \|f(M^{-j}\cdot)\|_p = m^{\frac{j}{p}-\frac{j}{2}}\|\varphi\|_q (2N+1)^{\frac{d}{p}} \|f\|_p.◊$$

Lemma 4.2.5 *Let* $f \in L_\infty(\mathbb{R}^d)$, $\varphi \in L_1(\mathbb{R}^d)$, φ *be compactly supported. Then*

$$\sup_{k\in\mathbb{Z}^d} |\langle f, \varphi_{jk}\rangle| \le C_{\infty,\varphi}\, m^{-\frac{j}{2}} \|f\|_\infty.$$

The proof of this lemma is obvious.

Lemma 4.2.6 *Let* $1 < p < \infty$, $\frac{1}{p} + \frac{1}{q} = 1$, $\varphi \in L_q(\mathbb{R}^d)$, $f \in L_p(\mathbb{R}^d)$ *and* φ *be compactly supported. Then*

$$\lim_{j\to-\infty} m^{\frac{j}{2}-\frac{j}{p}} \left(\sum_{k\in\mathbb{Z}^d} |\langle f, \varphi_{jk}\rangle|^p\right)^{\frac{1}{p}} = 0 \tag{4.14}$$

Proof First, we assume that f is continuous on $\mathbb{R}^d$ and supp $f \subset B_R$, $R > 0$. Using Hölder's inequality, we obtain

$$m^{\frac{j}{2}-\frac{j}{p}}|\langle f,\varphi_{jk}\rangle| = m^{\frac{j}{q}}|\langle f,\varphi(M^j\cdot+k)\rangle| \le m^{\frac{j}{q}}\int\limits_{B_R}|f(x)||\varphi(M^jx+k)|\,dx \le$$

$$m^{\frac{j}{q}}\|f\|_p\left(\int\limits_{B_R}|\varphi(M^jx+k)|^q\,dx\right)^{\frac{1}{q}} \le \|f\|_p\left(\int\limits_{M^jB_R+k}|\varphi(y)|^q\,dy\right)^{\frac{1}{q}}.$$

Since φ is compactly supported, there exists only a finite number of $k \in \mathbb{Z}^d$ such that the latter integral not equals zero for at least one $j < 0$. So, it suffices to check that

$$\lim_{j\to-\infty}\int\limits_{M^jB_R+k}|\varphi(y)|^q\,dy = 0,$$

for any $k \in \mathbb{Z}^d$. Let $\chi_{M^jB_R+k}$ denote the characteristic function of M^jB_R+k, then

$$\int\limits_{M^jB_R+k}|\varphi(y)|^q\,dy = \int\limits_{\mathbb{R}^d}\chi_{M^jB_R+k}(y)|\varphi(y)|^q\,dy.$$

If $y \ne k$, then $\lim\limits_{j\to-\infty}\chi_{M^jB_R+k}(y) = 0$. Hence, using Lebesgue's dominated convergence theorem, we have

$$\lim_{j\to-\infty}\int\limits_{\mathbb{R}^d}\chi_{M^jB_R+k}(y)|\varphi(y)|^q\,dy = 0,$$

and (4.14) is proved for compactly supported functions f.

Now, let $f \in L_p(\mathbb{R}^d)$. Given $\varepsilon > 0$, we find a compactly supported continuous $\widetilde{f}$ such that $\|f-\widetilde{f}\|_p < \varepsilon$. Using Minkowski's inequality and Lemma 4.2.4, we obtain

$$m^{\frac{j}{2}-\frac{j}{p}}\left(\sum_{k\in\mathbb{Z}^d}|\langle f,\varphi_{jk}\rangle|^p\right)^{\frac{1}{p}} \le m^{\frac{j}{2}-\frac{j}{p}}\left(\sum_{k\in\mathbb{Z}^d}|\langle\widetilde{f},\varphi_{jk}\rangle|^p\right)^{\frac{1}{p}} + C_{p,\varphi}\varepsilon,$$

which yields (4.14).$\Diamond$

Theorem 4.2.7 *Let $1 \le p < \infty$, $\frac{1}{p}+\frac{1}{q}=1$, $\widetilde{\varphi} \in L_q(\mathbb{R}^d)$ and $\varphi \in L_p(\mathbb{R}^d)$ be compactly supported functions, $f \in L_p(\mathbb{R}^d)$. Then*

(a) for every $j \in \mathbb{Z}$, the series $\sum_{k\in\mathbb{Z}^d}\langle f,\widetilde{\varphi}_{jk}\rangle\varphi_{jk}$ converges unconditionally in $L_p(\mathbb{R}^d)$, in particular, $Q_j(\varphi,\widetilde{\varphi},f) \in L_p(\mathbb{R}^d)$;

(b) $\left\|Q_j(\varphi,\widetilde{\varphi},f)\right\|_p \le C_{p,\varphi,\widetilde{\varphi}}\|f\|_p$;

(c) if $p > 1$, then $\lim\limits_{j\to-\infty}\left\|Q_j(\varphi,\widetilde{\varphi},f)\right\|_p = 0$.

Proof Let Ω be a finite subset of $\mathbb{Z}^d$. By the Riesz representation theorem,

$$\left\| \sum_{k\in\Omega} \langle f, \widetilde{\varphi}_{jk}\rangle \varphi_{jk} \right\|_p = \left| \sum_{k\in\Omega} \langle f, \widetilde{\varphi}_{jk}\rangle \langle \varphi_{jk}, g\rangle \right|,$$

where $g = g(\Omega) \in L_q(\mathbb{R}^d)$, $\|g\|_q \le 1$. Using Hölder's inequality and Lemma 4.2.4, we obtain

$$\left| \sum_{k\in\Omega} \langle f, \widetilde{\varphi}_{jk}\rangle \langle \varphi_{jk}, g\rangle \right| \le C_{q,\varphi} m^{\frac{j}{q}-\frac{j}{2}} \|g\|_q \left(\sum_{k\in\Omega} |\langle f, \widetilde{\varphi}_{jk}\rangle|^p \right)^{\frac{1}{p}} \le$$

$$C_{q,\varphi} m^{\frac{j}{q}-\frac{j}{2}} \left(\sum_{k\in\Omega} |\langle f, \widetilde{\varphi}_{jk}\rangle|^p \right)^{\frac{1}{p}}.$$

This implies (a) because, due to Lemma 4.2.4, the series $\sum_{k\in\mathbb{Z}^d} |\langle f, \widetilde{\varphi}_{jk}\rangle|^p$ is convergent.

Similarly,

$$\left\| \sum_{k\in\mathbb{Z}^d} \langle f, \widetilde{\varphi}_{jk}\rangle \varphi_{jk} \right\|_p \le C_{q,\varphi} m^{\frac{j}{q}-\frac{j}{2}} \left(\sum_{k\in\mathbb{Z}^d} |\langle f, \widetilde{\varphi}_{jk}\rangle|^p \right)^{\frac{1}{p}}.$$

To prove (b), it remains to apply Lemma 4.2.4; to prove (c), it remains to apply Lemma 4.2.6.◊

Now, we show that condition (c) from Theorem 4.2.7 does not hold for $p = 1$. Let $f, \varphi \in L_1(\mathbb{R}^d)$, $f, \varphi \ge 0$, $\operatorname{supp} f \subset B_r$, $\widetilde{\varphi} = \chi_{B_R}$. Choose the numbers R and r so that $M^j B_r \subset B_R$ for all $j < 0$. Then, for every $j < 0$

$$\left\| \sum_{k\in\mathbb{Z}^d} \langle f, \widetilde{\varphi}_{jk}\rangle \varphi_{jk} \right\|_1 \ge \int_{\mathbb{R}^d} \langle f, \widetilde{\varphi}_{j0}\rangle \varphi_{j0}(x)\, dx \ge$$

$$m^j \int_{\mathbb{R}^d} \varphi(M^j x)\, dx \int_{B_r} f(t) \widetilde{\varphi}(M^j t)\, dt = \|\varphi\|_1 \|f\|_1.$$

Condition (a) from Theorem 4.2.7 does not hold for $p = \infty$. Indeed, let $\varphi = \widetilde{\varphi} = \chi_{[-\frac{1}{2};\frac{1}{2}]^d}$, $f \equiv 1$. Then, $\langle f, \widetilde{\varphi}_{0k}\rangle = 1$ for all $k \in \mathbb{Z}^d$, and, evidently, the series $\sum\limits_{k\in\mathbb{Z}^d} \langle f, \widetilde{\varphi}_{0k}\rangle \varphi_{0k} = \sum\limits_{k\in\mathbb{Z}^d} \varphi_{0k}$ does not converge in $L_\infty(\mathbb{R}^d)$. A fortiori, we can not discuss whether conditions (b) and (c) are satisfied in this case. This loss may be fixed by restricting the class of functions f. The radial decay of f changes the situation.

Set

$$L_\infty^0(\mathbb{R}^d) := \{f \in L_\infty(\mathbb{R}^d) : \lim_{|x|\to\infty} f(x) = 0\}.$$

Almost the same arguments as for the proofs of Lemma 4.2.6 and Theorem 4.2.7, using Lemma 4.2.5 instead of Lemma 4.2.4, give the following statements.

Lemma 4.2.8 *Let $\varphi \in L_1(\mathbb{R}^d)$, $f \in L_\infty^0(\mathbb{R}^d)$ and φ be compactly supported. Then*

$$\lim_{j\to-\infty} m^{\frac{j}{2}} \sup_{k\in\mathbb{Z}^d} |\langle f, \varphi_{jk}\rangle| = 0.$$

Theorem 4.2.9 *Let $\widetilde{\varphi} \in L_1(\mathbb{R}^d)$ and $\varphi \in L_\infty(\mathbb{R}^d)$ be compactly supported functions, $f \in L_\infty^0(\mathbb{R}^d)$. Then*

(a) for every $j \in \mathbb{Z}$, the series $\sum_{k\in\mathbb{Z}^d}\langle f, \widetilde{\varphi}_{jk}\rangle\varphi_{jk}$ converges unconditionally in $L_\infty(\mathbb{R}^d)$, in particular, $Q_j(\varphi, \widetilde{\varphi}, f) \in L_\infty(\mathbb{R}^d)$;
(b) $\left\|Q_j(\varphi, \widetilde{\varphi}, f)\right\|_\infty \le C_{\infty,\varphi,\widetilde{\varphi}}\|f\|_\infty$;
(c) $\lim_{j\to-\infty} \left\|Q_j(\varphi, \widetilde{\varphi}, f)\right\|_\infty = 0$.

To formulate a similar statement for the case where $\widetilde{\varphi}$ is a tempered distribution, we need the following definition.

Definition 4.2.10 A compactly supported tempered distribution φ is said to satisfy the *Strang-Fix condition* of order n, $n \in \mathbb{N}$, if $D^\alpha\widehat{\varphi}(l) = 0$ for all $l \in \mathbb{Z}^d$, $l \ne \mathbf{0}$, $[\alpha] \le n - 1$.

Theorem 4.2.11 *Let $\varphi \in L_2(\mathbb{R}^d)$, $\widetilde{\varphi} \in S'$, $\varphi, \widetilde{\varphi}$ be compactly supported, $f \in S$. Then*

(a) for every $j \in \mathbb{Z}$, the series $\sum_{k\in\mathbb{Z}^d}\langle f, \widetilde{\varphi}_{jk}\rangle\varphi_{jk}$ converges unconditionally in $L_2(\mathbb{R}^d)$, in particular, $Q_j(\varphi, \widetilde{\varphi}, f) \in L_2(\mathbb{R}^d)$;
(b) if $\widehat{\varphi}(\mathbf{0}) = \widehat{\widetilde{\varphi}}(\mathbf{0}) = 1$, then the Strang-Fix condition of order 1 for φ is necessary and sufficient for the convergence of $Q_j(\varphi, \widetilde{\varphi}, f)$ to f in L_2-norm as $j \to +\infty$.

Proof Let Ω be a finite subset of $\mathbb{Z}^d$. By the Riesz representation theorem,

$$\left\|\sum_{k\in\Omega}\langle f, \widetilde{\varphi}_{jk}\rangle\varphi_{jk}\right\|_2 = \left|\sum_{k\in\Omega}\langle f, \widetilde{\varphi}_{jk}\rangle\langle\varphi_{jk}, g\rangle\right|,$$

where $g = g(\Omega) \in L_2(\mathbb{R}^d)$, $\|g\|_2 \le 1$. Using the Cauchy inequality and Lemma 4.2.4, we obtain

$$\left|\sum_{k\in\Omega}\langle f, \widetilde{\varphi}_{jk}\rangle\langle\varphi_{jk}, g\rangle\right| \le C_{2,\varphi}\|g\|_2\left(\sum_{k\in\Omega}|\langle f, \widetilde{\varphi}_{jk}\rangle|^2\right)^{\frac{1}{2}} \le C_{2,\varphi}\left(\sum_{k\in\Omega}|\langle f, \widetilde{\varphi}_{jk}\rangle|^2\right)^{\frac{1}{2}}.$$

This proves (a) because, due to Lemma 4.2.1, the series $\sum_{k\in\mathbb{Z}^d} |\langle f, \widetilde{\varphi}_{jk}\rangle|^2$ is convergent.

Assume now that $\widehat{\varphi}(\mathbf{0}) = \widehat{\widetilde{\varphi}}(\mathbf{0}) = 1$. To prove (b), first of all we note that $\sum_{l\in\mathbb{Z}^d} |\widehat{\varphi}(\xi + l)|^2$ is equivalent to a trigonometric polynomial by Lemma 2.2.1. Using notations of Lemma 4.2.1, set $G_j(\xi) = G_j(\widetilde{\varphi}, f, \xi)$. By the Plancherel theorem and Lemma 4.2.1,

$$\left\| f - \sum_{k\in\mathbb{Z}^d} \langle f, \widetilde{\varphi}_{jk}\rangle \varphi_{jk} \right\|_2 = \left\| \widehat{f} - \sum_{k\in\mathbb{Z}^d} \langle f, \widetilde{\varphi}_{jk}\rangle \widehat{\varphi_{jk}} \right\|_2 =$$

$$\left(\int\limits_{\mathbb{R}^d} \left| \widehat{f}(\xi) - \sum_{k\in\mathbb{Z}^d} \widehat{G_j}(k) e^{2\pi i\,(k,\, M^{*-j}\xi)} \widehat{\varphi}(M^{*-j}\xi) \right|^2 d\xi \right)^{\frac{1}{2}} =$$

$$\left(\int\limits_{\mathbb{R}^d} \left| \widehat{f}(\xi) - G_j(M^{*-i}\xi) \widehat{\varphi}(M^{*-j}\xi) \right|^2 d\xi \right)^{\frac{1}{2}} =$$

$$\left(m^j \int\limits_{\mathbb{R}^d} \left| \widehat{f}(M^{*j}\xi) - G_j(\xi)\widehat{\varphi}(\xi) \right|^2 d\xi \right)^{\frac{1}{2}} = \Bigg(m^j \int\limits_{\mathbb{R}^d} \Big[\left| \widehat{f}(M^{*j}\xi) \right|^2 -$$

$$\widehat{f}(M^{*j}\xi)\overline{G_j(\xi)\widehat{\varphi}(\xi)} - \overline{\widehat{f}(M^{*j}\xi)} G_j(\xi)\widehat{\varphi}(\xi) + \left| G_j(\xi)\widehat{\varphi}(\xi) \right|^2 \Big] d\xi \Bigg)^{\frac{1}{2}}. \qquad (4.15)$$

Let $j \geq 0$. Since $f \in L_1(\mathbb{R}^d) \cap L_2(\mathbb{R}^d)$, taking into account the boundedness of $\widehat{\varphi}$ and (4.7), we have

$$m^j \int\limits_{\mathbb{R}^d \setminus [-\frac{1}{2}; \frac{1}{2}]^d} \left[\left| \widehat{f}(M^{*j}\xi) \right|^2 - \widehat{f}(M^{*j}\xi)\overline{G_j(\xi)\widehat{\varphi}(\xi)} - \overline{\widehat{f}(M^{*j}\xi)} G_j(\xi)\widehat{\varphi}(\xi) \right] d\xi \underset{j\to+\infty}{\longrightarrow} 0.$$

Combining this with (4.15), using (4.8) and the essential boundedness of the functions $\widehat{f}, \widehat{\widetilde{\varphi}}$ and $\sum_{l\in\mathbb{Z}^d} |\widehat{\varphi}(\cdot + l)|^2$ on $[-\frac{1}{2}; \frac{1}{2}]^d$, we obtain

$$\left\| f - \sum_{k\in\mathbb{Z}^d} \langle f, \widetilde{\varphi}_{jk}\rangle \varphi_{jk} \right\|_2^2 = m^j \int\limits_{[-\frac{1}{2}; \frac{1}{2}]^d} \Big[\left| \widehat{f}(M^{*j}\xi) \right|^2 -$$

$$2Re\,\widehat{f}(M^{*j}\xi)\overline{G_j(\xi)\widehat{\varphi}(\xi)} + \left| G_j(\xi) \right|^2 \sum_{l\in\mathbb{Z}^d} |\widehat{\varphi}(\xi + l)|^2 \Big] d\xi + o(1) =$$

$$m^j \int\limits_{[-\frac{1}{2}; \frac{1}{2}]^d} \left| \widehat{f}(M^{*j}\xi) \right|^2 \left(1 - 2Re\,\widehat{\widetilde{\varphi}}(\xi)\overline{\widehat{\varphi}(\xi)} + |\widehat{\widetilde{\varphi}}(\xi)|^2 \sum_{l\in\mathbb{Z}^d} |\widehat{\varphi}(\xi + l)|^2 \right) d\xi + o(1). \qquad (4.16)$$

After the change of variable, the latter integral is reduced to

$$\int_{\mathbb{R}^d} \chi_{M^{*j}[-\frac{1}{2};\frac{1}{2}]^d} \left|\widehat{f}(\xi)\right|^2 \left(1 - 2Re\,\widehat{\widetilde{\varphi}}(M^{*-j}\xi)\overline{\widehat{\varphi}(M^{*-j}\xi)} + |\widehat{\widetilde{\varphi}}(M^{*-j}\xi)|^2 \sum_{l\in\mathbb{Z}^d} |\widehat{\varphi}(M^{*-j}\xi + l)|^2\right) d\xi.$$

The integrand tends to

$$|\widehat{f}(\xi)|^2 \sum_{l\in\mathbb{Z}^d, l\neq\mathbf{0}} |\widehat{\varphi}(l)|^2$$

as $j \to +\infty$ for each $\xi \in \mathbb{R}^d$. Repeating the arguments of the proof of (4.11) and taking into account that $\sum_{l\in\mathbb{Z}^d} |\widehat{\varphi}(\cdot + l)|^2 \in L_\infty(\mathbb{R}^d)$, it is not difficult to see that the integrand has a summable majorant. Thus, by Lebesgue's dominated convergence theorem,

$$m^j \int_{[-\frac{1}{2};\frac{1}{2}]^d} \left|\widehat{f}(M^{*j}\xi)\right|^2 \left(1 - 2Re\,\widehat{\widetilde{\varphi}}(\xi)\overline{\widehat{\varphi}(\xi)} + |\widehat{\widetilde{\varphi}}(\xi)|^2 \sum_{l\in\mathbb{Z}^d} |\widehat{\varphi}(\xi + l)|^2\right) d\xi \underset{j\to+\infty}{\longrightarrow} \|\widehat{f}\|_2^2 \sum_{l\in\mathbb{Z}^d, l\neq\mathbf{0}} |\widehat{\varphi}(l)|^2.$$

Combining this with (4.16) completes the proof of (b).◊

4.3 Wavelet Expansions

Above, we studied the scaling operators $Q_j : f \to \sum_{k\in\mathbb{Z}^d} \langle f, \widetilde{\varphi}_{jk}\rangle \varphi_{jk}$ with arbitrary compactly supported functions or distributions φ, $\widetilde{\varphi}$.

Assume that φ, $\widetilde{\varphi}$ are refinable, and dual wavelet systems $\{\psi^{(\nu)}_{j,k}\}_{j,k,\nu}$, $\{\widetilde{\psi}^{(\nu)}_{j,k}\}_{j,k,\nu}$ are generated from φ, $\widetilde{\varphi}$ by MEP. Recall that the associated wavelet functions $\psi^{(\nu)}$, $\widetilde{\psi}^{(\nu)}$, $\nu = 1, \dots, r$, are defined by (4.4), and their masks m_ν, $\widetilde{m}_\nu$ are trigonometric polynomials (see Sect. 4.1). It follows that $\psi^{(\nu)}$, $\widetilde{\psi}^{(\nu)}$ are in the same function spaces as φ, $\widetilde{\varphi}$, respectively. We are interested, if the system $\{\psi^{(\nu)}_{j,k}\}_{j,k,\nu}$ is frame-like, i.e., if (4.5) holds for every function f (from an appropriate class) or almost frame-like, i.e., (4.6) holds for every f. Because of the following lemma, which establishes the so-called perfect reconstruction property, (4.6) holds if and only if $\lim\limits_{j\to+\infty} Q_j(f) = f$, and (4.5) holds if and only if $\lim\limits_{j\to+\infty} Q_j(f) = f$ and $\lim\limits_{j\to-\infty} Q_j(f) = 0$, where the convergence is in the same sense as the convergence of the series in (4.5) and (4.6).

Lemma 4.3.1 *Let $\varphi, \widetilde{\varphi} \in S'$ be compactly supported refinable distributions $\psi^{(\nu)}$, $\widetilde{\psi}^{(\nu)}$, $\nu = 1, \dots, r$, be the associated wavelet functions, $f \in S$, j, $j' \in \mathbb{Z}$, $j' > j$. Then*

$$\sum_{k\in\mathbb{Z}^d}\langle f,\widetilde{\varphi}_{j'k}\rangle\varphi_{j'k}-\sum_{k\in\mathbb{Z}^d}\langle f,\widetilde{\varphi}_{jk}\rangle\varphi_{jk}=\sum_{i=j}^{j'-1}\sum_{\nu=1}^{r}\sum_{k\in\mathbb{Z}^d}\langle f,\widetilde{\psi}_{ik}^{(\nu)}\rangle\psi_{ik}^{(\nu)} \tag{4.17}$$

If, moreover, $\widetilde{\varphi}\in L_q(\mathbb{R}^d)$, $1\le q\le\infty$, *then (4.17) holds for any* $f\in L_p(\mathbb{R}^d)$, $\frac{1}{p}+\frac{1}{q}=1$.

Proof Evidently, it suffices to prove (4.17) for the case $j'=j+1$, i.e., to check that

$$\sum_{k\in\mathbb{Z}^d}\langle f,\widetilde{\varphi}_{j+1,k}\rangle\varphi_{j+1,k}=\sum_{k\in\mathbb{Z}^d}\langle f,\widetilde{\varphi}_{jk}\rangle\varphi_{jk}+\sum_{\nu=1}^{r}\sum_{k\in\mathbb{Z}^d}\langle f,\widetilde{\psi}_{jk}^{(\nu)}\rangle\psi_{jk}^{(\nu)}. \tag{4.18}$$

It follows from the refinement equations for φ, $\widetilde{\varphi}$ and (4.4) that

$$\varphi_{jk}=\sum_{l\in\mathbb{Z}^d}h_{l-Mk}^{(0)}\varphi_{j+1,l},\quad \psi_{jk}^{(\nu)}=\sum_{l\in\mathbb{Z}^d}h_{l-Mk}^{(\nu)}\varphi_{j+1,l}, \tag{4.19}$$

$$\widetilde{\varphi}_{jk}=\sum_{l\in\mathbb{Z}^d}\widetilde{h}_{l-Mk}^{(0)}\widetilde{\varphi}_{j+1,l},\quad \widetilde{\psi}_{jk}^{(\nu)}=\sum_{l\in\mathbb{Z}^d}\widetilde{h}_{l-Mk}^{(\nu)}\widetilde{\varphi}_{j+1,l}, \tag{4.20}$$

where $h_n^{(\nu)}$ and $\widetilde{h}_n^{(\nu)}$ are the coefficients of the masks

$$m_\nu(\xi)=\frac{1}{\sqrt{m}}\sum_{n\in\mathbb{Z}^d}h_n^{(\nu)}e^{2\pi i(n,\xi)},\quad \widetilde{m}_\nu(\xi)=\frac{1}{\sqrt{m}}\sum_{n\in\mathbb{Z}^d}\widetilde{h}_n^{(\nu)}e^{2\pi i(n,\xi)},\quad \nu=0,\dots,r.$$

Substituting (4.19) into the right-hand side of (4.18), we obtain

$$\sum_{k\in\mathbb{Z}^d}\langle f,\widetilde{\varphi}_{jk}\rangle\sum_{l\in\mathbb{Z}^d}h_{l-Mk}^{(0)}\varphi_{j+1,l}+\sum_{\nu=1}^{r}\sum_{k\in\mathbb{Z}^d}\langle f,\widetilde{\psi}_{jk}^{(\nu)}\rangle\sum_{l\in\mathbb{Z}^d}h_{l-Mk}^{(\nu)}\varphi_{j+1,l}.$$

Denote by A_l the coefficient at $\varphi_{j+1,l}$. Using (4.20), we have

$$A_{l_0}=\sum_{k\in\mathbb{Z}^d}h_{l_0-Mk}^{(0)}\langle f,\widetilde{\varphi}_{jk}\rangle+\sum_{\nu=1}^{r}\sum_{k\in\mathbb{Z}^d}h_{l_0-Mk}^{(\nu)}\langle f,\widetilde{\psi}_{jk}^{(\nu)}\rangle=$$

$$\sum_{\nu=0}^{r}\sum_{k\in\mathbb{Z}^d}h_{l_0-Mk}^{(\nu)}\sum_{l\in\mathbb{Z}^d}\overline{\widetilde{h}_{l-Mk}^{(\nu)}}\langle f,\widetilde{\varphi}_{j+1,l}\rangle.$$

It follows from (4.3) that

$$\sum_{\nu=0}^{r}\sum_{k\in\mathbb{Z}^d}h_{l_0-Mk}^{(\nu)}\overline{\widetilde{h}_{l-Mk}^{(\nu)}}=\delta_{l,l_0}.$$

Hence, we obtain $A_{l_0} = \langle f, \widetilde{\varphi}_{j+1,l_0}\rangle$ and this yields (4.18).$\diamond$

Lemma 4.3.1 together with Theorem 4.2.2 implies immediately the following statement.

Theorem 4.3.2 *Let $f \in S$, $\varphi, \widetilde{\varphi} \in S'$, $\varphi, \widetilde{\varphi}$ be compactly supported and refinable, $\widehat{\varphi}(\mathbf{0}) = \widehat{\widetilde{\varphi}}(\mathbf{0}) = 1$. $\psi^{(\nu)}, \widetilde{\psi}^{(\nu)}, \nu = 1, \dots, r$, be the associated wavelet functions. Then (4.6) holds with the series converging in S', i.e., the system $\{\psi_{jk}^{(\nu)}\}_{j,k,\nu}$ is almost frame-like in S'.*

Due to Lemma 4.3.1, under the assumptions of Theorem 4.3.2, expansion (4.6) can be replaced by

$$f = \sum_{k\in\mathbb{Z}^d} \langle f, \widetilde{\varphi}_{jk}\rangle \varphi_{jk} + \sum_{i=j}^{\infty}\sum_{\nu=1}^{r}\sum_{k\in\mathbb{Z}^d} \langle f, \widetilde{\psi}_{ik}^{(\nu)}\rangle \psi_{ik}^{(\nu)} \quad \forall j \in \mathbb{Z}, \tag{4.21}$$

but it can not be replaced by (4.5) because generally speaking the sum $\sum_{k\in\mathbb{Z}^d}\langle f, \widetilde{\varphi}_{jk}\rangle$ $\langle \varphi_{jk}, g\rangle$, $f, g \in S$, does not tend to zero as $j \to -\infty$. The following example illustrates this. Let $d = 1$, $M = 2$, $\widetilde{\varphi} = \varphi$ be the δ-function. It is not difficult to see that the δ-function is a compactly supported refinable distribution. If $f \in S$, $f \geq 0$, $f(0) \neq 0$, $g = f$, then

$$\sum_{k\in\mathbb{Z}^d} \langle f, \widetilde{\varphi}_{jk}\rangle\langle \varphi_{jk}, g\rangle = \sum_{k\in\mathbb{Z}^d} |\langle f, \widetilde{\varphi}_{jk}\rangle|^2 \geq |\langle f, \widetilde{\varphi}_{j0}\rangle|^2 = 2^{-j}|f(0)|^2 \underset{j\to-\infty}{\longrightarrow} \infty.$$

Next, we discuss how to construct wavelet functions associated with a pair of given refinable distributions. According to Matrix Extension Principle, polyphase matrices $\mathcal{M}$, $\widetilde{\mathcal{M}}$ satisfying (4.3) should be constructed. In Sect. 3.3, we discussed how to solve this problem for the construction of dual frames in $L_2(\mathbb{R}^d)$. It was a difficult problem because providing vanishing moments for both systems $\{\psi_{j,k}^{(\nu)}\}_{j,k,\nu}$, $\{\widetilde{\psi}_{j,k}^{(\nu)}\}_{j,k,\nu}$ was needed. To construct a frame-like or an almost frame-like system, we should not take care of this. However, later we will see that similar to frames, vanishing moments for $\{\widetilde{\psi}_{j,k}^{(\nu)}\}_{j,k,\nu}$ are required for desirable approximation order of frame-type decompositions.

As above, we say that a wavelet system $\{\psi_{jk}^{(\nu)}\}_{j,k,\nu}$ has VM^n property, if for every ν the equality $D^{\beta}\widehat{\psi^{(\nu)}}(\mathbf{0}) = 0$ holds for each $\beta \in \mathbb{Z}_+^d$, $[\beta] < n$.

Lemma 4.3.3 *Let $\varphi, \widetilde{\varphi}$ be compactly supported refinable distributions, μ_{0k}, $\widetilde{\mu}_{0k}$, $k = 0, \dots, m-1$, be polyphase components of their masks $m_0, \widetilde{m}_0$, respectively, $\widehat{\varphi}(\mathbf{0}) = \widehat{\widetilde{\varphi}}(\mathbf{0}) = 1$, $n \in \mathbb{N}$. If*

$$D^{\beta}\mu_{0k}(\mathbf{0}) = \frac{(2\pi i)^{[\beta]}}{\sqrt{m}} \sum_{\mathbf{0}\leq\gamma\leq\beta} \lambda_{\gamma} \binom{\beta}{\gamma} (-M^{-1}s_k)^{\beta-\gamma} \forall\beta \in \mathbb{Z}_+^d, [\beta] < n,$$

$$k = 0, \dots, m-1; \tag{4.22}$$

for some complex numbers λ_γ, $\gamma \in \mathbb{Z}_+^d$, $[\gamma] < n$, $\lambda_0 = 1$, *and*

$$D^\beta \left(1 - \sum_{k=0}^{m-1} \mu_{0k}(\xi)\overline{\widetilde{\mu}_{0k}(\xi)} \right)\Bigg|_{\xi=0} = 0 \; \forall \beta \in \mathbb{Z}_+^d, [\beta] < n. \tag{4.23}$$

Then there exist associated wavelet functions $\psi^{(\nu)}, \widetilde{\psi}^{(\nu)}$, $\nu = 1, \ldots, m$, *such that the wavelet system* $\{\widetilde{\psi}_{jk}^{(\nu)}\}_{j,k,\nu}$ *has* VM^n *property.*

Proof Set $\mu_{0,m} = 1 - \sum_{k=0}^{m-1} \mu_{0k}\overline{\widetilde{\mu}_{0k}}$, $\widetilde{\mu}_{0,m} \equiv 1$. Using Hausholder's transform (see Sect. 2.5), we can define $(m+1) \times (m+1)$ matrices $\mathcal{N}, \widetilde{\mathcal{N}}$ whose rows form a biorthonormal system, as follows

$$\mathcal{N} := \begin{pmatrix} \mu_{00} & \mu_{01} & \cdots & \mu_{0,m-1} & \mu_{0,m} \\ 0 & 0 & \ldots & 1 & -\overline{\widetilde{\mu}_{0,m-1}} \\ 0 & 0 & \ldots & 0 & -\overline{\widetilde{\mu}_{0,m-2}} \\ \vdots & \vdots & \ddots & \vdots & \vdots \\ 0 & 1 & \ldots & 0 & -\overline{\widetilde{\mu}_{01}} \\ 1 & 0 & \ldots & 0 & -\overline{\widetilde{\mu}_{00}} \end{pmatrix}, \tag{4.24}$$

$$\widetilde{\mathcal{N}} := \begin{pmatrix} \widetilde{\mu}_{00} & \widetilde{\mu}_{01} & \cdots & \widetilde{\mu}_{0m-1} & 1 \\ -\widetilde{\mu}_{00}\overline{\mu_{0,m-1}} & -\widetilde{\mu}_{01}\overline{\mu_{0,m-1}} & \cdots & 1-\widetilde{\mu}_{0,m-1}\overline{\mu_{0,m-1}} & -\overline{\mu_{0,m-1}} \\ -\widetilde{\mu}_{00}\overline{\mu_{0,m-2}} & -\widetilde{\mu}_{01}\overline{\mu_{0,m-2}} & \cdots & -\widetilde{\mu}_{0,m-1}\overline{\mu_{0,m-2}} & -\overline{\mu_{0,m-2}} \\ \vdots & \vdots & \ddots & \vdots & \vdots \\ -\widetilde{\mu}_{00}\overline{\mu_{01}} & 1-\widetilde{\mu}_{01}\overline{\mu_{01}} & \cdots & -\widetilde{\mu}_{0,m-1}\overline{\mu_{01}} & -\overline{\mu_{01}} \\ 1-\widetilde{\mu}_{00}\overline{\mu_{00}} & -\widetilde{\mu}_{01}\overline{\mu_{00}} & \cdots & -\widetilde{\mu}_{0,m-1}\overline{\mu_{00}} & -\overline{\mu_{00}} \end{pmatrix}. \tag{4.25}$$

If now $m_\nu, \widetilde{m}_\nu$, $\nu = 1, \ldots, m$, are trigonometric polynomials whose polyphase components are $\mu_{0k}, \widetilde{\mu}_{0k}$, $k = 1, \ldots, m-1$, respectively (see Sect. 3.3), and $\psi^{(\nu)}, \widetilde{\psi}^{(\nu)}$ are defined by (4.4), then, due to Theorem 3.3.8 (with changing the roles of m_0 and $\widetilde{m}_0$), the wavelet system $\{\widetilde{\psi}_{jk}^{(\nu)}\}$ has VM^n property.$\diamondsuit$

Remark 4.3.4 Due to Proposition 3.3.2 and Lemma 2.5.1, conditions (4.22), (4.23) in Lemma 4.3.3 may be replaced by

$$(i)\; D^\beta m_0(M^{*-1}\xi)|_{\xi=q} = 0, \quad \forall q \in D(M^*) \setminus \{\mathbf{0}\}, \forall \beta \in \mathbb{Z}_+^d, [\beta] < n,$$
$$(ii)\; D^\beta \left(1 - m_0(\xi)\overline{\widetilde{m}_0(\xi)}\right)\Big|_{\xi=0} = 0 \; \forall \beta \in \mathbb{Z}_+^d, [\beta] < n.$$

Thus, for every pair of refinable masks $m_0, \widetilde{m}_0$ satisfying (i), (ii), Lemma 4.3.3 gives an explicit construction of associated wavelet functions $\psi^{(\nu)}, \widetilde{\psi}^{(\nu)}$ such that the system $\{\widetilde{\psi}_{jk}^{(\nu)}\}_{j,k,\nu}$ has VM^n property.

In Theorem 4.3.2, we proved that dual wavelet systems generated from $\varphi, \widetilde{\varphi} \in S'$ by MEP are almost frame-like in S'. Generally speaking, associated wavelet functions

$\psi^{(\nu)}, \widetilde{\psi}^{(\nu)}$ are tempered distributions in this case. It is very doubtful that such wavelet decomposition can be useful for applications. Next, we are interested what happens, if φ is in $L_p(\mathbb{R}^d)$, $1 \le p \le \infty$.

Let λ denote the minimal (in modulus) eigenvalue of M. Recall (see comments to (2.2) in Sect. 2.1) that $|\lambda|^{-1}$ is the spectral radius of the matrix M^{-1}. It follows that

$$\|M^{-i}\| \le \frac{C(\varepsilon)}{(|\lambda| - \varepsilon)^i} \tag{4.26}$$

for every $\varepsilon \in (0, 1 - |\lambda|)$ and $i \in \mathbb{N}$.

Lemma 4.3.5 *Let $\varphi \in L_1(\mathbb{R}^d)$ be a refinable function with a polynomial mask, $\widehat{\varphi}(\mathbf{0}) = 1$. Then φ satisfies the Strang-Fix condition of order 1.*

Proof Let φ satisfy refinable Eq. (4.1), where m_0 is a trigonometric polynomial. Assume that $\widehat{\varphi}(n) \ne 0$ for some $n \in \mathbb{Z}^d, n \ne \mathbf{0}$. Since $\widehat{\varphi}$ is continuous and $\widehat{\varphi}(\mathbf{0}) = 1$, there exists $\varepsilon > 0$ such that $\widehat{\varphi}$ does not vanish on the set $\Omega_\varepsilon := \{\xi \in \mathbb{R}^d : |\xi| \le \varepsilon\}$. It follows from the boundedness of $\widehat{\varphi}$ and (4.1) that there exists a constant C_ε such that

$$\inf_{\xi \in \Omega_\varepsilon} \left| \prod_{j=1}^{k} m_0(M^{*-j}\xi) \right| \ge C_\varepsilon \quad \forall k \in \mathbb{N}.$$

Let $\eta \in \Omega_\varepsilon$ and $k \in \mathbb{N}$. Iterating (4.1) with $\xi = M^{*k}n + \eta$, we obtain

$$\widehat{\varphi}(M^{*k}n + \eta) = \widehat{\varphi}(n + M^{*-k}\eta) \prod_{j=1}^{k} m_0(M^{*-j}\eta).$$

Hence,

$$|\widehat{\varphi}(M^{*k}n + \eta)| \ge C_\varepsilon |\widehat{\varphi}(n + M^{*-k}\eta)| \quad \forall k \in \mathbb{N}.$$

Due to continuity of $\widehat{\varphi}$ and (2.1), $\widehat{\varphi}(n + M^{*-k}\eta) \to \widehat{\varphi}(n)$ as $|k| \to \infty$. On the other hand, by (2.3) and the Riemann–Lebesgue theorem, $\widehat{\varphi}(M^{*k}n + \eta)$ tends to zero as $|k| \to \infty$, which contradicts to our assumption.◊

Theorem 4.3.6 *Let $f \in S$, $\varphi \in L_2(\mathbb{R}^d)$, $\widetilde{\varphi} \in S'$, $\varphi, \widetilde{\varphi}$ be compactly supported and refinable, $\widehat{\varphi}(\mathbf{0}) = \widehat{\widetilde{\varphi}}(\mathbf{0}) = 1$, and let $\psi^{(\nu)}, \widetilde{\psi}^{(\nu)}, \nu = 1, \dots, r$, be associated wavelet functions. Then (4.6) holds with the series converging in L_2-norm, i.e., the system $\{\psi_{jk}^{(\nu)}\}_{j,k,\nu}$ is almost frame-like in $L_2(\mathbb{R}^d)$. If, moreover, $\varphi, \widetilde{\varphi}$ are as in Lemma 4.3.3, then*

$$\left\| f - \sum_{k \in \mathbb{Z}^d} \langle f, \widetilde{\varphi}_{0k} \rangle \varphi_{0k} - \sum_{i=0}^{j-1} \sum_{\nu=1}^{r} \sum_{k \in \mathbb{Z}^d} \langle f, \widetilde{\psi}_{ik}^{(\nu)} \rangle \psi_{ik}^{(\nu)} \right\|_2 \le \frac{C \|f\|_{W_2^{n^*}}}{(|\lambda| - \varepsilon)^{jn}}, \tag{4.27}$$

for every $\varepsilon \in (0, 1 - |\lambda|)$ and some $n^* \ge n$; C and n^* do not depend on f and j.

Proof First of all, we note that due to Theorem 4.2.11, the series $\sum_{k\in\mathbb{Z}^d}\langle f,\widetilde{\varphi}_{0k}\rangle\varphi_{0k}$ and $\sum_{k\in\mathbb{Z}^d}\langle f,\widetilde{\psi}_{ik}^{(\nu)}\rangle\psi_{ik}^{(\nu)}$, $i\in\mathbb{Z}$, $\nu=1,\ldots,r$, converge unconditionally in L_2-norm. By Lemma 4.3.1,

$$\left\|f-\sum_{k\in\mathbb{Z}^d}\langle f,\widetilde{\varphi}_{0k}\rangle\varphi_{0k}-\sum_{i=0}^{j-1}\sum_{\nu=1}^{r}\sum_{k\in\mathbb{Z}^d}\langle f,\widetilde{\psi}_{ik}^{(\nu)}\rangle\psi_{ik}^{(\nu)}\right\|_2= \left\|f-\sum_{k\in\mathbb{Z}^d}\langle f,\widetilde{\varphi}_{jk}\rangle\varphi_{jk}\right\|_2. \quad (4.28)$$

This relation, together with Theorem 4.2.11 and Lemma 4.3.5, yields (4.6).

Now, we assume that φ, $\widetilde{\varphi}$ satisfy all assumptions of Lemma 4.3.3. Since the right-hand side of (4.28) does not depend on the choice of associated wavelet functions, it suffices to check (4.27) for at least one selection of $\psi^{(\nu)}$, $\widetilde{\psi}^{(\nu)}$, $\nu=1,\ldots,r$. Due to Lemma 4.3.3, without loss of generality, we can consider that $\{\widetilde{\psi}_{jk}^{(\nu)}\}$ has VM^n property. It follows from (4.6) that

$$\left\|f-\sum_{k\in\mathbb{Z}^d}\langle f,\widetilde{\varphi}_{0k}\rangle\varphi_{0k}-\sum_{i=0}^{j-1}\sum_{\nu=1}^{r}\sum_{k\in\mathbb{Z}^d}\langle f,\widetilde{\psi}_{ik}^{(\nu)}\rangle\psi_{ik}^{(\nu)}\right\|_2= \left\|\sum_{i=j}^{\infty}\sum_{\nu=1}^{r}\sum_{k\in\mathbb{Z}^d}\langle f,\widetilde{\psi}_{ik}^{(\nu)}\rangle\psi_{ik}^{(\nu)}\right\|_2. \quad (4.29)$$

Due to the Paley–Wiener theorem for tempered distributions, there exist $N\in\mathbb{N}$ and $C_{\widetilde{\varphi}}>0$ so that $\left|\widehat{\widetilde{\varphi}}(\xi)\right|\le C_{\widetilde{\varphi}}\,|\xi|^N$ for all $\xi\notin[-\frac{1}{2};\frac{1}{2}]^d$. Let $n^*\ge n$ and $n^*>N+d/2$. Using notations of Lemma 4.2.1, set $G_i(\xi)=G_i(\widetilde{\psi}^{(\nu)},f,\xi)$. Similar to (4.15), using the Plancherel theorem and Lemma 4.2.1, we have

$$\left\|\sum_{k\in\mathbb{Z}^d}\langle f,\widetilde{\psi}_{ik}^{(\nu)}\rangle\psi_{ik}^{(\nu)}\right\|_2=\left(m^i\int_{\mathbb{R}^d}\left|G_i(\xi)\widehat{\psi^{(\nu)}}(\xi)\right|^2 d\xi\right)^{\frac{1}{2}}= \left(m^i\int_{[-\frac{1}{2};\frac{1}{2}]^d}|G_i(\xi)|^2\sum_{l\in\mathbb{Z}}|\widehat{\psi^{(\nu)}}(\xi+l)|^2d\xi\right)^{\frac{1}{2}}. \quad (4.30)$$

Since $\psi^{(\nu)}$ is in $L_2(\mathbb{R}^d)$ and compactly supported, we have

$$\sum_{l\in\mathbb{Z}^d}|\widehat{\psi^{(\nu)}}(\xi+l)|^2\le C_1. \quad (4.31)$$

Since

$$|\widehat{\widetilde{\psi}^{(\nu)}}(\xi)| \leq C_2|\xi|^n, \xi \in \left[-\frac{1}{2};\frac{1}{2}\right]^d, \quad |\widehat{\widetilde{\psi}^{(\nu)}}(\xi)| \leq C_3|\xi|^N, \xi \notin \left[-\frac{1}{2};\frac{1}{2}\right]^d,$$

we obtain

$$\begin{aligned} m^i \int\limits_{[-\frac{1}{2};\frac{1}{2}]^d} |\widehat{f}(M^{*i}\xi)\widehat{\widetilde{\psi}^{(\nu)}}(\xi)|^2 d\xi \leq C_2^2 m^i \int\limits_{[-\frac{1}{2};\frac{1}{2}]^d} |\xi|^{2n}|\widehat{f}(M^{*i}\xi)|^2 d\xi = \\ C_2^2 \int\limits_{M^{*i}[-\frac{1}{2};\frac{1}{2}]^d} |M^{*-i}\xi|^{2n}|\widehat{f}(\xi)|^2 d\xi \leq C_2^2\|M^{-i}\|^{2n} \int\limits_{\mathbb{R}^d} |\xi|^{2n}|\widehat{f}(\xi)|^2\, d\xi \leq \\ C_2^2\|M^{-i}\|^{2n}\left(\int\limits_{B_1} |\widehat{f}(\xi)|^2\, d\xi + \int\limits_{\mathbb{R}^d\setminus B_1} |\xi|^{2n^*}|\widehat{f}(\xi)|^2\, d\xi\right) \leq \\ C_2^2\|f\|^2_{W_2^{n^*}}\|M^{-i}\|^{2n}; \end{aligned} \tag{4.32}$$

$$\begin{aligned} m^i \int\limits_{[-\frac{1}{2};\frac{1}{2}]^d} \left|\sum_{l\in\mathbb{Z},\, l\neq\mathbf{0}} \widehat{f}(M^{*i}(\xi+l))\widehat{\widetilde{\psi}^{(\nu)}}(\xi+l)\right|^2 d\xi \leq \\ C_3^2 m^i \int\limits_{[-\frac{1}{2};\frac{1}{2}]^d} \left|\sum_{\substack{l\in\mathbb{Z}^d\\ l\neq\mathbf{0}}} \frac{|\xi+l|^{n^*}|\widehat{f}(M^{*i}(\xi+l))|}{|\xi+l|^{n^*-N}}\right|^2 d\xi \leq \\ \sup_{\xi\in[-\frac{1}{2};\frac{1}{2}]^d} \sum_{\substack{l\in\mathbb{Z}^d\\ l\neq\mathbf{0}}} \frac{C_3^2}{|\xi+l|^{2(n^*-N)}} m^i \sum_{\substack{l\in\mathbb{Z}^d\\ l\neq\mathbf{0}}} \int\limits_{[-\frac{1}{2};\frac{1}{2}]^d} |\xi+l|^{2n^*}|\widehat{f}(M^{*i}(\xi+l))|^2 d\xi \leq \\ C_4 m^i \int\limits_{\mathbb{R}^d} |\xi|^{2n^*}|\widehat{f}(M^{*i}\xi)|^2 d\xi \leq C_4\|f\|^2_{W_2^{n^*}}\|M^{-i}\|^{2n^*} \leq \\ C_5\|f\|^2_{W_2^{n^*}}\|M^{-i}\|^{2n}. \end{aligned} \tag{4.33}$$

Combining (4.31), (4.32), (4.33) with (4.30), we have

$$\left\|\sum_{k\in\mathbb{Z}^d} \langle f, \widetilde{\psi}_{ik}^{(\nu)}\rangle \psi_{ik}^{(\nu)}\right\|_2 \leq C_6\|f\|_{W_2^{n^*}}\|M^{-i}\|^n,$$

which together with (4.26) yields that

$$\left\|\sum_{i=j}^{\infty}\sum_{\nu=1}^{r}\sum_{k\in\mathbb{Z}^d} \langle f, \widetilde{\psi}_{ik}^{(\nu)}\rangle \psi_{ik}^{(\nu)}\right\|_2 \leq \sum_{i=j}^{\infty}\sum_{\nu=1}^{r}\left\|\sum_{k\in\mathbb{Z}^d} \langle f, \widetilde{\psi}_{ik}^{(\nu)}\rangle \psi_{ik}^{(\nu)}\right\|_2 \leq \frac{C\|f\|_{W_2^{n^*}}}{(|\lambda|-\varepsilon)^{jn}}. \tag{4.34}$$

It remains to combine (4.34) with (4.29).◊

Due to Lemma 4.3.1, under the assumptions of Theorem 4.3.6 (a), (4.6) can be replaced by (4.21), but it can not be replaced by (4.5) because, generally speaking, $\|Q_j(\varphi, \widetilde{\varphi}, f)\|_2$ does not tend to zero as $j \to -\infty$. The following example illustrates this. Let $d = 1$, $M = 2$, $\widetilde{\varphi}$ be the δ-function, $\varphi = \chi_{[0,1]}$. It is not difficult to see that $\widetilde{\varphi}$ and φ are refinable. If $f \in S$, $f \ge 0$, $f(0) \ne 0$, then

$$\left\|\sum_{k\in\mathbb{Z}^d} \langle f, \widetilde{\varphi}_{jk}\rangle \varphi_{jk}\right\|_2^2 = \int_{\mathbb{R}} \left|\sum_{k\in\mathbb{Z}} f(-2^{-j}k)\varphi(2^j x + k)\right|^2 dx \ge$$

$$\int_{\mathbb{R}} \left|f(0)\varphi(2^j x)\right|^2 dx = 2^{-j}|f(0)|^2 \underset{j\to-\infty}{\longrightarrow} \infty.$$

Theorem 4.3.7 *Let $f \in L_2(\mathbb{R}^d)$, $\varphi, \widetilde{\varphi} \in L_2(\mathbb{R}^d)$ be compactly supported refinable functions, $\widehat{\varphi}(\mathbf{0}) = \widehat{\widetilde{\varphi}}(\mathbf{0}) = 1$, $\psi^{(\nu)}, \widetilde{\psi}^{(\nu)}$, $\nu = 1, \dots, r$, be associated wavelet functions. Then (4.5) holds with the series converging in L_2-norm, i.e., the system $\{\psi_{jk}^{(\nu)}\}_{j,k,\nu}$ is frame-like in $L_2(\mathbb{R}^d)$. If, moreover, $\varphi, \widetilde{\varphi}$ are as in Lemma 4.3.3, then for every $f \in W_2^n(\mathbb{R}^d)$*

$$\left\|f - \sum_{i=-\infty}^{j-1} \sum_{\nu=1}^{r} \sum_{k\in\mathbb{Z}^d} \langle f, \widetilde{\psi}_{ik}^{(\nu)}\rangle \psi_{ik}^{(\nu)}\right\|_2 \le \frac{C\|f\|_{W_2^n}}{(|\lambda| - \varepsilon)^{jn}}, \tag{4.35}$$

where λ is a minimal (in module) eigenvalue of M, $\varepsilon > 0$, $|\lambda| - \varepsilon > 1$, C does not depend on f and j.

Proof Due to Lemma 4.3.1 and Theorem 4.2.7 (c),

$$\left\|f - \sum_{i=-\infty}^{j-1} \sum_{\nu=1}^{r} \sum_{k\in\mathbb{Z}^d} \langle f, \widetilde{\psi}_{ik}^{(\nu)}\rangle \psi_{ik}^{(\nu)}\right\|_2 = \left\|f - \sum_{k\in\mathbb{Z}^d} \langle f, \widetilde{\varphi}_{jk}\rangle \varphi_{jk}\right\|_2. \tag{4.36}$$

If $f \in S$, then the right-hand side tends to zero because of Theorem 4.3.6, which yields (4.5). Approximating $f \in L_2(\mathbb{R}^d)$ by $\widetilde{f} \in S$ in $L_2(\mathbb{R}^d)$ and taking into account Theorem 4.2.7 (b), we obtain (4.5) for arbitrary $f \in L_2(\mathbb{R}^d)$.

Now, we assume that $\varphi, \widetilde{\varphi}$ satisfy all assumptions of Lemma 4.3.3. Since the right-hand side of (4.36) does not depend of the choice of associated wavelet functions, it suffices to check (4.35) for at least one selection of $\psi^{(\nu)}, \widetilde{\psi}^{(\nu)}$, $\nu = 1, \dots, r$. Due to Lemma 4.3.3, without loss of generality, we can consider that $\{\widetilde{\psi}_{jk}^{(\nu)}\}$ has VM^n property.

First, we assume that $f \in S$. To prove that

$$\left\|f - \sum_{k\in\mathbb{Z}^d} \langle f_j, \widetilde{\varphi}_{jk}\rangle \varphi_{jk}\right\|_2 \le C\frac{\|f\|_{W_2^n}}{(|\lambda| - \varepsilon)^{jn}} \tag{4.37}$$

we repeat the proof of the Theorem 4.3.6 with the following modifications. Since

$$\sum_{l\in\mathbb{Z}^d} |\widehat{\widetilde{\psi}^{(\nu)}}(\xi+l)|^2 \le \widetilde{C}_2,$$

which holds true because $\widetilde{\psi}^{(\nu)}$ is in $L_2(\mathbb{R}^d)$ and compactly supported, (4.33) can be replaced by

$$\begin{aligned} m^i \int\limits_{[-\frac{1}{2};\frac{1}{2}]^d} \left| \sum_{\substack{l\in\mathbb{Z}\\ l\neq 0}} \widehat{f}(M^{*i}(\xi+l))\widehat{\widetilde{\psi}^{(\nu)}}(\xi+l) \right|^2 d\xi \le \\ m^i \int\limits_{[-\frac{1}{2};\frac{1}{2}]^d} \left| \sum_{\substack{l\in\mathbb{Z}\\ l\neq 0}} |\xi+l|^n |\widehat{f}(M^{*i}(\xi+l))| \frac{|\widehat{\widetilde{\psi}^{(\nu)}}(\xi+l)|}{|\xi+l|^n} \right|^2 d\xi \le \\ \sup_{\xi\in[-\frac{1}{2};\frac{1}{2}]^d} \sum_{\substack{l\in\mathbb{Z}^d\\ l\neq 0}} \frac{|\widehat{\widetilde{\psi}^{(\nu)}}(\xi+l)|^2}{|\xi+l|^{2n}} \left(m^i \sum_{\substack{l\in\mathbb{Z}^d\\ l\neq 0}} \int\limits_{[-\frac{1}{2};\frac{1}{2}]^d} |\xi+l|^{2n} |\widehat{f}(M^{*i}(\xi+l))|^2 d\xi \right) \le \\ \widetilde{C}_2 2^{2n} m^i \int\limits_{\mathbb{R}^d} |\xi|^{2n} |\widehat{f}(M^{*i}\xi)|^2 d\xi \le \widetilde{C}_2 2^{2n} \|f\|^2_{W_2^n} \|M^{-i}\|^{2n}. \end{aligned} \tag{4.38}$$

Evidently, n^* can be replaced by n in (4.32). After these changes, we obtain (4.34) with n instead of n^*, which, due to (4.36), yields (4.37) for $f \in S$. Approximating $f \in W_2^n(\mathbb{R}^d)$ by $\widetilde{f} \in S$ in W_2^n-norm and taking into account Theorem 4.2.7 (b), we obtain (4.37) for arbitrary $f \in W_2^n(\mathbb{R}^d)$. Using (4.36) again, we complete the proof.$\Diamond$

Relation (4.35) provides approximation order n for the frame-like expansions. The technique used for the proof of this fact does not work for the spaces $L_p(\mathbb{R}^d)$, $p \neq 2$. To establish approximation order in L_p-norm, we shall use another technique. The following lemma will be very useful for us.

Lemma 4.3.8 *Let* $1 \le p \le \infty$, $\varphi \in L_p(\mathbb{R}^d)$ *be compactly supported,* $c = \{c_k\}_{k\in\mathbb{Z}^d} \in \ell_p(\mathbb{Z}^d)$, $i \in \mathbb{Z}$. *Then*

$$\left\| \sum_{k\in\mathbb{Z}^d} c_k \varphi_{ik} \right\|_p \le C_\varphi m^{\frac{i}{2}-\frac{i}{p}} \|c\|_{\ell_p}, \tag{4.39}$$

where C_φ *depends only on* φ *and* p.

Proof Let first $i = 0$. We have

$$\left\|\sum_{k\in\mathbb{Z}^d} c_k\varphi_{0k}\right\|_p^p = \int_{\mathbb{R}^d}\left|\sum_{k\in\mathbb{Z}^d} c_k\varphi(x+k)\right|^p dx = \int_{[0,1]^d}\sum_{l\in\mathbb{Z}^d}\left|\sum_{k\in\mathbb{Z}^d} c_k\varphi(x+l+k)\right|^p dx.$$

For every fixed x, by a discrete version of Young's inequality,

$$\left(\sum_{l\in\mathbb{Z}^d}\left|\sum_{k\in\mathbb{Z}^d} c_k\varphi(x+l+k)\right|^p\right)^{1/p} \le \|c\|_{\ell_p}\sum_{k\in\mathbb{Z}^d}|\varphi(x+k)|,$$

which yields

$$\left\|\sum_{k\in\mathbb{Z}^d} c_k\varphi_{0k}\right\|_p^p \le \|c\|_{\ell_p}^p\int_{[0,1]^d}\left(\sum_{k\in\mathbb{Z}^d}|\varphi(x+k)|\right)^p dx.$$

Denoting the latter integral by C_φ^p, we obtain (4.39) for $i=0$. To prove (4.39) for arbitrary i, it remains to note that

$$\left\|\sum_{k\in\mathbb{Z}^d} c_k\varphi_{ik}\right\|_p = m^{\frac{i}{2}-\frac{i}{p}}\left\|\sum_{k\in\mathbb{Z}^d} c_k\varphi_{0k}\right\|_p. \diamondsuit$$

Theorem 4.3.9 *Let* $1<p\le\infty$, $\frac{1}{p}+\frac{1}{q}=1$, $f,\varphi\in L_p(\mathbb{R}^d)$, $\widetilde{\varphi}\in L_q(\mathbb{R}^d)$, $\varphi,\widetilde{\varphi}$ *be compactly supported refinable functions satisfying all conditions of Lemma 4.3.3, and let* $\psi^{(\nu)},\widetilde{\psi}^{(\nu)}$, $\nu=1,\dots,r$, *be associated wavelet functions. Then (4.5) holds with the series converging in* L_p*-norm, i.e., the system* $\{\psi_{jk}^{(\nu)}\}_{j,k,\nu}$ *is frame-like in* $L_p(\mathbb{R}^d)$ *(in* $L_\infty(\mathbb{R}^d)\cap L_\infty^0(\mathbb{R}^d)$ *for* $p=\infty$*). If, moreover,* $f\in W_p^n(\mathbb{R}^d)$ *(*$f\in W_\infty^n\cap L_\infty^0(\mathbb{R}^d)$ *for* $p=\infty$*), then*

$$\left\|f-\sum_{i=-\infty}^{j-1}\sum_{\nu=1}^{r}\sum_{k\in\mathbb{Z}^d}\langle f,\widetilde{\psi}_{ik}^{(\nu)}\rangle\psi_{ik}^{(\nu)}\right\|_p \le \frac{C\|f\|_{W_p^n}}{(|\lambda|-\varepsilon)^{jn}}, \tag{4.40}$$

for every $\varepsilon\in(0,1-|\lambda|)$*;* C *does not depend on* f *and* j.

Proof By Theorems 4.2.7 (c) and 4.2.9 (c), and Lemma 4.3.1,

$$\sum_{i=-\infty}^{j-1}\sum_{\nu=1}^{r}\sum_{k\in\mathbb{Z}^d}\langle f,\widetilde{\psi}_{ik}^{(\nu)}\rangle\psi_{ik}^{(\nu)} = \sum_{k\in\mathbb{Z}^d}\langle f,\widetilde{\varphi}_{jk}\rangle\varphi_{jk}. \tag{4.41}$$

Since the right-hand side of (4.41) does not depend on the choice of associated wavelet functions, it suffices to check (4.40) for at least one selection of $\psi^{(\nu)},\widetilde{\psi}^{(\nu)}$, $\nu=1,\dots,r$. Due to Lemma 4.3.3, without loss of generality, we can consider that $\{\widetilde{\psi}_{jk}^{(\nu)}\}$ has VM^n property.

First of all, we note that due to Theorems 4.2.7 (a) and 4.2.9 (a), the series $\sum_{k\in\mathbb{Z}^d}\langle f,\widetilde{\psi}_{ik}^{(\nu)}\rangle\psi_{ik}^{(\nu)}$, $i\in\mathbb{Z}$, $\nu=1,\dots,r$, converges unconditionally in L_p-norm. Suppose $p<\infty$, $f\in W_p^n(\mathbb{R}^d)$ and prove that for every $i\in\mathbb{N}$

$$\left\|\sum_{k\in\mathbb{Z}^d}\langle f,\widetilde{\psi}_{ik}^{(\nu)}\rangle\psi_{ik}^{(\nu)}\right\|_p\le\widetilde{C}\|M^{-i}\|^n\,\|f\|_{W_p^n},\tag{4.42}$$

where $\widetilde{C}$ does not depend on f and i.

Let $k\in\mathbb{Z}^d$, $z\in[1/2,1/2]^d-k$, $y=M^{-i}z$. Since VM^n property is equivalent to

$$\int\limits_{\mathbb{R}^d}x^\alpha\widetilde{\psi}_{ik}^{(\nu)}(x)dx=0,\qquad\nu=1,\dots,r,\ \forall\alpha\in\mathbb{Z}^d,\ [\alpha]<n,$$

using Taylor's formula with the integral remainder, we obtain

$$\left|\langle f,\widetilde{\psi}_{ik}^{(\nu)}\rangle\right|=\left|\int\limits_{\mathbb{R}^d}f(x)\overline{\widetilde{\psi}_{ik}^{(\nu)}(x)}\,dx\right|=$$

$$\left|\int\limits_{\mathbb{R}^d}\overline{\widetilde{\psi}_{ik}^{(\nu)}(x)}\left(\sum_{k=0}^{n-1}\frac{1}{k!}((x_1-y_1)\partial_1+\dots+(x_d-y_d)\partial_d)^k f(y)+\right.\right.$$

$$\left.\left.\int\limits_0^1\frac{(1-t)^{n-1}}{(n-1)!}((x_1-y_1)\partial_1+\dots+(x_d-y_d)\partial_d)^n f(y+t(x-y))\,dt\right)dx\right|\le$$

$$C_1\int\limits_{\mathbb{R}^d}dx\,|x-y|^n|\widetilde{\psi}_{ik}^{(\nu)}(x)|\int\limits_0^1\sum_{[\beta]=n}|D^\beta f(y+t(x-y))|\,dt.$$

Suppose $supp\,\widetilde{\psi}^{(\nu)}\subset B_R$, $R\ge\sqrt{d}$. If $\widetilde{\psi}_{ik}^{(\nu)}(x)\ne0$, then

$$|y-x|^n\le\|M^{-i}\|^n|z-M^ix|^n\le(2R)^n\|M^{-i}\|^n,$$

and if $y-x\notin M^{-i}(B_{2R})$, then $\widetilde{\psi}_{ik}^{(\nu)}(x)=0$. It follows that

$$\left|\langle f,\widetilde{\psi}_{ik}^{(\nu)}\rangle\right|\le C_2\|M^{-i}\|^n\int\limits_{\mathbb{R}^d}dx|\widetilde{\psi}_{ik}^{(\nu)}(x)|\int\limits_0^1\sum_{[\beta]=n}|D^\beta f(y+t(x-y))|\,dt=$$

$$C_2m^{i/2}\|M^{-i}\|^n\int\limits_{y-x\in M^{-i}(B_{2R})}dx|\widetilde{\psi}^{(\nu)}(M^ix+k)|\int\limits_0^1\sum_{[\beta]=n}|D^\beta f(y+t(x-y))|\,dt.$$

Using Hölder's inequality, we have

$$\left|\langle f, \widetilde{\psi}_{ik}^{(\nu)}\rangle\right| \le C_2 m^{i/2}\|M^{-i}\|^n \left(\int\limits_{\mathbb{R}^d} |\widetilde{\psi}^{(\nu)}(M^i x + k)|^q\, dx\right)^{1/q} \cdot$$

$$\left(\int\limits_{x-y\in M^{-i}(B_{2R})} dx \left(\int\limits_0^1 \sum_{[\beta]=n} |D^\beta f(y+t(x-y))|\, dt\right)^p\right)^{\frac{1}{p}} \le$$

$$C_2 m^{\frac{i}{2}-\frac{i}{q}}\|M^{-i}\|^n \|\widetilde{\psi}^{(\nu)}\|_q \left(\int\limits_{M^{-i}(B_{2R})} du \int\limits_0^1 \sum_{[\beta]=n} |D^\beta f(y+tu)|^p\, dt\right)^{\frac{1}{p}}.$$

Combining this with Lemma 4.3.8, we obtain

$$\left\|\sum_{k\in\mathbb{Z}^d} \langle f, \widetilde{\psi}_{ik}^{(\nu)}\rangle \psi_{ik}^{(\nu)}\right\|_p^p \le C_\varphi^p m^{p(\frac{i}{2}-\frac{i}{p})} \sum_{k\in\mathbb{Z}^d} \left|\langle f, \widetilde{\psi}_{ik}^{(\nu)}\rangle\right|^p =$$

$$C_\varphi^p m^{p(\frac{i}{2}-\frac{i}{p})} \sum_{k\in\mathbb{Z}^d} \int\limits_{[1/2,1/2]^d+k} dz \left|\langle f, \widetilde{\psi}_{ik}^{(\nu)}\rangle\right|^p \le$$

$$C_3\|M^{-i}\|^{pn} \int\limits_{\mathbb{R}^d} dz \int\limits_{M^{-i}(B_{2R})} du \int\limits_0^1 \sum_{[\beta]=n} |D^\beta f(M^{-i}z+tu)|^p\, dt =$$

$$C_3\|M^{-i}\|^{pn} \int\limits_{\mathbb{R}^d} dz \int\limits_{B_{2R}} du \int\limits_0^1 \sum_{[\beta]=n} |D^\beta f(z+tu)|^p\, dt =$$

$$C_3\|M^{-i}\|^{pn} \int\limits_{B_{2R}} du \int\limits_0^1 dt \int\limits_{\mathbb{R}^d} dz \sum_{[\beta]=n} |D^\beta f(z)|^p\, dz \le C_4\|M^{-i}\|^{pn}\|f\|_{W_p^n}^p.$$

This yields (4.42) for the case $p < \infty$.

Similarly, if $p = \infty$, $f \in W_\infty^n \cap L_\infty^0(\mathbb{R}^d)$, then we have

$$\left|\langle f, \widetilde{\psi}_{ik}^{(\nu)}\rangle\right| \le C_2 m^{-i/2}\|M^{-i}\|^n \operatorname{vraisup}_{\tau\in\mathbb{R}^d} \sum_{[\beta]=n} |D^\beta f(\tau)| \int\limits_{\mathbb{R}^d} |\widetilde{\psi}^{(\nu)}(x)|dx \le$$

$$C_2 m^{-i/2}\|M^{-i}\|^n \|\widetilde{\psi}^{(\nu)}\|_1 \|f\|_{W_\infty^n}.$$

Combining this with Lemma 4.3.8, we obtain

$$\left\| \sum_{k\in\mathbb{Z}^d} \langle f, \widetilde{\psi}_{ik}^{(\nu)} \rangle \psi_{ik}^{(\nu)} \right\|_\infty \le C_\varphi m^{i/2} \sup_{k\in\mathbb{Z}^d} \left| \langle f, \widetilde{\psi}_{ik}^{(\nu)} \rangle \right| \le C_5 \|M^{-i}\|^n \|f\|_{W_\infty^n},$$

which completes the proof of (4.42) for $p = \infty$.

Let us prove that $\{\psi_{jk}^{(\nu)}\}_{j,k,\nu}$ is frame-like in $L_p(\mathbb{R}^d)$. Assume first that $f \in S'$. It follows from (4.42) and (4.26) that the series

$$\sum_{i=1}^{j} \sum_{\nu=1}^{r} \sum_{k\in\mathbb{Z}^d} \langle f, \widetilde{\psi}_{ik}^{(\nu)} \rangle \psi_{ik}^{(\nu)}$$

converges in L_p-norm. By Lemma 4.3.1, this yields that $Q_j(\varphi, \widetilde{\varphi}, f)$ converges in L_p-norm as $j \to +\infty$. On the other hand, due to Theorem 4.2.2, we know that $Q_j(\varphi, \widetilde{\varphi}, f)$ tends to f in S' as $j \to +\infty$. Hence, $Q_j(\varphi, \widetilde{\varphi}, f)$ tends to f in L_p-norm as $j \to +\infty$. If now $f \in L_p(\mathbb{R}^d)$, approximating f by $\widetilde{f} \in S$ in $L_p(\mathbb{R}^d)$ and using Theorem 4.2.7 (b), we conclude that $Q_j(\varphi, \widetilde{\varphi}, f)$ tends to f in L_p-norm as $j \to +\infty$. Combining this with Theorems 4.2.7 (c) and 4.2.9 (c), and Lemma 4.3.1, we obtain (4.5), i.e., the system $\{\psi_{jk}^{(\nu)}\}_{j,k,\nu}$ is frame-like in $L_p(\mathbb{R}^d)$ (in $L_\infty(\mathbb{R}^d) \cap L_\infty^0(\mathbb{R}^d)$ for $p = \infty$). To prove (4.40), it remains to note that

$$f - \sum_{i=-\infty}^{j} \sum_{\nu=1}^{r} \sum_{k\in\mathbb{Z}^d} \langle f, \widetilde{\psi}_{ik}^{(\nu)} \rangle \psi_{ik}^{(\nu)} = \sum_{i=j+1}^{\infty} \sum_{\nu=1}^{r} \sum_{k\in\mathbb{Z}^d} \langle f, \widetilde{\psi}_{ik}^{(\nu)} \rangle \psi_{ik}^{(\nu)}$$

and apply (4.42) and (4.26).◊

Theorem 4.3.9 is not true for $p = 1$ because (4.41) does not hold. Using

$$\sum_{k\in\mathbb{Z}^d} \langle f, \widetilde{\varphi}_{0k} \rangle \varphi_{0k} + \sum_{i=0}^{j-1} \sum_{\nu=1}^{r} \sum_{k\in\mathbb{Z}^d} \langle f, \widetilde{\psi}_{ik}^{(\nu)} \rangle \psi_{ik}^{(\nu)} = \sum_{k\in\mathbb{Z}^d} \langle f, \widetilde{\varphi}_{jk} \rangle \varphi_{jk}.$$

instead of (4.41) and repeating all other arguments of the proof of Theorem 4.3.9 and Lemma 4.3.8, we obtain

Theorem 4.3.10 *Let $f, \varphi \in L_1(\mathbb{R}^d)$, $\widetilde{\varphi} \in L_\infty(\mathbb{R}^d)$, $\varphi, \widetilde{\varphi}$ be compactly supported refinable functions satisfying all conditions of Lemma 4.3.3, and let $\psi^{(\nu)}, \widetilde{\psi}^{(\nu)}$, $\nu = 1, \dots, r$, be associated wavelet functions. Then (4.6) holds with the series converging in L_p-norm, i.e., the system $\{\psi_{jk}^{(\nu)}\}_{j,k,\nu}$ is almost frame-like in $L_1(\mathbb{R}^d)$. If, moreover, $f \in W_1^n(\mathbb{R}^d)$, then*

$$\left\| f - \sum_{k\in\mathbb{Z}^d} \langle f, \widetilde{\varphi}_{0k} \rangle \varphi_{0k} - \sum_{i=0}^{j-1} \sum_{\nu=1}^{r} \sum_{k\in\mathbb{Z}^d} \langle f, \widetilde{\psi}_{ik}^{(\nu)} \rangle \psi_{ik}^{(\nu)} \right\|_1 \le \frac{C\|f\|_{W_1^n}}{(|\lambda| - \varepsilon)^{jn}}, \qquad (4.43)$$

for every $\varepsilon \in (0, 1 - |\lambda|)$; C does not depend on f and j.

4.4 Examples

In this section, we shall give several examples of frame-like and almost frame-like wavelet systems to illustrate the results of this chapter. All examples are based on the construction of wavelet functions $\psi^{(\nu)}, \widetilde{\psi}^{(\nu)}$ associated with φ, $\widetilde{\varphi}$ according to the scheme described in Sect. 4.3.

1. Let $M = \begin{pmatrix} 1 & 1 \\ 1 & -1 \end{pmatrix}$ be the quincunx dilation matrix. For this matrix $m = 2$, $D(M) = \{s_0 = (0, 0), s_1 = (1, 0)\}$ is a set of digits.

Let φ be a box spline whose Fourier transform is defined by

$$\widehat{\varphi}(\xi) = \left(\frac{1 - e^{-2\pi i\xi_1}}{2\pi i\xi_1}\right)\left(\frac{1 - e^{-2\pi i\xi_2}}{2\pi i\xi_2}\right)^2\left(\frac{1 - e^{-2\pi i(\xi_1+\xi_2)}}{2\pi i(\xi_1 + \xi_2)}\right)\left(\frac{1 - e^{-2\pi i(\xi_1-\xi_2)}}{2\pi i(\xi_1 - \xi_2)}\right)^2.$$

It is not difficult to check that this box spline is in C^2 (see [6]) and satisfies the refinement equation with the mask

$$m_0(\xi) = \left(\frac{1 + e^{-2\pi i\xi_1}}{2}\right)\left(\frac{1 + e^{-2\pi i\xi_2}}{2}\right)^2.$$

The polyphase components

$$\mu_{00} = \frac{1}{\sqrt{2}}\left(\frac{1}{4} + \frac{1}{4}e^{-2\pi i(\xi_1-\xi_2)} + \frac{1}{2}e^{-2\pi i\xi_1}\right),$$

$$\mu_{01} = \frac{1}{\sqrt{2}}\left(\frac{1}{2}e^{-2\pi i\xi_1} + \frac{1}{4}e^{-2\pi i(\xi+\xi_2)} + \frac{1}{4}e^{-4\pi i\xi_1}\right)$$

satisfy condition (4.22) in Lemma 4.3.3 with $n = 2$ and $\lambda_{00} = 1$, $\lambda_{10} = -\frac{3}{4}$, $\lambda_{01} = \frac{1}{4}$. Since $e^{2\pi i(c,\xi)}m_0(\xi)$ with $c = (-\frac{1}{2}, -1)$ is real and even, the refinable function φ is real and symmetric with respect to the point $C = (M - E)^{-1}c = (-2, -\frac{1}{2})$.

Let $\widetilde{\varphi}$ be the δ-function. Then, $\widetilde{m}_0 \equiv 1$ and the corresponding polyphase components are $\widetilde{\mu}_{00} \equiv \sqrt{2}$, $\widetilde{\mu}_{01} \equiv 0$. We construct associated wavelet functions $\psi^{(\nu)}, \widetilde{\psi}^{(\nu)}$, $\nu = 1, 2$, using the method given in the proof of Lemma 4.3.3. The polyphase matrices $\mathcal{N}, \widetilde{\mathcal{N}}$ defined by (4.24), (4.25) look as

$$\mathcal{N} = \begin{pmatrix} \mu_{00} & \mu_{01} & 1 - \sqrt{2}\mu_{00} \\ 0 & 1 & 0 \\ 1 & 0 & -\sqrt{2} \end{pmatrix}, \quad \widetilde{\mathcal{N}} = \begin{pmatrix} \sqrt{2} & 0 & 1 \\ -\sqrt{2}\overline{\mu}_{01} & 1 & -\overline{\mu}_{01} \\ 1 - \sqrt{2}\overline{\mu}_{00} & 0 & -\overline{\mu}_{00} \end{pmatrix}.$$

This leads to the wavelet masks $m_1(\xi) = \frac{1}{\sqrt{2}}e^{2\pi i\xi_1}$, $m_2(\xi) \equiv 1$ and the corresponding wavelet functions $\psi^{(1)}(x) = \sqrt{2}\varphi(Mx + \begin{pmatrix} 1 \\ 0 \end{pmatrix})$, $\psi^{(2)}(x) = \sqrt{2}\varphi(Mx)$ which are in C^2 and symmetric with respect to the points $(-\frac{7}{4}, -\frac{5}{4})$, $(-\frac{5}{4}, -\frac{3}{4})$, respectively.

The dual wavelet masks and the corresponding wavelet distributions are

$$
\begin{aligned}
\widetilde{m}_1(\xi) &= \frac{1}{\sqrt{2}}\left(-\frac{1}{2}e^{2\pi i(\xi_1+\xi_2)} - \frac{1}{4}e^{4\pi i(\xi_1+\xi_2)} - \frac{1}{4}e^{4\pi i\xi_1} + e^{2\pi i\xi_1}\right),\\
\widetilde{m}_2(\xi) &= \frac{1}{\sqrt{2}}\left(\frac{3}{4} - \frac{1}{4}e^{4\pi i\xi_2} - \frac{1}{2}e^{2\pi i(\xi_1+\xi_2)}\right),\\
\widetilde{\psi}^{(1)}(x) &= \sqrt{2}\left[-\frac{1}{2}\delta\left(Mx+\begin{pmatrix}1\\1\end{pmatrix}\right) - \frac{1}{4}\delta\left(Mx+\begin{pmatrix}2\\2\end{pmatrix}\right) - \right.\\
&\qquad\left.\frac{1}{4}\delta\left(Mx+\begin{pmatrix}2\\0\end{pmatrix}\right) + \delta\left(Mx+\begin{pmatrix}1\\0\end{pmatrix}\right)\right],\\
\widetilde{\psi}^{(2)}(x) &= \sqrt{2}\left[\frac{3}{4}\delta(Mx) - \frac{1}{4}\delta\left(Mx+\begin{pmatrix}0\\2\end{pmatrix}\right) - \frac{1}{2}\delta\left(Mx+\begin{pmatrix}1\\1\end{pmatrix}\right)\right].
\end{aligned}
$$

By Theorem 4.3.6, for any $f \in S$, expansion (4.6) holds in L_2-norm and looks as follows

$$
\begin{aligned}
f(x) = \sum_{k\in\mathbb{Z}^2} f(-k)\varphi(x+k) + \sum_{j=0}^{+\infty}\sum_{k\in\mathbb{Z}^2}&\left[-\frac{1}{2}f\left(-M^{-j-1}\left(Mk+\begin{pmatrix}1\\1\end{pmatrix}\right)\right) - \right.\\
&\frac{1}{4}f\left(-M^{-j-1}\left(Mk+\begin{pmatrix}2\\2\end{pmatrix}\right)\right) - \frac{1}{4}f\left(-M^{-j-1}\left(Mk+\begin{pmatrix}2\\0\end{pmatrix}\right)\right) + \\
&\left. f\left(-M^{-j-1}\left(Mk+\begin{pmatrix}1\\0\end{pmatrix}\right)\right)\right]\varphi\left(M^{j+1}x+Mk+\begin{pmatrix}1\\0\end{pmatrix}\right) + \\
&\sum_{j=0}^{+\infty}\sum_{k\in\mathbb{Z}^2}\left[\frac{3}{4}f\left(-M^{-j}k\right) - \frac{1}{4}f\left(-M^{-j-1}\left(Mk+\begin{pmatrix}0\\2\end{pmatrix}\right)\right) - \right.\\
&\left.\frac{1}{2}f\left(-M^{-j-1}\left(Mk+\begin{pmatrix}1\\1\end{pmatrix}\right)\right)\right]\varphi(M^{j+1}x+Mk).
\end{aligned}
$$

Since all assumptions of Lemma 4.3.3 are satisfied with $n = 2$, this expansion has approximation order 2.

Now, using the same refinable function φ, we construct an almost frame-like wavelet system providing approximation order 3. Choosing

$$
\widetilde{\mu}_{00} = \sqrt{2} - \frac{3i}{2\sqrt{2}}\sin 2\pi\xi_1 + \frac{i}{2\sqrt{2}}\sin 2\pi\xi_2, \quad \widetilde{\mu}_{01} \equiv 0,
$$

we have (4.23) satisfied with $n = 3$. Again we construct $\psi^{(\nu)}$, $\widetilde{\psi}^{(\nu)}$, $\nu = 1, 2$, using the method given in the proof of Lemma 4.3.3. The polyphase matrices $\mathcal{N}, \widetilde{\mathcal{N}}$ defined by (4.24), (4.25) look as

$$
\mathcal{N} = \begin{pmatrix} \mu_{00} & \mu_{01} & \mu_{02}\\ 0 & 1 & 0\\ 1 & 0 & -\widetilde{\mu}_{00}\end{pmatrix}, \quad \widetilde{\mathcal{N}} = \begin{pmatrix} \widetilde{\mu}_{00} & 0 & 1\\ -\widetilde{\mu}_{00}\overline{\mu}_{01} & 1 & -\overline{\mu}_{01}\\ 1-\widetilde{\mu}_{00}\overline{\mu}_{00} & 0 & -\overline{\mu}_{00}\end{pmatrix},
$$

where $\mu_{02}(\xi) = 1 - \mu_{00}(\xi)\overline{\widetilde{\mu}_{00}}(\xi)$. This leads to the same wavelet masks m_1, m_2 and wavelet functions $\psi^{(1)}$, $\psi^{(2)}$ as above. The dual refinable mask is

$$\widetilde{m}_0(\xi) = 1 - \frac{3}{8}e^{2\pi i(\xi_1+\xi_2)} + \frac{3}{8}e^{-2\pi i(\xi_1+\xi_2)} + \frac{1}{8}e^{2\pi i(\xi_1-\xi_2)} - \frac{1}{8}e^{-2\pi i(\xi_1-\xi_2)}.$$

Since $\widetilde{m}_0$ is a trigonometric polynomial, $\widetilde{\varphi}$ is a compactly supported tempered distribution. The dual wavelet masks $\widetilde{m}_1, \widetilde{m}_2$ are trigonometric polynomials whose Fourier coefficients are given by the following tables

$$\frac{1}{32\sqrt{2}}\begin{pmatrix} 0&0&0&0&1&0&3\\ 0&0&0&2&0&-2&0\\ 0&0&0&0&-18&0&2\\ 0&0&0&-6&32&-10&0\\ 0&0&0&0&-3&0&-1\\ 0&0&0&0&0&0&0\\ 0&0&0&0&0&0&0 \end{pmatrix}, \quad \frac{1}{32\sqrt{2}}\begin{pmatrix} 0&1&0&3&0\\ 0&0&-6&0&6\\ 0&-2&0&-14&0\\ 0&0&18&0&-2\\ 0&-3&0&-1&0\\ 0&0&0&0&0\\ 0&0&0&0&0 \end{pmatrix},$$

and the corresponding dual wavelet distributions $\widetilde{\psi}^{(1)}, \widetilde{\psi}^{(2)}$ are defined by (4.4). By Theorem 4.3.6, for any $f \in S$, expansion (4.6) holds in L_2-norm and has approximation order 2.

2. Let $M = \begin{pmatrix} 1 & -2\\ 2 & -1 \end{pmatrix}$. For this matrix,

$$D(M) = \{s_0 = (0,0), s_1 = (0,-1), s_2 = (0,1)\}$$

is a set of digits, $m = 3$. Let the mask m_0 be defined by the table of its Fourier coefficients

$$\frac{1}{2187}\begin{pmatrix} 0&0&0&0&7&4&0&4&7\\ 0&0&0&4&0&-32&-32&0&4\\ 0&0&0&-32&-20&0&-20&-32&0\\ 0&4&-32&0&312&312&0&-32&4\\ 7&0&-20&312&729&312&-20&0&7\\ 4&-32&0&312&312&0&-32&4&0\\ 0&-32&-20&0&-20&-32&0&0&0\\ 4&0&-32&-32&0&4&0&0&0\\ 7&4&0&4&7&0&0&0&0 \end{pmatrix}.$$

The mask is interpolatory, i.e., $\mu_{00} \equiv \frac{1}{\sqrt{3}}$, and the corresponding refinable function φ is in C^2, supported on $[-4,4]^2$, symmetric with respect to the origin and with respect to the axes. Condition (4.22) from Lemma 4.3.3 is satisfied with $n = 6$ and $\lambda_0 = 1$, $\lambda_k = 0$, $k \in \mathbb{Z}^2_+, 0 < [k] < 6$.

Let $\widetilde{\varphi}$ be the δ-function. Then, the corresponding polyphase components are $\widetilde{\mu}_{00} \equiv \sqrt{3}$, $\widetilde{\mu}_{01} \equiv 0$, $\widetilde{\mu}_{02} \equiv 0$. So, $\sum_{k=0}^{m-1} \mu_{0k}\overline{\widetilde{\mu}_{0k}} \equiv 1$, and matrix extension may be realized with the minimal number of wavelet functions in this case. Moreover, to provide symmetric/antisymmetric wavelet functions, we can modify polyphase matrices constructed in Lemma 4.3.3 as follows

$$\mathcal{N} = \begin{pmatrix} \frac{1}{\sqrt{3}} & \mu_{01} & \mu_{02} \\ 0 & 1 & 1 \\ 0 & 1 & -1 \end{pmatrix}, \quad \widetilde{\mathcal{N}} = \begin{pmatrix} \sqrt{3} & 0 & 0 \\ -\frac{\sqrt{3}}{2}(\overline{\mu}_{02} + \overline{\mu}_{01}) & \frac{1}{2} & \frac{1}{2} \\ \frac{\sqrt{3}}{2}(\overline{\mu}_{01} - \overline{\mu}_{02}) & \frac{1}{2} & -\frac{1}{2} \end{pmatrix}.$$

The corresponding wavelet functions are

$$\psi^{(1)} = \sqrt{3}\left[\varphi\left(Mx + \begin{pmatrix} 0 \\ -1 \end{pmatrix}\right) + \varphi\left(Mx + \begin{pmatrix} 0 \\ 1 \end{pmatrix}\right)\right],$$

$$\psi^{(2)} = \sqrt{3}\left[\varphi\left(Mx + \begin{pmatrix} 0 \\ -1 \end{pmatrix}\right) - \varphi\left(Mx + \begin{pmatrix} 0 \\ 1 \end{pmatrix}\right)\right].$$

The dual wavelet masks $\widetilde{m}_1, \widetilde{m}_2$ are trigonometric polynomials whose Fourier coefficients are given by the following tables

$$\frac{1}{1458\sqrt{3}} \begin{pmatrix} 0 & 0 & 0 & 0 & 0 & -4 & 0 & 0 & -7 \\ 0 & 0 & 0 & -4 & 0 & 0 & 32 & 0 & 0 \\ 0 & 0 & 0 & 0 & 13 & 0 & 0 & 28 & 0 \\ 0 & 0 & 32 & 0 & 0 & -280 & 0 & 0 & -8 \\ -7 & 0 & 0 & -280 & 729 & 0 & 40 & 0 & 0 \\ 0 & 28 & 0 & 0 & -624 & 0 & 0 & 28 & 0 \\ 0 & 0 & 40 & 0 & 729 & -280 & 0 & 0 & -7 \\ -8 & 0 & 0 & -280 & 0 & 0 & 32 & 0 & 0 \\ 0 & 28 & 0 & 0 & 13 & 0 & 0 & 0 & 0 \\ 0 & 0 & 32 & 0 & 0 & -4 & 0 & 0 & 0 \\ -7 & 0 & 0 & -4 & 0 & 0 & 0 & 0 & 0 \end{pmatrix},$$

$$\frac{1}{1458\sqrt{3}} \begin{pmatrix} 0 & 0 & 0 & 0 & 0 & -4 & 0 & 0 & -7 \\ 0 & 0 & 0 & -4 & 0 & 0 & 32 & 0 & 0 \\ 0 & 0 & 0 & 0 & 27 & 0 & 0 & 36 & 0 \\ 0 & 0 & 32 & 0 & 0 & -344 & 0 & 0 & 0 \\ -7 & 0 & 0 & -344 & -729 & 0 & 0 & 0 & 0 \\ 0 & 36 & 0 & 0 & 0 & 0 & 0 & -36 & 0 \\ 0 & 0 & 0 & 0 & 729 & 344 & 0 & 0 & 7 \\ 0 & 0 & 0 & 344 & 0 & 0 & -32 & 0 & 0 \\ 0 & -36 & 0 & 0 & -27 & 0 & 0 & 0 & 0 \\ 0 & 0 & -32 & 0 & 0 & 4 & 0 & 0 & 0 \\ 7 & 0 & 0 & 4 & 0 & 0 & 0 & 0 & 0 \end{pmatrix}.$$

The corresponding dual wavelet distributions $\widetilde{\psi}^{(1)}, \widetilde{\psi}^{(2)}$ defined by (4.4) are finite linear combinations of the M-scales and integer shifts of the δ-function.

By Theorem 4.3.6, for any $f \in S$, expansion (4.6) holds in L_2-norm and has approximation order 6; the wavelet functions $\psi^{(1)}, \psi^{(2)}$ are in C^2, $\psi^{(1)}, \widetilde{\psi}^{(1)}$ are symmetric with respect to the origin, and $\psi^{(2)}, \widetilde{\psi}^{(2)}$, are antisymmetric with respect to the origin.

3. The following example is intended to illustrate Theorems 4.3.9 and 4.3.10. Let $d = 1$, $M = 2$, φ be the B-spline of order 3 whose Fourier transform is given by

$$\widehat{\varphi}(\xi) = \left(\frac{\sin \pi\xi}{\pi\xi}\right)^4.$$

The function φ is refinable with the mask $m_0(\xi) = \cos^4 \pi\xi$; φ is in C^2 and supported on $[-2, 2]$. The polyphase components of the mask

$$\mu_{00}(\xi) = \sqrt{2}\left(\frac{1}{16}e^{2\pi i\xi} + \frac{1}{16}e^{-2\pi i\xi} + \frac{3}{8}\right), \quad \mu_{01}(\xi) = \sqrt{2}\left(\frac{1}{4}e^{-2\pi i\xi} + \frac{1}{4}\right)$$

satisfy condition (4.22) of Lemma 4.3.3 with $n = 4$ and $\lambda_0 = 1$, $\lambda_1 = 0$, $\lambda_2 = -\pi^2$, $\lambda_3 = 0$. To provide approximation order 2, we have to satisfy (4.23) from Lemma 4.3.3 with $n = 2$. This condition is satisfied for the B-spline of order 1 as $\widetilde{\varphi}$. Its Fourier transform is

$$\widehat{\widetilde{\varphi}}(\xi) = \left(\frac{\sin \pi\xi}{\pi\xi}\right)^2,$$

the function $\widetilde{\varphi}$ is continuous, supported on $[-1, 1]$ and refinable with the mask $\widetilde{m}_0(\xi) = \cos^2 \pi\xi$. The polyphase components of the mask are

$$\widetilde{\mu}_{00}(\xi) = \frac{\sqrt{2}}{2}, \quad \widetilde{\mu}_{01}(\xi) = \sqrt{2}\left(\frac{1}{4}e^{-2\pi i\xi} + \frac{1}{4}\right).$$

Next, we construct polyphase matrices $\mathcal{N}$, $\widetilde{\mathcal{N}}$ by (4.24), (4.25), which leads to the wavelet functions

$$\begin{aligned}
\psi^{(1)}(x) &= \sqrt{2}\varphi(2x+1),\\
\psi^{(2)}(x) &= \sqrt{2}\varphi(2x),\\
\widetilde{\psi}^{(1)}(x) &= \frac{1}{2\sqrt{2}}\left(-\frac{1}{2}\widetilde{\varphi}(2x-1) - \widetilde{\varphi}(2x) + 3\widetilde{\varphi}(2x+1) - \widetilde{\varphi}(2x+2) - \frac{1}{2}\widetilde{\varphi}(2x+3)\right),\\
\widetilde{\psi}^{(2)}(x) &= \frac{1}{8\sqrt{2}}\left(-\frac{1}{2}\widetilde{\varphi}(2x-3) - \widetilde{\varphi}(2x-2) - \frac{7}{2}\widetilde{\varphi}(2x-1) + 10\widetilde{\varphi}(2x) - \right.\\
&\qquad \left. \frac{7}{2}\widetilde{\varphi}(2x+1) - \widetilde{\varphi}(2x+2) - \frac{1}{2}\widetilde{\varphi}(2x+3)\right).
\end{aligned}$$

Since $\varphi, \widetilde{\varphi}$ are bounded functions, they are in $L_p(\mathbb{R})$ for any $p \in [1, \infty]$. So all assumptions of Theorems 4.3.9 and 4.3.10 are fulfilled. Thus, $\{\psi_{jk}^{(\nu)}\}$ is frame-like with approximation order 2 in $L_p(\mathbb{R}^d)$ ($L_\infty(\mathbb{R}^d) \cap L_\infty^0(\mathbb{R}^d)$ for $p = \infty$) and almost frame-like with approximation order 2 in $L_1(\mathbb{R})$.

References

1. Averbuch, A.Z., Zheludev, V.A., Cohen, T.: Interpolatory frames in signal space. IEEE Trans. Sig. Proc. **54**(6), 2126–2139 (2006)
2. Han, B., Shen, Z.: Dual wavelet frames and Riesz bases in Sobolev spaces. Constr. Approx. **29**, 369–406 (2009)
3. Ehler, M.: The multiresolution structure of pairs of dual wavelet frames for a pair of Sobolev spaces. Jaen J. Approx. **2**(2), 193–214 (2010)
4. Vladimirov, V.S.: Generalized Functions in Mathematical Physics. MIR (1979). (Translated from Russian)
5. Gel'fand, I.M. Ramanujan, M.S., Shilov G.E.: Generalized functions. Volume 1. Prop. Oper. Amer. Math. Monthly **74**(8), 1026 (1967)
6. Ron, A., Shen, Z.: Compactly supported tight affine spline frames in $L_2(R^d)$. Math. Comp. **67**, 191–207 (1998)

Chapter 5
Symmetric Wavelets

Abstract Different kinds of symmetry of wavelets are desirable in many applications, since they obey linear-phase properties. We discuss the construction of symmetric refinable masks in general settings. A variety of methods for the construction of symmetric wavelet frames and frame-like wavelet systems are also analysed.

5.1 Symmetric Refinable Masks

In the previous chapters, we gave a variety of methods for the construction of dual wavelet bases, frames, and frame-like wavelet systems in the multivariate case. We also discussed how to provide good approximation properties for the constructed wavelet systems, which are demanded in applications. But among the other desirable for applications, properties are symmetry properties of refinable and wavelet functions together with the minimum possible number of nonzero coefficients of the constructed masks.

In the previous chapters, the starting points of our constructions of wavelet systems were refinable masks. In this section, we study the general framework for the construction of symmetric refinable masks in the multivariate case. Throughout this chapter, we assume that refinable mask m_0 is a trigonometric polynomial and $m_0(\mathbf{0}) = 1$. The symmetry properties of functions are described using the notion of a symmetry group.

Definition 5.1.1 A finite set $\mathcal{H}$ of $d \times d$ integer matrices with determinant ± 1 is a *symmetry group* on $\mathbb{Z}^d$, if $\mathcal{H}$ forms a group under the matrix multiplication.

A compactly supported distribution f is called $\mathcal{H}$*-symmetric* with respect to a center $C \in \mathbb{R}^d$, if

$$\widehat{f}(\xi) = \widehat{f}(E^*\xi)e^{2\pi i(EC-C,\xi)}, \quad \forall \xi \in \mathbb{R}^d, \quad \forall E \in \mathcal{H}. \tag{5.1}$$

For a continuous compactly supported function f, condition (5.1) is equivalent to

$$f(x) = f(E(x - C) + C), \quad \forall x \in \mathbb{R}^d, \quad \forall E \in \mathcal{H}.$$

A. Krivoshein et al., *Multivariate Wavelet Frames*,
Industrial and Applied Mathematics, DOI 10.1007/978-981-10-3205-9_5

This condition means that function f is invariant to some rotations or reflections of its support around some center.

For trigonometric polynomials, it is convenient to use a bit different definition of symmetry which is compatible with the above definition of $\mathcal{H}$-symmetric functions in a sense of Lemma 5.1.2 below. A symmetry center $c \in \mathbb{R}^d$ is called *appropriate* for a symmetry group $\mathcal{H}$ on $\mathbb{Z}^d$, if $c - Ec \in \mathbb{Z}^d$ for all $E \in \mathcal{H}$. A trigonometric polynomial $t(\xi) = \sum_{k\in\mathbb{Z}^d} h_k e^{2\pi i(k,\xi)}$ is called *$\mathcal{H}$-symmetric* with respect to an appropriate center c, if

$$t(\xi) = e^{2\pi i(c-Ec,\xi)} t(E^*\xi), \qquad \forall \xi \in \mathbb{R}^d, \quad \forall E \in \mathcal{H}. \tag{5.2}$$

Note that condition (5.2) is equivalent to $h_k = h_{E(k-c)+c}$, $\forall k \in \mathbb{Z}^d$, $\forall E \in \mathcal{H}$.

The $\mathcal{H}$-symmetry of refinable mask does not always carry over to its refinable function. In fact, a dilation matrix should be appropriately chosen. A dilation matrix M is called *appropriate* for a symmetry group $\mathcal{H}$ on $\mathbb{Z}^d$, if

$$M^{-1}EM \in \mathcal{H}, \quad \forall E \in \mathcal{H}.$$

This property can be also interpreted as follows: For each $E \in \mathcal{H}$, there exists $E' \in \mathcal{H}$ such that

$$EM = ME' \quad \text{or} \quad M^{-1}E = E'M^{-1}. \tag{5.3}$$

The following Lemma states the correspondence between symmetric refinable mask and its refinable function.

Lemma 5.1.2 *Let a dilation matrix M and a symmetry center c be appropriate for a symmetry group $\mathcal{H}$. Suppose $\varphi \in S'$ is a compactly supported refinable distribution with a refinable mask m_0. Then, m_0 is $\mathcal{H}$-symmetric with respect to the center c if and only if φ is $\mathcal{H}$-symmetric with respect to the center $C \in \mathbb{R}^d$, where $C = (I_d - M)^{-1}c$.*

Proof Necessity. Suppose m_0 is $\mathcal{H}$-symmetric with respect to the center c. Since $\widehat{\varphi}(\xi) = \prod_{j=1}^{\infty} m_0(M^{*-j}\xi)$, then

$$\begin{aligned}
\widehat{\varphi}(E^*\xi) &= \prod_{j=1}^{\infty} m_0(M^{*-j}E^*\xi) = \prod_{j=1}^{\infty} m_0((M^jEM^{-j})^*M^{*-j}\xi) \\
&= \prod_{j=1}^{\infty} e^{2\pi i(M^jEM^{-j}c-c,M^{*-j}\xi)} \prod_{j=1}^{\infty} m_0(M^{*-j}\xi) \\
&= \widehat{\varphi}(\xi) \prod_{j=1}^{\infty} e^{2\pi i(EM^{-j}c-M^{-j}c,\xi)} \\
&= \widehat{\varphi}(\xi) e^{2\pi i((E-I_d)\sum_{j=1}^{\infty} M^{-j}c,\xi)} = \widehat{\varphi}(\xi) e^{2\pi i(C-EC,\xi)},
\end{aligned}$$

where $C = -\sum_{j=1}^{\infty} M^{-j}c$. Thus, φ is $\mathcal{H}$-symmetric with respect to the center C. Note that

$$C - MC = \sum_{j=1}^{\infty} M^{-j}c + \sum_{j=0}^{\infty} M^{-j}c = c \quad \text{or} \quad C = (I_d - M)^{-1}c.$$

Sufficiency. Conversely, suppose φ is $\mathcal{H}$-symmetric with respect to the center C. From the refinement equation $\widehat{\varphi}(M^*\xi) = m_0(\xi)\widehat{\varphi}(\xi)$, it follows that

$$\widehat{\varphi}(M^*E^*\xi) = m_0(E^*\xi)\widehat{\varphi}(E^*\xi) = m_0(E^*\xi)\widehat{\varphi}(\xi)e^{2\pi i(C-EC,\xi)}.$$

Since $M^{-1}EM \in \mathcal{H}$, then

$$\begin{aligned}\widehat{\varphi}(M^*E^*\xi) &= \widehat{\varphi}((M^{-1}EM)^*M^*\xi)\\ &= \widehat{\varphi}(M^*\xi)e^{2\pi i(C-M^{-1}EMC,M\xi)} = \widehat{\varphi}(M^*\xi)e^{2\pi i(MC-EMC,\xi)}.\end{aligned}$$

This implies that

$$\begin{aligned}\widehat{\varphi}(M^*\xi) &= e^{2\pi i(EMC-MC,\xi)}\widehat{\varphi}(M^*E^*\xi)\\ &= e^{2\pi i(EMC-MC,\xi)}m_0(E^*\xi)\widehat{\varphi}(\xi)e^{2\pi i(C-EC,\xi)}\\ &= e^{2\pi i(c-Ec,\xi)}m_0(E^*\xi)\widehat{\varphi}(\xi),\end{aligned}$$

since $(I_d - M)C = c$ and

$$EMC - MC + C - EC = (I_d - M)C - E(I_d - M)C = c - Ec.$$

Since φ is compactly supported and $\widehat{\varphi}(\mathbf{0}) \neq 0$, therefore $\widehat{\varphi}$ is holomorphic and $\widehat{\varphi}(\xi) \neq 0$ almost everywhere.

From the above considerations, it follows that

$$e^{2\pi i(c-Ec,\xi)}m_0(E^*\xi) = \widehat{\varphi}(M^*\xi)/\widehat{\varphi}(\xi) = m_0(\xi), \quad \text{a.e. on } \mathbb{R}^d.$$

Since m_0 is a trigonometric polynomial, then m_0 is $\mathcal{H}$-symmetric with respect to the center c.$\Diamond$

Before we proceed further, let us introduce several examples of symmetry groups. Assume that $\mathcal{H} = \{I_d, -I_d\}$ and $c \in \mathbb{Z}^d$. When $c = \mathbf{0}$, the $\mathcal{H}$-symmetry of a continuous function f means that the function is even. For trigonometric polynomial $t(\xi) = \sum_{k\in\mathbb{Z}^d} h_k e^{2\pi i(k,\xi)}$, the $\mathcal{H}$-symmetry condition (5.2) reduces to

$$t(\xi) = e^{2\pi i(2c,\xi)}t(-\xi) \quad \text{or} \quad h_k = h_{2c-k}, \ \forall k \in \mathbb{Z}^d. \tag{5.4}$$

Such trigonometric polynomial t is called *symmetric with respect to the point* c (or just point symmetric). In this case, any dilation matrix M is appropriate for $\mathcal{H}$.

Point symmetry is closely related to the notion of linear-phase moments of trigonometric polynomial. We say that a trigonometric polynomial t has *linear-phase moments of order* n with phase $c \in \mathbb{R}^d$ if

$$D^{\beta}t(\mathbf{0}) = D^{\beta}e^{2\pi i(c,\xi)}\Big|_{\xi=\mathbf{0}} = (2\pi i)^{[\beta]}c^{\beta}, \quad \forall \beta \in \mathbb{Z}_+^d, [\beta] < n.$$

The linear-phase moments play important role in the setting of polynomial reproduction and subdivision schemes. If trigonometric polynomial t is point symmetric with respect to c, then

$$D^{e_j}t(\mathbf{0}) = D^{e_j}\left(e^{2\pi i(2c,\xi)}t(-\xi)\right)\Big|_{\xi=\mathbf{0}} = 2\pi i2(c)_j - D^{e_j}t(\mathbf{0}),$$

where for $j = 1, \dots, d$, vectors $e_j \in \mathbb{Z}^d$ are the standard basis for $\mathbb{R}^d$. Therefore, $D^{e_j}t(\mathbf{0}) = 2\pi i c^{e_j}$, $j = 1, \dots, d$, i.e., point symmetric trigonometric polynomial has linear-phase moments at least of order 2 and the phase must match with the symmetry center c.

A symmetry group defined by

$$\mathcal{H}^{axis} := \left\{\mathtt{diag}(u_1, \dots, u_d) : u_j = \pm 1, j = 1, \dots, d\right\}$$

is called *the axial symmetry group on* $\mathbb{Z}^d$. It is easy to see that $\mathcal{H}^{axis}$ is an abelian group, $\#\mathcal{H}^{axis} = 2^d$ and for all $E \in \mathcal{H}^{axis}$ we have $E^2 = I_d$ and $E = E^*$. The set of all appropriate dilation matrices in this case is given in the following Lemma Axial symmetry group.

Lemma 5.1.3 *A dilation matrix M is appropriate for $\mathcal{H}^{axis}$ if and only if $M = \mathtt{diag}(m_1, \dots, m_d)P$, where $m_1, \dots, m_d \in \mathbb{Z}$ and P is a $d \times d$ permutation matrix (i.e., a square matrix that has exactly one entry equal to 1 in each row and each column and zeros elsewhere).*

Proof Necessity. Assume that M is appropriate for $\mathcal{H}^{axis}$. For $j = 1, \dots, d$, matrix $I_d - 2e_je_j^T$ is in $\mathcal{H}^{axis}$, where e_j is the unit jth coordinate column vector in $\mathbb{R}^d$. Then, matrix $M(I_d - 2e_je_j^T)M^{-1} = I_d - 2(Me_j)(e_j^TM^{-1})$ is also in $\mathcal{H}^{axis}$. Since all elements in $\mathcal{H}^{axis}$ are diagonal, then matrix $(Me_j)(e_j^TM^{-1})$ also should be a diagonal matrix. This is possible only if vectors Me_j and $e_j^TM^{-1}$ have exactly one nonzero entry for any $j = 1, \dots, d$. Hence, there exists a $d \times d$ permutation matrix P such that $M = \mathtt{diag}(m_1, \dots, m_d)P$, where $m_1, \dots, m_d \in \mathbb{Z}$.

Sufficiency. Let $M = \mathtt{diag}(m_1, \dots, m_d)P$, where $m_1, \dots, m_d \in \mathbb{Z}$ and P is a $d \times d$ permutation matrix. Then for any $E \in \mathcal{H}^{axis}$, $MEM^{-1} = \mathtt{diag}(m_1, \dots, m_d)$ $PEP^{-1}\mathtt{diag}(1/m_1, \dots, 1/m_d) \in \mathcal{H}^{axis}$, since PEP^{-1} is again a diagonal matrix where the diagonal elements are the same as on the diagonal of E but possibly permuted.$\diamondsuit$

A symmetry group defined by

$$\mathcal{H} := \left\{\pm I_2, \pm\begin{pmatrix}-1 & 0\\ 0 & 1\end{pmatrix}, \pm\begin{pmatrix}0 & 1\\ -1 & 0\end{pmatrix}, \pm\begin{pmatrix}0 & 1\\ 1 & 0\end{pmatrix}\right\}$$

is called the *fourfold (or full) symmetry group on* $\mathbb{Z}^2$.

A symmetry group defined by

$$\mathcal{H} = \left\{ \pm I_2, \pm \begin{pmatrix} 0 & 1 \\ 1 & 0 \end{pmatrix}, \pm \begin{pmatrix} 1 & 0 \\ 1 & -1 \end{pmatrix}, \right.$$
$$\left. \pm \begin{pmatrix} 1 & -1 \\ 1 & 0 \end{pmatrix}, \pm \begin{pmatrix} 0 & 1 \\ -1 & 1 \end{pmatrix}, \pm \begin{pmatrix} -1 & 1 \\ 0 & 1 \end{pmatrix} \right\}$$

is called the *sixfold (or hexagonal) symmetry group on* $\mathbb{Z}^2$. The following group

$$\mathcal{H} = \left\{ I_2, \begin{pmatrix} 0 & -1 \\ 1 & -1 \end{pmatrix}, \begin{pmatrix} -1 & 1 \\ -1 & 0 \end{pmatrix} \right\}$$

is called the *hexagonal abelian group symmetry group on* $\mathbb{Z}^2$.

Further, we need some results from the theory of finite groups. Let $\mathcal{H}$ be a finite group (with a binary operation "$\cdot$") and let Ω be a finite set. A group action of $\mathcal{H}$ on Ω is a map $\chi : \mathcal{H} \times \Omega \longrightarrow \Omega$ such that for any element $\omega \in \Omega$ the following conditions hold

1. the identity, i.e., $\chi(I, \omega) = \omega$, where I is the identity element of $\mathcal{H}$;
2. the associativity, i.e., $\chi(E, \chi(\widetilde{E}, \omega)) = \chi(E \cdot \widetilde{E}, \omega)$ for all $E, \widetilde{E} \in \mathcal{H}$.

A finite set Ω is called an $\mathcal{H}$-space, if there exists a group action χ of $\mathcal{H}$ on Ω. In fact, a group action of $\mathcal{H}$ on Ω permutes the elements of Ω.

Suppose Ω is an $\mathcal{H}$-space, χ is a group action of $\mathcal{H}$ on Ω. For $E \in \mathcal{H}$ and $\omega \in \Omega$, denote $\chi(E, \omega)$ by $E\omega$, for convenience. Fix $\omega \in \Omega$. The *orbit* of ω is a subset of Ω defined by $\mathcal{H}\omega := \{E\omega, E \in \mathcal{H}\}$. It is not hard to see that two orbits are either equal or disjoint. Thus, Ω can be represented as the union of disjoint orbits.

Lemma 5.1.4 *Suppose Ω is an $\mathcal{H}$-space. Then, there exists a set $\Lambda \subseteq \Omega$ such that $\Omega = \bigcup_{\omega \in \Lambda} \mathcal{H}\omega$ and the orbits $\mathcal{H}\omega$, $\omega \in \Lambda$, are mutually disjoint.*

For $\omega \in \Omega$, the *stabilizer* of ω is a subgroup $\mathcal{H}_\omega$ of $\mathcal{H}$ defined by $\mathcal{H}_\omega := \{E \in \mathcal{H} : E\omega = \omega\}$. $\mathcal{H}/\mathcal{H}_\omega$ is the quotient group of $\mathcal{H}$ modulo $\mathcal{H}_\omega$. Denote by $\mathcal{E}_\omega$ a complete set of representatives of the cosets $\mathcal{H}/\mathcal{H}_\omega$.

Lemma 5.1.5 *Suppose Ω is an $\mathcal{H}$-space, $\omega \in \Omega$ and $\mathcal{E}_\omega$ is a complete set of representatives of the cosets $\mathcal{H}/\mathcal{H}_\omega$. Then, $E^{-1}\widetilde{E} \notin \mathcal{H}_\omega$ for any $E, \widetilde{E} \in \mathcal{E}_\omega$ such that $E \neq \widetilde{E}$ and for any $K \in \mathcal{H}$ there exists a unique pair of elements $F \in \mathcal{H}_\omega$ and $E \in \mathcal{E}_\omega$ such that $K = EF$.*

Note that the orbit of $\omega \in \Omega$ can be represented as $\mathcal{H}\omega = \bigcup_{E \in \mathcal{E}_\omega} \{E\omega\}$. Indeed, obviously, $\bigcup_{E \in \mathcal{E}_\omega} \{E\omega\} \subseteq \mathcal{H}\omega$. Also, $\mathcal{H}\omega \subseteq \bigcup_{E \in \mathcal{E}_\omega} \{E\omega\}$, since any $\eta \in \mathcal{H}\omega$ is given by $\eta = K\omega$ for some $K \in \mathcal{H}$. Therefore, by Lemma 5.1.5 $\eta = EF\omega = E\omega$ for some $E \in \mathcal{E}_\omega$, $F \in \mathcal{H}_\omega$. Hence, Ω can be decomposed as

$$\Omega = \bigcup_{\omega \in \Lambda} \bigcup_{E \in \mathcal{E}_\omega} \{E\omega\}, \tag{5.5}$$

where Λ is defined in Lemma 5.1.4. Also note that $\mathcal{E}_\omega$ and $\mathcal{H}_\omega$ are subsets of $\mathcal{H}$, $\#\mathcal{E}_\omega = \#\mathcal{H}\omega$.

Suppose $\omega \in \Omega$, $\eta \in \mathcal{H}\omega$, i.e., there exists $K \in \mathcal{H}$ such that $\eta = K\omega$. Then, the stabilizer of η and the stabilizer of ω are conjugate subgroups, namely $K\mathcal{H}_\omega K^{-1} = \mathcal{H}_\eta$.

Now, we study how a symmetry group acts on a set of digits of an appropriate dilation matrix. This information will help to characterize a symmetric mask in terms of its polyphase components. Let $\mathcal{H}$ be a symmetry group on $\mathbb{Z}^d$, M be an appropriate for $\mathcal{H}$ dilation matrix, c be an appropriate for $\mathcal{H}$ symmetry center. By Lemma 2.1.2 any $\alpha \in \mathbb{Z}^d$ can be uniquely represented as $\alpha = M\beta + s$, where $\beta \in \mathbb{Z}^d$, $s \in D(M)$. This fact yields that for each digit $s \in D(M)$ and matrix $E \in \mathcal{H}$ there exist a unique digit $q \in D(M)$ and a unique vector $r_s^E \in \mathbb{Z}^d$ such that

$$Es = Mr_s^E + q + Ec - c. \tag{5.6}$$

The indices of r_s^E mean that vector r_s^E depends on digit s and matrix E.

The coset corresponding to digit $s \in D(M)$ we denote by $\langle s\rangle$, i.e., $\langle s\rangle = \{M\beta + s, \beta \in \mathbb{Z}^d\} = M\mathbb{Z}^d + s$. Let $\mathcal{D} := \{\langle s\rangle, s \in D(M)\}$. In other words, $\mathcal{D}$ is the collection of cosets $\mathbb{Z}^d/M\mathbb{Z}^d$. Define a group action χ from the set $\mathcal{H} \times \mathcal{D}$ to $\mathcal{D}$ by

$$\chi(E, \langle s\rangle) = E\langle s\rangle := \{EM\beta + Es + c - Ec, \beta \in \mathbb{Z}^d\}, \quad E \in H, \quad s \in D(M). \tag{5.7}$$

By (5.3) and (5.6), it is easy to see that $E\langle s\rangle$ is an element of $\mathcal{D}$, i.e., $\exists q \in D(M)$ such that $E\langle s\rangle = \langle q\rangle$.

Proposition 5.1.6 *The set $\mathcal{D}$ is an $\mathcal{H}$-space, where the group action is defined by (5.7).*

Proof Proposition 5.1.6 can be proved by direct computations. Let us fix $E \in \mathcal{H}$ and show that $E\langle s\rangle \in \mathcal{D}$ for all $s \in D(M)$. Due to (5.6) there always exists a unique digit $q \in D(M)$ such that $Es + c - Ec = M\gamma + q$, $\gamma \in \mathbb{Z}^d$. Therefore,

$$E\langle s\rangle = \{EM\beta + Es + c - Ec, \beta \in \mathbb{Z}^d\} = \{M(E'\beta + \gamma) + q, \beta \in \mathbb{Z}^d\} = \langle q\rangle,$$

where E' is such that $EM = ME'$. The identity condition in the definition of group action is obviously valid. It remains to show the associativity. Suppose $E, \widetilde{E} \in \mathcal{H}$. Then

$$\widetilde{E}E\langle s\rangle = \{\widetilde{E}EM\beta + \widetilde{E}Es + c - \widetilde{E}Ec, \beta \in \mathbb{Z}^d\} = \{M\widetilde{E}'E'\beta + \widetilde{E}Es + c - \widetilde{E}Ec, \beta \in \mathbb{Z}^d\},$$

where E', $\widetilde{E}'$ are such that $EM = ME'$ and $\widetilde{E}M = M\widetilde{E}'$. On the other hand, $\widetilde{E}(E\langle s\rangle) = \widetilde{E}\langle q\rangle = \{\widetilde{E}M\beta + \widetilde{E}q + c - \widetilde{E}c, \beta \in \mathbb{Z}^d\}$. Since $q = Es + c - Ec - M\gamma$, we see that

$$\begin{aligned}\widetilde{E}(E\langle s\rangle) &= \{\widetilde{E}M\beta + \widetilde{E}(Es + c - Ec - M\gamma) + c - \widetilde{E}c, \beta \in \mathbb{Z}^d\} \\ &= \{M(\widetilde{E}'\beta - \gamma) + \widetilde{E}Es + c - \widetilde{E}Ec, \beta \in \mathbb{Z}^d\}.\end{aligned}$$

Therefore, $\widetilde{E}(E\langle s\rangle) = \widetilde{E}E\langle s\rangle$.$\Diamond$

Since $\mathcal{D}$ is an $\mathcal{H}$-space, Lemmas 5.1.4 and 5.1.5 are valid for the set $\mathcal{D}$. Now we introduce suitable notations that will be used throughout the chapter. These notations are illustrated by Example 5.1.7 below.

- $\mathcal{H}\langle s\rangle = \{E\langle s\rangle, E \in \mathcal{H}\}$ is the *orbit* of the coset $\langle s\rangle \in \mathcal{D}$. In fact, $\mathcal{H}\langle s\rangle \subseteq \mathcal{D}$. Two orbits are either equal or disjoint.
- A set $\Lambda \subseteq \mathcal{D}$ contains representatives from each orbit. Therefore, $\mathcal{D} = \cup_{\langle s\rangle \in \Lambda}\mathcal{H}\langle s\rangle$ by Lemma 5.1.4.
- For convenience, redenote the elements of the set Λ by $\langle s_{p,0}\rangle$, where $p = 0, \ldots, \#\Lambda - 1$.
- $\mathcal{H}_{p,0} := \mathcal{H}_{\langle s_{p,0}\rangle} = \{F \in \mathcal{H} : F\langle s_{p,0}\rangle = \langle s_{p,0}\rangle\}$ is the *stabilizer* of $\langle s_{p,0}\rangle$; $\mathcal{H}_{p,0} \subseteq \mathcal{H}$. $\mathcal{H}_{p,0}$ is a subgroup of $\mathcal{H}$.
- A set $\mathcal{E}_p := \mathcal{E}_{\langle s_{p,0}\rangle}$ contains a complete set of representatives of $\mathcal{H}/\mathcal{H}_{p,0}$; $\mathcal{E}_p \subseteq \mathcal{H}$. The orbit $\mathcal{H}\langle s_{p,0}\rangle$ can be represented as

$$\mathcal{H}\langle s_{p,0}\rangle = \bigcup_{E \in \mathcal{E}_p} \{E\langle s_{p,0}\rangle\}.$$

- The elements of the orbit $\mathcal{H}\langle s_{p,0}\rangle$ we denote by $\langle s_{p,i}\rangle$, $i = 0, \ldots, \#\mathcal{E}_p - 1$.
- For a fixed index p, the matrices of the set $\mathcal{E}_p$ we denote by $E^{(i)}$ such that $E^{(i)}\langle s_{p,0}\rangle = \langle s_{p,i}\rangle$, $i = 0, \ldots, \#\mathcal{E}_p - 1$. Note that $E^{(0)} = I_d$.
- The digit corresponding to the coset $\langle s_{p,i}\rangle$ we denote by $s_{p,i}$.

By Lemma 5.1.5 for a fixed p, $p = 0, \ldots, \#\Lambda - 1$, $\#\mathcal{H} = \#\mathcal{H}_{p,0} \cdot \#\mathcal{E}_p$ and symmetry group $\mathcal{H}$ can be uniquely represented as follows $\mathcal{H} = \mathcal{E}_p \times \mathcal{H}_{p,0}$, i.e., for each matrix $\widetilde{E} \in \mathcal{H}$ there exist matrices $E \in \mathcal{E}_p$ and $F \in \mathcal{H}_{p,0}$ such that $\widetilde{E} = EF$. These sets $\mathcal{E}_p$, $\mathcal{H}_{p,1}$ can be considered as the "*coordinate axes*" of the symmetry group $\mathcal{H}$. For each p, $p = 0, \ldots, \#\Lambda - 1$, these "coordinate axes" of $\mathcal{H}$ can be different. The stabilizers $\mathcal{H}_{p,0}$ and $\mathcal{H}_{p,i}$ are conjugate subgroups, namely $E^{(i)}\mathcal{H}_{p,0}(E^{(i)})^{-1} = \mathcal{H}_{p,i}$, $E^{(i)} \in \mathcal{E}_p$.

Now we point out some features of the choice of digits and indicate how the matrices from the symmetry group $\mathcal{H}$ act on the digits. This information helps to characterize the inner structure of $\mathcal{H}$-symmetric masks. Let us fix a digit $s_{p,0} \in D(M)$ and a matrix $E^{(i)} \in \mathcal{E}_p$. Since $E^{(i)}\langle s_{p,0}\rangle = \langle s_{p,i}\rangle$, by (5.6) we get $E^{(i)}s_{p,0} + c - E^{(i)}c = Mr_{p,0}^{E^{(i)}} + s_{p,i}$, where $r_{p,0}^{E^{(i)}} \in \mathbb{Z}^d$. Let us rechoose the digits $s_{p,i}$ such that $r_{p,0}^{E^{(i)}} = 0$, i.e.,

$$E^{(i)}s_{p,0} + c - E^{(i)}c =: s_{p,i}, \quad i = 1, \ldots, \#\mathcal{E}_p - 1. \tag{5.8}$$

Throughout the text, we assume that the digits $s_{p,i}$ are chosen in that way.

Suppose F is a matrix from the stabilizer $\mathcal{H}_{p,0}$. Then, $F\langle s_{p,0}\rangle = \langle s_{p,0}\rangle$. Hence, by (5.6)

$$Fs_{p,0} = Mr^F_{p,0} + s_{p,0} + Fc - c, \tag{5.9}$$

where $r^F_{p,0} \in \mathbb{Z}^d$. Notice that $r^F_{p,0} = M^{-1}(c - s_{p,0}) - M^{-1}F(c - s_{p,0})$.

For a matrix $K \in \mathcal{H}$, there exist matrices $E^{(i)} \in \mathcal{E}_p$ and $F \in \mathcal{H}_{p,0}$ such that $K = E^{(i)}F$. Therefore, $K\langle s_{p,0}\rangle = E^{(i)}F\langle s_{p,0}\rangle = \langle s_{p,i}\rangle$. Together with (5.8) and (5.9) we obtain that

$$Ks_{p,0} = E^{(i)}Fs_{p,0} = E^{(i)}(Mr^F_{p,0} + s_{p,0} + Fc - c) = Mr^K_{p,0} + s_{p,i} + Kc - c,$$

where $r^K_{p,0} = M^{-1}E^{(i)}Mr^F_{p,0}$.

Analogously, we can represent $Ks_{p,i}$. Note that $K\langle s_{p,i}\rangle = KE^{(i)}\langle s_{p,0}\rangle$. There exist matrices $E^{(j(p,i,K))} \in \mathcal{E}_p$ and $F \in \mathcal{H}_{p,0}$ such that $KE^{(i)} = E^{(j(p,i,K))}F$. Therefore,

$$K\langle s_{p,i}\rangle = E^{(j(p,i,K))}F\langle s_{p,0}\rangle = \langle s_{p,j(p,i,K)}\rangle. \tag{5.10}$$

Here the notation $j(p,\cdot,K)$ means the map from the set of indices $\{0,\dots,\#\mathcal{E}_p - 1\}$ to itself and the index $j(p,i,K)$ is uniquely defined by the index i and the matrix $K \in \mathcal{H}$ for each p. Thus, together with (5.8) and (5.9) we obtain

$$\begin{aligned} Ks_{p,i} &= KE^{(i)}s_{p,0} + Kc - KE^{(i)}c \\ &= E^{(j)}Mr^F_{p,0} + E^{(j)}s_{p,1} + E^{(j)}Fc - E^{(j)}c + Kc - KE^{(i)}c \\ &= E^{(j)}Mr^F_{p,0} + s_{p,j} + E^{(j)}Fc - c + Kc - KE^{(i)}c \\ &= Mr^K_{p,i} + s_{p,j} + Kc - c, \end{aligned} \tag{5.11}$$

where $r^K_{p,i} = M^{-1}E^{(j)}Mr^F_{p,0}$ and $j = j(p,i,K)$.

Example 5.1.7 Suppose $d = 2$, $M = 2I_2$, $c = \binom{0}{0}$ and $\mathcal{H}$ is the full symmetry group on $\mathbb{Z}^2$

$$\mathcal{H} = \mathcal{H}^{full} := \left\{\pm I_2, \pm\begin{pmatrix}-1 & 0\\ 0 & 1\end{pmatrix}, \pm\begin{pmatrix}0 & 1\\ -1 & 0\end{pmatrix}, \pm\begin{pmatrix}0 & 1\\ 1 & 0\end{pmatrix}\right\}.$$

The set of digits is $D(M) = \{\binom{0}{0}, \binom{0}{1}, \binom{1}{0}, \binom{1}{1}\}$. The set of cosets is defined as $\mathcal{D} := \{\langle\binom{0}{0}\rangle, \langle\binom{1}{0}\rangle, \langle\binom{0}{1}\rangle, \langle\binom{1}{1}\rangle\}$. The set of cosets is split into three disjoint orbits

$$\mathcal{H}\langle\tbinom{0}{0}\rangle = \{\langle\tbinom{0}{0}\rangle\},\ \mathcal{H}\langle\tbinom{1}{1}\rangle = \{\langle\tbinom{1}{1}\rangle\},\ \mathcal{H}\langle\tbinom{1}{0}\rangle = \{\langle\tbinom{1}{0}\rangle, \langle\tbinom{0}{1}\rangle\}.$$

Therefore, $\Lambda = \{\langle\binom{0}{0}\rangle, \langle\binom{1}{1}\rangle, \langle\binom{1}{0}\rangle\}$. Let us reindex the digits according to the above considerations: $s_{0,0} = \binom{0}{0}$, $s_{1,0} = \binom{1}{1}$, $s_{2,0} = \binom{1}{0}$, $s_{2,1} = \binom{0}{1}$. The sets $\mathcal{E}_p$, $\mathcal{H}_{p,0}$ are

$$\mathcal{H}_{0,0} = \mathcal{H}^{full},\ \mathcal{E}_0 = \{I_2\},\ \mathcal{H}_{1,0} = \mathcal{H}^{full},\ \mathcal{E}_1 = \{I_2\},$$

$$\mathcal{H}_{2,0} = \left\{ \pm I_2, \pm \begin{pmatrix} 1 & 0 \\ 0 & -1 \end{pmatrix} \right\}, \quad \mathcal{E}_2 = \left\{ I_2, \begin{pmatrix} 0 & 1 \\ 1 & 0 \end{pmatrix} \right\} = \{E^{(1)}, E^{(2)}\}.$$

In what follows in the chapter, we will use two types of notations for the polyphase components of an $\mathcal{H}$-symmetric trigonometric polynomial $t(\xi)$. The first type is with one index as it was defined in Sect. 2.5: $\tau_k(\xi)$. The second type is with double index. Namely, the polyphase components of an $\mathcal{H}$-symmetric trigonometric polynomial $t(\xi)$ we enumerate as the corresponding digits using double index: $\tau_{p,i}(\xi)$, $i = 0, \ldots, \#\mathcal{E}_p - 1$, $p = 0, \ldots, \#\Lambda - 1$.

Now we reformulate the $\mathcal{H}$-symmetry conditions for a trigonometric polynomial (5.2) in terms of its polyphase components.

Lemma 5.1.8 *Let a dilation matrix M and a symmetry center c be appropriate for a symmetry group $\mathcal{H}$. A trigonometric polynomial t is $\mathcal{H}$-symmetric with respect to c if and only if for each $p \in \{0, \ldots, \#\Lambda - 1\}$ its polyphase components $\tau_{p,i}$, $i = 0, \ldots, \#\mathcal{E}_p - 1$, satisfy*

$$\tau_{p,i}((M^{-1}KM)^*\xi)e^{2\pi i (r^K_{p,i}, \xi)} = \tau_{p,j}(\xi), \quad \forall K \in \mathcal{H} \tag{5.12}$$

where index $j = j(p, i, K)$ is defined as above, i.e., $KE^{(i)} = E^{(j)}F$ with $E^{(i)}, E^{(j)} \in \mathcal{E}_p$, $F \in \mathcal{H}_{p,0}$, vector $r^K_{p,i} = M^{-1}E^{(j)}Mr^F_{p,0}$.

Proof Necessity. Suppose t is an $\mathcal{H}$-symmetric with respect to the center c trigonometric polynomial. Then, using the polyphase representation of t together with (5.2) and (5.11) we obtain that for any $K \in \mathcal{H}$

$$\begin{aligned} t(K^*\xi) &= \frac{1}{\sqrt{m}} \sum_{p=0}^{\#\Lambda-1} \sum_{i=0}^{\#\mathcal{E}_p-1} e^{2\pi i (Ks_{p,i}, \xi)} \tau_{p,i}(M^*K^*\xi) \\ &= \frac{e^{2\pi i (Kc-c, \xi)}}{\sqrt{m}} \sum_{p=0}^{\#\Lambda-1} \sum_{i=0}^{\#\mathcal{E}_p-1} e^{2\pi i (s_{p,j}, \xi)} \tau_{p,i}(M^*K^*\xi) e^{2\pi i (Mr^K_{p,i}, \xi)}, \\ e^{2\pi i (Kc-c, \xi)} t(\xi) &= \frac{e^{2\pi i (Kc-c, \xi)}}{\sqrt{m}} \sum_{p=0}^{\#\Lambda-1} \sum_{i=0}^{\#\mathcal{E}_p-1} e^{2\pi i (s_{p,j}, \xi)} \tau_{p,j}(M^*\xi), \end{aligned}$$

where $j = j(p, i, K)$ is defined as above, i.e., for each $p \in \{0, \ldots, \#\Lambda - 1\}$ there exist matrices $E^{(j(p,i,K))} \in \mathcal{E}_p$ and $F \in \mathcal{H}_{p,0}$ such that $KE^{(i)} = E^{(j(p,i,K))}F$. Since the polyphase representation is unique with respect to the chosen digits, we get (5.12). The converse statement can be checked by the similar direct computations.◊

Condition (5.12) has much more simple form when matrix K is from the "coordinate axis" of $\mathcal{H}$, i.e., from $\mathcal{H}_{p,0}$ and $\mathcal{E}_p$. Namely, if $F \in \mathcal{H}_{p,0}$, then from (5.12) with $K = F$ it follows that

$$\tau_{p,0}(\xi) = e^{2\pi i (r_{p,0}^F, \xi)} \tau_{p,0}((M^{-1}FM)^*\xi), \tag{5.13}$$

Thus, $\tau_{p,0}(\xi)$ should be $M^{-1}\mathcal{H}_{p,0}M-$symmetric with respect to the center $M^{-1}(c - s_{p,0})$. If $E^{(i)} \in \mathcal{E}_p$, $i = 0, \ldots, \#\mathcal{E}_p - 1$, then from (5.12) with $K = E^{(i)}$ it follows that

$$\tau_{p,i}(\xi) = \tau_{p,0}((M^{-1}E^{(i)}M)^*\xi). \tag{5.14}$$

Conversely, if conditions (5.13) and (5.14) are valid for all $i = 0, \ldots, \#\mathcal{E}_p - 1$, $p = 0, \ldots, \#\Lambda - 1$, then condition (5.12) is valid for all $i = 0, \ldots, \#\mathcal{E}_p - 1$, $p = 0, \ldots, \#\Lambda - 1$. Indeed, for fixed indices p, i and a matrix $K \in \mathcal{H}$ and for index j such that $KE^{(i)} = E^{(j)}F$ with $E^{(i)}, E^{(j)} \in \mathcal{E}_p$, $F \in \mathcal{H}_{p,0}$. Thus, (5.13) and (5.14) yield that

$$\begin{aligned}
\tau_{p,i}((M^{-1}KM)^*\xi) &= \tau_{p,0}(M^*E^{(i)*}K^*M^{*-1}\xi) = \tau_{p,0}(M^*F^*E^{(j)*}M^{*-1}\xi) \\
&= \tau_{p,0}(M^*E^{(j)*}M^{*-1}\xi)e^{-2\pi i (r_{p,0}^F, M^*E^{(j)*}M^{*-1}\xi)} \\
&= \tau_{p,j}(\xi)e^{\;2\pi i (r_{p,i}^K, \xi)},
\end{aligned}$$

where $r_{p,i}^K = M^{-1}E^{(j)}Mr_{p,0}^F$.

Notice that conditions (5.13) and (5.14) imply that the polyphase component $\tau_{p,i}$ is $M^{-1}\mathcal{H}_{p,i}M-$symmetric with respect to the center $M^{-1}(c - s_{p,i})$. It can be checked by direct computations using the fact that $\mathcal{H}_{p,0}$ and $\mathcal{H}_{p,i}$ are conjugate subgroups. Indeed, since the group $\mathcal{H}_{p,i}$ is conjugate to $\mathcal{H}_{p,0}$, i.e., $\mathcal{H}_{p,i} = E^{(i)}\mathcal{H}_{p,0}(E^{(i)})^{-1}$, for any $K \in \mathcal{H}_{p,i}$ there exists $F \in \mathcal{H}_{p,0}$ such that $K = E^{(i)}F(E^{(i)})^{-1}$. Let us show that $\tau_{p,i}((M^{-1}KM)^*\xi)e^{2\pi i (r_{p,i}^K, \xi)} = \tau_{p,i}(\xi)$, where $r_{p,i}^K = M^{-1}E^{(j)}Mr_{p,0}^F$. By (5.13) and (5.14) we obtain

$$\begin{aligned}
\tau_{p,i}((M^{-1}KM)^*\xi)e^{2\pi i (r_{p,i}^K, \xi)} &= \tau_{p,0}(M^*F^*E^{(i)*}M^{*-1}\xi)e^{2\pi i (r_{p,i}^K, \xi)} \\
&= \tau_{p,0}(M^{-1}E^{(i)}M)^*\xi)e^{-2\pi i (r_{p,i}^K - M^{-1}E^{(i)}Mr_{p,0}^F, \xi)} \\
&= \tau_{p,i}(\xi).
\end{aligned}$$

This establishes the $M^{-1}\mathcal{H}_{p,i}M-$symmetry with respect to the center $M^{-1}(c - s_{p,i})$, since

$$\begin{aligned}
r_{p,i}^K &= M^{-1}E^{(i)}Mr_{p,0}^F \\
&= M^{-1}E^{(i)}(c - s_{p,0} - F(c - s_{p,0})) \\
&= M^{-1}(c - s_{p,i}) - (M^{-1}KM)M^{-1}(c - s_{p,i}).
\end{aligned}$$

since $E^{(i)}(c - s_{p,0}) = c - s_{p,i}$ by (5.8) and $KE^{(i)} = E^{(i)}F$.

Next, we move to the construction of symmetric refinable masks. But firstly, we state two simple utility lemmas.

Lemma 5.1.9 *Let $t(\xi)$ be a trigonometric polynomial, $n \in \mathbb{N}$, $v \in \mathbb{R}^d$, and $t'(\xi) := e^{-2\pi i(v,\xi)}t(\xi)$. Then $D^\beta t'(\mathbf{0}) = (2\pi i)^{[\beta]}\kappa'_\beta$, $\forall \beta \in \mathbb{Z}^d_+$, $[\beta] < n$, if and only if $D^\beta t(\mathbf{0}) = (2\pi i)^{[\beta]} \sum_{\alpha \le \beta} \kappa'_\alpha \binom{\beta}{\alpha} v^{\beta-\alpha}$, $\forall \beta \in \mathbb{Z}^d_+$, $[\beta] < n$.*

Proof Let $D^\beta t'(\mathbf{0}) = (2\pi i)^{|\beta|}\kappa'_\beta$, $\forall \beta \in \mathbb{Z}^d_+$, $[\beta] < n$. Then, the statement follows from the Leibniz differentiation rule. Conversely, let us denote $\kappa_\beta := \sum_{\alpha \le \beta} \kappa'_\alpha \binom{\beta}{\alpha} v^{\beta-\alpha}$, $\forall \beta \in \mathbb{Z}^d_+$, $[\beta] < n$. So $D^\beta t(\mathbf{0}) = (2\pi i)^{[\beta]}\kappa_\beta$. Then, by the Leibniz differentiation rule

$$\begin{aligned} D^\beta t'(\mathbf{0}) &= (2\pi i)^{[\beta]} \sum_{\alpha \le \beta} (-v)^{\beta-\alpha} \binom{\beta}{\alpha} \kappa_\alpha \\ &= (2\pi i)^{[\beta]} \sum_{\alpha \le \beta} (-v)^{\beta-\alpha} \binom{\beta}{\alpha} \sum_{\gamma \le \alpha} \kappa'_\gamma \binom{\alpha}{\gamma} v^{\alpha-\gamma}, \quad \forall \beta \in \mathbb{Z}^d_+, [\beta] < n. \end{aligned}$$

After changing the order of summation and taking into account that $\binom{\alpha}{\gamma}\binom{\beta}{\alpha} = \binom{\beta}{\gamma}\binom{\beta-\gamma}{\alpha-\gamma}$ it follows that for all $\forall \beta \in \mathbb{Z}^d_+$, $[\beta] < n$

$$\begin{aligned} D^\beta t'(\mathbf{0}) &= (2\pi i)^{[\beta]} \sum_{\gamma \le \beta} \kappa'_\gamma \sum_{\alpha \in \mathbb{Z}^d_+,\, \gamma \le \alpha \le \beta} (-v)^{\beta-\alpha} \binom{\beta}{\alpha}\binom{\alpha}{\gamma} v^{\alpha-\gamma} \\ &= (2\pi i)^{[\beta]} \sum_{\gamma \le \beta} \kappa'_\gamma \binom{\beta}{\gamma} \sum_{\alpha \in \mathbb{Z}^d_+,\, \gamma \le \alpha \le \beta} (-v)^{\beta-\alpha} \binom{\beta-\gamma}{\alpha-\gamma} v^{\alpha-\gamma} \\ &= (2\pi i)^{[\beta]} \sum_{\gamma \le \beta} \kappa'_\gamma \binom{\beta}{\gamma} (v-v)^{\beta-\gamma} = (2\pi i)^{[\beta]}\kappa'_\beta. \Diamond \end{aligned}$$

The construction of mask m_0 will be carried out using its polyphase components μ_{0k}, $k = 0, \ldots, m-1$. So, we will frequently use the sum rule condition (3.22) in the polyphase form (3.20), i.e., for all $k = 0, \ldots, m-1$,

$$D^\beta \mu_{0k}(\mathbf{0}) = \frac{(2\pi i)^{[\beta]}}{\sqrt{m}} \sum_{\gamma \le \beta} \lambda_\gamma \binom{\beta}{\gamma} (-M^{-1}s_k)^{\beta-\gamma}, \ \forall \beta \in \mathbb{Z}^d_+, [\beta] < n, \tag{5.15}$$

for some complex numbers λ_γ, $\gamma \in \mathbb{Z}^d_+$, $[\gamma] < n$. Following traditional terminology, condition (5.15) will be referred as the *sum rule of order n* in the sequel. And additionally, we need the following notion that will be useful for the construction of $\mathcal{H}$-symmetric masks.

Definition 5.1.10 Suppose $n \in \mathbb{N}$. The numbers $\nu_\beta \in \mathbb{C}$, for all $\beta \in \mathbb{Z}^d_+$, $[\beta] < n$, are called *admissible* for a symmetry group $\mathcal{H}$ if for any smooth enough function

$f(\xi)$ such that $D^{\beta} f(\xi)\Big|_{\xi=\mathbf{0}} = (2\pi i)^{|\beta|}\nu_{\beta}$, for all $\beta \in \mathbb{Z}_{+}^{d}$, $[\beta] < n$. for any $E \in \mathcal{H}$ holds $D^{\beta} f(E^{*}\xi)\Big|_{\xi=\mathbf{0}} = (2\pi i)^{|\beta|}\nu_{\beta}$, for all $\beta \in \mathbb{Z}_{+}^{d}$, $[\beta] < n$.

Next, we reformulate sum rule conditions (3.22) with respect to the $\mathcal{H}-$symmetric mask m_0.

Lemma 5.1.11 *Let a dilation matrix M and a symmetry center c be appropriate for a symmetry group $\mathcal{H}$, $n \in \mathbb{N}$. Suppose m_0 is an $\mathcal{H}-$symmetric with respect to the center c mask satisfying condition (3.20) with some numbers $\lambda_{\alpha} \in \mathbb{C}$, $\alpha \in \mathbb{Z}_{+}^{d}$, $[\alpha] < n$. Then the numbers $\lambda_{\alpha} \in \mathbb{C}$ can be represented as*

$$\lambda_{\alpha} = \sum_{\gamma \le \alpha} \lambda'_{\gamma} \binom{\alpha}{\gamma} (M^{-1}c)^{\alpha-\gamma}, \quad \forall \alpha \in \mathbb{Z}_{+}^{d}, [\alpha] < n. \tag{5.16}$$

where $\lambda'_{\alpha} \in \mathbb{C}$, $\forall \alpha \in \mathbb{Z}_{+}^{d}$, $[\alpha] < n$, are some admissible numbers for $\mathcal{H}$. Also, condition (3.20) for μ_{0k}, $k = 0, \dots, m-1$, is equivalent to

$$D^{\beta}\mu_{0k}(\mathbf{0}) = \frac{(2\pi i)^{[\beta]}}{\sqrt{m}} \sum_{\gamma \le \beta} \lambda'_{\gamma} \binom{\beta}{\gamma} (M^{-1}c - M^{-1}s_k)^{\beta-\gamma}, \ \forall \beta \in \mathbb{Z}_{+}^{d}, [\beta] < n. \tag{5.17}$$

Proof Suppose $m'_0(\xi) := e^{-2\pi i (c,\xi)} m_0(\xi)$. Therefore, $m'_0(\xi) = m'_0(E^{*}\xi)$, for all $E \in \mathcal{H}$. Define the numbers λ'_{α}, $\alpha \in \mathbb{Z}_{+}^{d}$, $[\alpha] < n$, by

$$\lambda'_{\alpha} := \frac{1}{(2\pi i)^{[\alpha]}} D^{\alpha} m'_0(M^{*-1}\xi)\Big|_{\xi=\mathbf{0}}. \tag{5.18}$$

They are admissible for $\mathcal{H}$. Due to the Leibniz differentiation rule, we obtain (5.16). After combining (3.20) with (5.16) we establish the equivalence of conditions (3.20) and (5.17).$\Diamond$

Now we suggest an algorithm for the construction of refinable masks which are $\mathcal{H}$-symmetric with respect to an appropriate center $c \in \mathbb{R}^{d}$ and which have arbitrary order of sum rule.

Theorem 5.1.12 *Let a dilation matrix M and a symmetry center c be appropriate for a symmetry group $\mathcal{H}$, $n \in \mathbb{N}$, for $\gamma \in \mathbb{Z}_{+}^{d}$, $[\gamma] < n$, $\lambda'_{\gamma} \in \mathbb{C}$ be admissible numbers for $\mathcal{H}$. Then, there exists a mask m_0 that is $\mathcal{H}$-symmetric with respect to the center c and has sum rule of order n. This mask m_0 can be represented by*

$$m_0(\xi) = \frac{1}{\sqrt{m}} \sum_{p=0}^{\#\Lambda-1} \sum_{i=0}^{\#\mathcal{E}_p-1} e^{2\pi i (s_{p,i},\xi)} \mu_{0,p,i}(M^{*}\xi); \tag{5.19}$$

here for each $p = 0, \dots, \#\Lambda - 1$

$$\mu_{0,p,i}(M^*\xi) = \frac{1}{\#\mathcal{H}_{p,0}} \sum_{F \in \mathcal{H}_{p,0}} G_p(M^* F^* E^{(i)*}\xi) e^{2\pi i (r^F_{p,0}, M^* E^{(i)*}\xi)},$$

where $i = 0, \dots, \#\mathcal{E}_p - 1$, $E^{(i)} \in \mathcal{E}_p$ *and* G_p *are arbitrary trigonometric polynomials such that*

$$D^\beta G_p(\mathbf{0}) = \frac{(2\pi i)^{[\beta]}}{\sqrt{m}} \sum_{\gamma \le \beta} \lambda'_\gamma \binom{\beta}{\gamma} (M^{-1}c - M^{-1}s_{p,0})^{\beta-\gamma} \quad \forall \beta \in \mathbb{Z}^d_+, [\beta] < n. \tag{5.20}$$

Proof Let us construct the polyphase components $\mu_{0,p,i}$, $i = 0, \dots, \#\mathcal{E}_p - 1$, $p = 0, \dots, \#\Lambda - 1$ such that they satisfy conditions (5.12) and (5.17). For a fixed p, the algorithm for the construction of $\mu_{0,p,i}$ is the following. Firstly, define trigonometric polynomial G_p such that (5.20) is valid. Thus, the required $\mu_{0,p,0}$ can be obtained as follows:

$$\mu_{0,p,0}(\xi) = \frac{1}{\#\mathcal{H}_{p,0}} \sum_{F \in \mathcal{H}_{p,0}} G_p((M^{-1}FM)^*\xi) e^{2\pi i (r^F_{p,0}, \xi)}. \tag{5.21}$$

Next, for $i = 1, \dots, \#\mathcal{E}_p - 1$, define $\mu_{0,p,i}$ as $\mu_{0,p,i}(\xi) = \mu_{0,p,0}((M^{-1}E^{(i)}M)^*\xi)$, where $E^{(i)} \in \mathcal{E}_p$. Finally, mask m_0 is combined with the polyphase representation formula (3.16).

Let us check that we construct mask m_0 with the necessary properties. Firstly, for a fixed p we show that conditions (5.13) and (5.17) are valid for $\mu_{0,p,0}$. For any $\widetilde{F} \in \mathcal{H}_{p,0}$, we get

$$\begin{aligned}
\mu_{0,p,0}((M^{-1}\widetilde{F}M)^*\xi) &= \frac{1}{\#\mathcal{H}_{p,0}} \sum_{F \in \mathcal{H}_{p,0}} G_p((M^{-1}\widetilde{F}FM)^*\xi) e^{2\pi i (M^{-1}\widetilde{F}Mr^F_{p,0}), \xi)} = \\
&= \frac{1}{\#\mathcal{H}_{p,0}} \sum_{F \in \mathcal{H}_{p,0}} G_p((M^{-1}\widetilde{F}FM)^*\xi) e^{2\pi i (r^{\widetilde{F}F}_{p,0} - r^{\widetilde{F}}_{p,0}, \xi)} \\
&= \mu_{0,p,0}(\xi) e^{-2\pi i (r^{\widetilde{F}}_{p,0}, \xi)},
\end{aligned}$$

since $M^{-1}\widetilde{F}Mr^F_{p,0} = r^{\widetilde{F}F}_{p,0} - r^{\widetilde{F}}_{p,0}$. Thus, (5.13) is valid for $\mu_{0,p,0}$. By Lemma 5.1.9, condition (5.20) is equivalent to

$$D^\beta \left(G_p(\xi) e^{-2\pi i (M^{-1}(c - s_{p,0}), \xi)} \right) \Big|_{\xi = \mathbf{0}} = \frac{(2\pi i)^{[\beta]}}{\sqrt{m}} \lambda'_\beta \quad \forall \beta \in \mathbb{Z}^d_+, [\beta] < n.$$

Since numbers λ'_β are admissible and since for any $F \in \mathcal{H}_{p,0}$ matrix $M^{-1}FM$ is in $\mathcal{H}$, we obtain that $\forall \beta \in \mathbb{Z}^d_+$, $[\beta] < n$

$$D^{\beta}\left(G_p((M^{-1}FM)^*\xi)e^{-2\pi i(M^{-1}F(c-s_{p,0}),\xi)}\right)\Big|_{\xi=\mathbf{0}} = \frac{(2\pi i)^{[\beta]}}{\sqrt{m}}\lambda'_{\beta} \quad \forall F \in \mathcal{H}_{p,0}.$$

Then, by Lemma 5.1.9 and (5.9) we conclude that $\forall F \in \mathcal{H}_{p,0}$ and $\forall \beta \in \mathbb{Z}^d_+$, $[\beta] < n$

$$D^{\beta}\left(G_p((M^{-1}FM)^*\xi)e^{2\pi i(r^F_{p,0},\xi)}\right)\Big|_{\xi=\mathbf{0}} = \frac{(2\pi i)^{[\beta]}}{\sqrt{m}}\sum_{\gamma\le\beta}\lambda'_{\gamma}\binom{\beta}{\gamma}(M^{-1}c - M^{-1}s_{p,0})^{\beta-\gamma}.$$

Hence, condition (5.17) for $\mu_{0,p,0}$ is valid.

Next, for a fixed $i = 0, \dots, \#\mathcal{E}_p - 1$, we show that condition (5.17) is valid for $\mu_{0,p,i}$. Since (5.17) is true for $\mu_{0,p,0}$, then

$$D^{\beta}\left(\mu_{0,p,0}(\xi)e^{-2\pi i(M^{-1}(c-s_{p,0}),\xi)}\right)\Big|_{\xi=\mathbf{0}} = \frac{(2\pi i)^{[\beta]}}{\sqrt{m}}\lambda'_{\beta} \quad \forall \beta \in \mathbb{Z}^d_+,\ [\beta] < n.$$

Again, since the numbers λ'_{β} are admissible and for $E^{(i)} \in \mathcal{E}_p$ the matrix $M^{-1}E^{(i)}M$ is in $\mathcal{H}$, we obtain that $\forall \beta \in \mathbb{Z}^d_+$, $[\beta] < n$, and $\forall E^{(i)} \in \mathcal{E}_p$

$$D^{\beta}\left(\mu_{0,p,0}((M^{-1}E^{(i)}M)^*\xi)e^{-2\pi i(M^{-1}E^{(i)}(c-s_{p,0}),\xi)}\right)\Big|_{\xi=\mathbf{0}} = \frac{(2\pi i)^{[\beta]}}{\sqrt{m}}\lambda'_{\beta}.$$

It remains to note that $E^{(i)}(c - s_{p,0}) = c - s_{p,i}$. Therefore, $\forall \beta \in \mathbb{Z}^d_+$, $[\beta] < n$

$$D^{\beta}\mu_{0,p,i}(\mathbf{0}) = \frac{(2\pi i)^{[\beta]}}{\sqrt{m}}\sum_{\gamma\le\beta}\lambda'_{\gamma}\binom{\beta}{\gamma}(M^{-1}c - M^{-1}s_{p,i})^{\beta-\gamma}.$$

Hence, the polyphase components $\mu_{0,p,i}$, $i = 0, \dots, \#\mathcal{E}_p - 1$, $p = 0, \dots, \#\Lambda - 1$ satisfy conditions (5.17) and symmetry conditions (5.13) and (5.14) and therefore (5.12). Thus, by Lemmas 5.1.8 and 5.1.11 the constructed mask m_0 is $\mathcal{H}$-symmetric and has sum rule of order n.$\diamondsuit$

Note that trigonometric polynomials G_p satisfying (5.20) can be found by solving the corresponding linear system of equations. Alternatively, using a trigonometric analogue of Taylor's formula with trigonometric polynomials T_{β} such that $D^{\gamma}T_{\beta}(\mathbf{0}) = \delta_{\beta\gamma}$ it is easy to construct G_p analytically. Formulas for T_{β} are at the end of Sect. 3.4.

Remark 5.1.13 To provide linear-phase moments of order n, $n \in \mathbb{N}$, for refinable masks constructed by Theorem 5.1.12, we have to take $\lambda'_{\beta} = \delta_{\mathbf{0}\beta}$ for all $\beta \in \mathbb{Z}^d_+$, $[\beta] < n$ in Theorem 5.1.12. Then, due to Lemmas 5.1.9 and 5.1.11, relation $D^{\beta}m_0(M^{*-1}\xi)\Big|_{\xi=\mathbf{0}} = (2\pi i)^{[\beta]}(M^{-1}c)^{\beta}$ is valid for all $\beta \in \mathbb{Z}^d_+$, $[\beta] < n$. Hence, m_0 has linear-phase moments of order n.

Remark 5.1.14 For applications, it is important to get refinable masks with the minimal number of nonzero coefficients or with the minimal number of elements in the spectrum (recall that the spectrum of trigonometric polynomial $t(\xi) = \sum_k h_k e^{2\pi i (k,\xi)}$ is defined by $\mathrm{spec}(t) = \{k \in \mathbb{Z}^d : h_k \neq 0\}$). This feature can be provided by Theorem 5.1.12. Note that for a fixed initial parameters the $\mathcal{H}$-symmetric mask m_0 given by (5.19) depends only on the trigonometric polynomials G_p. Thus, for fixed p and for any vector $k \in \mathrm{spec}(G_p)$, the set of integer points $\{M^{-1}FMk + r^F_{p,0} : F \in \mathcal{H}_{p,0}\}$ is a subset of $\mathrm{spec}(\mu_{0,p,0})$ (see (5.21)). So, the general recommendation for choosing the minimal number of nonzero coefficients of mask is the following. For some fixed $k_1 \in \mathbb{Z}^d$, one should check whether the system of linear equations (5.20) is solvable with $\mathrm{spec}(G_p) = K_1 := \{M^{-1}FMk_1 + r^F_{p,0} : F \in \mathcal{H}_{p,0}\}$. If yes, then one solves the system. If not, one should add another point $k_2 \in \mathbb{Z}^d$ and check the solvability of the system with $\mathrm{spec}(G_p) = K_1 \cup \{M^{-1}FMk_2 + r^F_{p,0} : F \in \mathcal{H}_{p,0}\}$ and so on.

Using this technique of construction of symmetric mask, it is possible to describe the whole class of $\mathcal{H}$-symmetric masks with sum rule of order n.

Theorem 5.1.15 *Let a dilation matrix M and a symmetry center c be appropriate for a symmetry group $\mathcal{H}$, $n \in \mathbb{N}$, for $\gamma \in \mathbb{Z}^d_+$, $[\gamma] < n$, $\lambda'_\gamma \in \mathbb{C}$ be admissible numbers for $\mathcal{H}$. A general form for all masks m_0 that are $\mathcal{H}$-symmetric with respect to the center c and have sum rule of order n is given by*

$$m_0(\xi) = m_0^\dagger(\xi) + \sum_{p=0}^{\#\Lambda-1} \sum_{i=0}^{\#\mathcal{E}_p-1} \frac{e^{2\pi i (s_{p,i},\xi)}}{\#\mathcal{H}_{p,0}\cdot\sqrt{m}} \sum_{F\in\mathcal{H}_{p,0}} e^{2\pi i (r^F_{p,0}, M^*E^{(i)*}\xi)} T_p(M^*F^*E^{(i)*}\xi), \quad (5.22)$$

where T_p are trigonometric polynomials such that $D^\beta T_p(\mathbf{0}) = 0$, for all $\beta \in \mathbb{Z}^d_+$, $[\beta] < n$, mask $m_0^\dagger$ is constructed by Theorem 5.1.12 with the admissible numbers $\lambda'_\gamma \in \mathbb{C}$, $\gamma \in \mathbb{Z}^d_+$, $[\gamma] < n$.

Proof Let m_0 be an $\mathcal{H}$-symmetric mask with respect to the center c that has sum rule of order n. Let us set $m'_0(\xi) = e^{-2\pi i (c,\xi)} m_0(\xi)$ and define numbers λ'_γ for all $\gamma \in \mathbb{Z}^d_+$, $[\gamma] < n$, by (5.18). They are admissible. For the polyphase components μ_{0k} of mask m_0, $k = 0, \dots, m-1$, condition (5.17) holds with λ'_γ, $\gamma \in \mathbb{Z}^d_+$, $[\gamma] < n$.

Due to Theorem 5.1.12, there exists a mask $m_0^\dagger$ that is $\mathcal{H}$-symmetric with respect to the point c and satisfies condition (5.17) with λ'_γ. The polyphase components of $m_0 - m_0^\dagger$ are $\mu_{0,p,i} - \mu^\dagger_{0,p,i}$, $i = 0, \dots, \#\mathcal{E}_p - 1$, $p = 0, \dots, \#\Lambda - 1$. where $\mu^\dagger_{0,p,i}$ are the polyphase components of $m_0^\dagger$. Then, due to the fact that condition (5.17) holds for m_0 and $m_0^\dagger$ with the same numbers λ'_γ, we have $D^\beta(\mu_{0,p,i}(\xi) - \mu^\dagger_{0,p,i}(\xi))\Big|_{\xi=\mathbf{0}} = 0$, $i = 0, \dots, \#\mathcal{E}_p - 1$, $p = 0, \dots, \#\Lambda - 1$ for all $\beta \in \mathbb{Z}^d_+$, $[\beta] < n$,

It remains to show that there exist trigonometric polynomials T_p such that $\mu_{0,p,i}(\xi) - \mu^\dagger_{0,p,i}(\xi)$ can be represented as

$$\mu_{0,p,i}(\xi) - \mu^{\dagger}_{0,p,i}(\xi) = \sum_{F \in \mathcal{H}_{p,0}} \frac{e^{2\pi i (r^F_{p,0}, M^* E^{(i)*} M^{-1*}\xi)}}{\#\mathcal{H}_{p,0}} T_p(M^* F^* E^{(i)*} M^{-1*}\xi), \quad (5.23)$$

where $D^{\beta}T_p(\mathbf{0}) = 0$, for all $\beta \in \mathbb{Z}^d_+$, $[\beta] < n$, $E^{(i)} \in \mathcal{E}_p$. If we assume that $T_p = \mu_{0,p,0} - \mu^{\dagger}_{0,p,0}$, then for $i = 0$ equality (5.23) is valid. Since $\mu_{0,p,i}(\xi) = \mu_{0,p,0}((M^{-1}E^{(i)}M)^*\xi)$, we get the required representation.

In the converse case, we assume that m_0 is defined by (5.22). Trigonometric polynomial $m^{\dagger}_0$ is $\mathcal{H}$-symmetric with respect to the center c and has sum rule of order n. Repeating the elements of the proof of Theorem 5.1.12 and by Lemma 5.1.8, it can be shown that m_0 is $\mathcal{H}$-symmetric. The polyphase components of $m_0 - m^{\dagger}_0$ are given by (5.23). Since $D^{\beta}T_p(\mathbf{0}) = 0$, for all $\beta \in \mathbb{Z}^d_+$, $[\beta] < n$, we get $D^{\beta}(\mu_{0k}(\xi) - \mu^{\dagger}_{0k}(\xi))\Big|_{\xi=\mathbf{0}} = 0$, $k = 0, \dots, m-1$, for all $\beta \in \mathbb{Z}^d_+$, $[\beta] < n$. Therefore, conditions (5.17) are valid for the polyphase components μ_{0k} of mask m_0 with numbers λ'_{γ}, because they are valid for $\mu^{\dagger}_{0k}$. Therefore, m_0 has sum rule of order n.◊

5.2 Interpolatory Symmetric Wavelets

In this section, we discuss the construction of dual wavelet frames with symmetry properties for the case, when matrix extension can be explicitly realized. Let t be a trigonometric polynomial. We say that t is *interpolatory* if

$$\sum_{s \in D(M^*)} t(\xi + M^{*-1}s) \equiv 1.$$

It follows from (3.17) that for the interpolatory trigonometric polynomial t its polyphase component corresponding to the digit $s_0 = \mathbf{0}$ is a constant, namely $\tau_0 \equiv \frac{1}{\sqrt{m}}$.

For an interpolatory trigonometric polynomial t conditions (3.20) (which are equivalent to sum rule conditions) can be simplified. Since $\tau_0 = \frac{1}{\sqrt{m}}$, then from condition (3.20) for τ_0 we get that $\lambda_{\alpha} = \delta_{\mathbf{0}\alpha}$, for all $\alpha \in \mathbb{Z}^d_+$, $[\alpha] < n$. Thus, the conditions on the polyphase components of t are

$$D^{\beta}\tau_k(\mathbf{0}) = \frac{(2\pi i)^{[\beta]}}{\sqrt{m}}(-M^{-1}s_k)^{\beta} \quad \forall \beta \in \mathbb{Z}^d_+, [\beta] < n, \quad (5.24)$$

where $k = 0, \dots, m-1$ and

$$D^{\beta}t(M^{*-1}\xi)\Big|_{\xi=\mathbf{0}} = \delta_{\mathbf{0}\beta}, \quad \forall \beta \in \mathbb{Z}^d_+, [\beta] < n. \quad (5.25)$$

The following Lemma shows that an $\mathcal{H}$-symmetric interpolatory trigonometric polynomial t has some restrictions on the symmetry center.

Lemma 5.2.1 *Let a dilation matrix M and a symmetry center c be appropriate for a symmetry group $\mathcal{H}$. Suppose t is an interpolatory trigonometric polynomial that is $\mathcal{H}$-symmetric with respect to c and has sum rule of order n, $n \geq 2$. Then, $c = Ec$ for all $E \in \mathcal{H}$.*

Proof From (5.25) and the higher chain formula for the linear change of variables we get that $\forall E \in \mathcal{H}$

$$D^{\beta}t(M^{*-1}\xi)\Big|_{\xi=\mathbf{0}} = D^{\beta}t(E^{*}M^{*-1}\xi)\Big|_{\xi=\mathbf{0}} = (2\pi i)^{[\beta]}\delta_{\mathbf{0}\beta}, \quad \forall \beta \in \mathbb{Z}_{+}^{d}, [\beta] < n.$$

From the other hand, since t is $\mathcal{H}$-symmetric, we get that $\forall E \in \mathcal{H}$ and $\forall \beta \in \mathbb{Z}_{+}^{d}$, $[\beta] < n$

$$\begin{aligned} D^{\beta}t(M^{*-1}\xi)\Big|_{\xi=\mathbf{0}} &= D^{\beta}e^{2\pi i(c-Ec,M^{*-1}\xi)}t(E^{*}M^{*-1}\xi)\Big|_{\xi=\mathbf{0}} \\ &= (2\pi i)^{[\beta]}(M^{-1}(c-Ec))^{\beta}. \end{aligned}$$

Thus, $M^{-1}(c-Ec) = \mathbf{0}$, or, equivalently, $c = Ec$, $\forall E \in \mathcal{H}$.◊

The condition $c = Ec$ for all $E \in \mathcal{H}$ means that either $c = \mathbf{0}$ or $\det(I_d - E) = 0$ for all $E \in \mathcal{H}$. In the latter case, c belongs to the intersection of the null spaces of matrices $I_d - E$, $\forall E \in \mathcal{H}$, and condition (5.2) reduces to $t(\xi) = t(E^{*}\xi)$, $\forall E \in \mathcal{H}$. Actually, we can say that t is $\mathcal{H}$-symmetric with respect to any point from the intersection of the null spaces of matrices $I_d - E$, $\forall E \in \mathcal{H}$. So, it is does not matter how the symmetry center will be chosen. For example, if $\mathcal{H} := \left\{I_2, \begin{pmatrix} -1 & 0 \\ 0 & 1 \end{pmatrix}\right\}$, then the symmetry center of trigonometric polynomial defined in Lemma 5.2.1 can be at any point on the $y-$axis. For convenience, we assume that the symmetry center for $\mathcal{H}$-symmetric interpolatory trigonometric polynomials is $c = \mathbf{0}$. Due to this fact, $\mathcal{H}_{0,0} = \mathcal{H}$, $\mathcal{E}_0 = \{I_d\}$.

As it was described in Sect. 2.5, the key problem during the construction of wavelets is the matrix extension problem. For an interpolatory refinable mask m_0, the matrix extension can be done explicitly. But firstly we need to construct a dual refinable mask that will be appropriate in some sense. Let m_0 is constructed by Theorem 5.1.12, $\mu_{00}, \ldots, \mu_{0,m-1}$ are the polyphase components of m_0. We say that $\widetilde{m}_0$ is admissible for m_0 as a dual refinable mask if $\widetilde{m}_0$ is $\mathcal{H}$-symmetric with respect to the origin, $\widetilde{m}_0$ has sum rule at least of order 1 and

$$\sum_{k=0}^{m-1} \mu_{0k}\overline{\widetilde{\mu}_{0k}} = 1, \tag{5.26}$$

where $\widetilde{\mu}_{00}, \widetilde{\mu}_{01}, \dots, \widetilde{\mu}_{0,m-1}$ are the polyphase components of $\widetilde{m}_0$. In order to construct symmetric wavelets, we require the same symmetry properties for $\widetilde{m}_0$ as for m_0. Sum rule of order 1 for dual refinable mask $\widetilde{m}_0$ leads to vanishing moments of order 1 for wavelet masks m_ν. And this is sufficient for obtaining a dual wavelet frame (see Sect. 3.1). Equality (5.26) is required for the matrix extension.

An admissible dual mask $\widetilde{m}_0$ can be explicitly defined in the following form via its polyphase components

$$(\widetilde{\mu}_{00}, \widetilde{\mu}_{01}, \dots, \widetilde{\mu}_{0,m-1}) := \left(\sqrt{m}\left(1 - \sum_{k=1}^{m-1} \mu_{0k}\overline{\widetilde{\mu}_{0k}}\right), \mu_{01}, \dots, \mu_{0,m-1}\right). \quad (5.27)$$

The properties of this mask are given in the following Lemma.

Lemma 5.2.2 *Let m_0 be an interpolatory $\mathcal{H}$-symmetric refinable mask that has sum rule of order n. Then, dual refinable mask $\widetilde{m}_0$ defined by (5.27) is $\mathcal{H}$-symmetric, has sum rule of order n and equality (5.26) is valid.*

Proof Equality (5.26) is obviously valid since $\mu_{00} = \frac{1}{\sqrt{m}}$. Since each function μ_{0k}, $k = 1, \dots, m-1$ coincides with $\widetilde{\mu}_{0k}$, it satisfies (5.24) and symmetry conditions (5.12). It remains to check that these conditions are valid for $\widetilde{\mu}_{00}$. Evidently, $\widetilde{\mu}_{00}(\mathbf{0}) = \frac{1}{\sqrt{m}}$. Let $\beta \in \mathbb{Z}_+^d$, $0 < [\beta] < n$. For any $k = 1, \dots, m-1$, using the Leibniz differentiation rule and (5.24) we have

$$\begin{aligned} D^\beta \left|\widetilde{\mu}_{0k}(\xi)\right|^2 \Big|_{\xi=\mathbf{0}} &= D^\beta \left(\widetilde{\mu}_{0k}(\xi)\overline{\widetilde{\mu}_{0k}(\xi)}\right) \Big|_{\xi=\mathbf{0}} \\ &= \sum_{\mathbf{0}\le\gamma\le\beta} \binom{\beta}{\gamma} D^\gamma \widetilde{\mu}_{0k}(\mathbf{0}) \overline{D^{\beta-\gamma}\widetilde{\mu}_{0k}(\mathbf{0})} \\ &= \frac{(2\pi i)^{[\beta]}}{m} \sum_{\mathbf{0}\le\gamma\le\beta} \binom{\beta}{\gamma} (-M^{-1}s_k)^\gamma (M^{-1}s_k)^{\beta-\gamma} \\ &= \frac{(2\pi i)^{[\beta]}}{m} (M^{-1}s_k - M^{-1}s_k)^\beta = 0. \end{aligned}$$

It follows that $D^\beta \widetilde{\mu}_{00}(\mathbf{0}) = 0$ for all $\beta \in \mathbb{Z}_+^d$, $0 < [\beta] < n$. So (5.24) is valid for $\widetilde{\mu}_{00}$. The symmetry conditions (5.12) for $\widetilde{\mu}_{00}$ means that $\widetilde{\mu}_{00}$ should be $\mathcal{H}$-symmetric with respect to the origin. This is valid since

$$\begin{aligned} \widetilde{\mu}_{00}(K^*\xi) &= \sqrt{m}\left(1 - \sum_{p=1}^{\#\Lambda-1} \sum_{i=0}^{\#\mathcal{E}_p-1} \left|\widetilde{\mu}_{0,p,i}(K^*\xi)\right|^2\right) \\ &= \sqrt{m}\left(1 - \sum_{p=1}^{\#\Lambda-1} \sum_{i=0}^{\#\mathcal{E}_p-1} \left|\widetilde{\mu}_{0,p,j(p,i,K)}(\xi)\right|^2\right), \end{aligned}$$

where $j = j(p, i, K)$ is the mapping defined in Sect. 5.1.$\Diamond$

More flexible way for choosing a dual refinable mask is the following. Suppose $\widetilde{m}_0$ is an interpolatory $\mathcal{H}$-symmetric mask constructed using Theorem 5.1.12 with $n = 1$, $c = \mathbf{0}$. And $\widetilde{\mu}_{0k}$, $k = 0, \dots, m-1$, are its polyphase components. Now we modify mask $\widetilde{m}_0$ by replacing its first polyphase component $\widetilde{\mu}_{00} = \frac{1}{\sqrt{m}}$ with $\widetilde{\mu}_{00} = \sqrt{m}(1 - \sum_{k=1}^{m-1} \overline{\mu_{0k}}\widetilde{\mu}_{0k})$, where μ_{0k} are the polyphase components of the mask m_0. It is easy to see that the modified mask $\widetilde{m}_0$ is admissible for m_0. This flexibility gives a possibility for choosing a dual refinable mask with a smaller spectrum. Note that the smaller order of sum rule for $\widetilde{m}_0$ we demand, the smaller number of nonzero coefficients of $\widetilde{m}_0$ we obtain.

For a refinable mask and an admissible dual mask, the matrix extension can be realized as follows. Let $P = (\mu_{01}, \dots, \mu_{0,m-1})$, $\widetilde{P} = (\widetilde{\mu}_{01}, \dots, \widetilde{\mu}_{0,m-1})$. Then,

$$\mathcal{N} = \left(\begin{array}{c|c} \frac{1}{\sqrt{m}} & P \\ \hline -U\widetilde{P}^* & \sqrt{m}(U - U\widetilde{P}^*P) \end{array}\right), \quad \widetilde{\mathcal{N}} = \left(\begin{array}{c|c} \sqrt{m}(1 - \widetilde{P}P^*) & \widetilde{P} \\ \hline -\widetilde{U}P^* & \frac{1}{\sqrt{m}}\widetilde{U} \end{array}\right), \tag{5.28}$$

Here $U, \widetilde{U}$ are $(m-1) \times (m-1)$ matrices consisting of trigonometric polynomials such that $U\widetilde{U}^* \equiv I_{m-1}$. The matrices U, $\widetilde{U}$, for example, can be taken as follows $U = \widetilde{U} = I_{m-1}$. It can be checked easily that $\mathcal{N}\widetilde{\mathcal{N}}^* \equiv I_m$.

Theorem 5.2.3 *Let a dilation matrix M and a symmetry center c be appropriate for a symmetry group $\mathcal{H}$, $n \in \mathbb{N}$, and let m_0 and $\widetilde{m}_0$ be $\mathcal{H}$-symmetric with respect to the origin masks such that m_0 is interpolatory and has sum rule of order n, $\widetilde{m}_0$ is an admissible dual refinable mask. Then, wavelet masks $m_{(p,i)}$ and $\widetilde{m}_{(p,i)}$, $i = 0, \dots, \#\mathcal{E}_p - 1$, $p = 1, \dots, \#\Lambda - 1$, constructed using the matrix extension (5.28) with $U = \widetilde{U} = I_{m-1}$ have vanishing moments of order n and the symmetry properties. Namely, for a fixed number $p = 1, \dots, \#\Lambda - 1$,*

- *$m_{(p,i)}$ and $\widetilde{m}_{(p,i)}$ are $\mathcal{H}_{p,i}$-symmetric with respect to the center $s_{p,i}$;*
- *$m_{(p,i)} = m_{(p,0)}(E^{(i)*}\xi)$, $\widetilde{m}_{(p,i)} = \widetilde{m}_{(p,0)}(E^{(i)*}\xi)$, where $E^{(i)} \in \mathcal{E}_p$, $i = 0, \dots, \#\mathcal{E}_p - 1$.*

If refinable functions φ, $\widetilde{\varphi}$ corresponding to refinable masks m_0 and $\widetilde{m}_0$ are in $L_2(\mathbb{R}^d)$, then the corresponding wavelet system is a dual wavelet frame.

Proof Let $\mu_{0,p,i}$ be the polyphase components of m_0, $\widetilde{\mu}_{0,p,i}$ be the polyphase components of $\widetilde{m}_0$, $i = 0, \dots, \#\mathcal{E}_p - 1$, $p = 1, \dots, \#\Lambda - 1$, and let $T_p = (\mu_{0,p,0}, \dots, \mu_{0,p,N_p-1})$, $\widetilde{T}_p = (\widetilde{\mu}_{0,p,0}, \dots, \widetilde{\mu}_{0,p,N_p-1})$, where $N_p := \#\mathcal{E}_p$, $p = 1, \dots, \#\Lambda - 1$. Thus, we can set $P = (T_1, \dots, T_{\#\Lambda-1})$ and $\widetilde{P} = (\widetilde{T}_1, \dots, \widetilde{T}_{\#\Lambda-1})$. Consider the submatrix $I_m - \widetilde{P}^*P$ in (5.28). It is a block matrix and

$$I_m - \widetilde{P}^*P = \begin{pmatrix} I_{N_1-1} - \widetilde{T}_1^*T_1 & -\widetilde{T}_1^*T_2 & \dots & -\widetilde{T}_1^*T_{\#\Lambda-1} \\ -\widetilde{T}_2^*T_1 & I_{N_2-1} - \widetilde{T}_2^*T_2 & \dots & -\widetilde{T}_2^*T_{\#\Lambda-1} \\ \vdots & & \ddots & \vdots \\ -\widetilde{T}_{\#\Lambda-1}^*T_1 & \dots & \dots & I_{N_{(\#\Lambda-1)}-1} - \widetilde{T}_{\#\Lambda-1}^*T_{\#\Lambda-1} \end{pmatrix}.$$

The rows of the submatrix we number by the double indices (p, i). The (p, i)th row is the ith row in the pth block. The wavelet mask corresponding to the (p, i)th row is denoted by $m_{(p,i)}$. Thus,

$$m_{(p,i)}(\xi) = e^{2\pi i(s_{p,i},\xi)} - \overline{\widetilde{\mu}_{0,p,i}(M^*\xi)}m_0(\xi).$$

Let us show that mask $m_{(p,0)}$ is $\mathcal{H}_{p,0}$-symmetric with respect to the center $s_{p,0}$. Indeed, due to (5.9) and (5.13) for all $F \in \mathcal{H}_{p,0}$ we have

$$m_{(p,0)}(F^*\xi) = e^{2\pi i(Fs_{p,0},\xi)} - \overline{\widetilde{\mu}_{0,p,0}(M^*F^*\xi)}m_0(F^*\xi) = m_{(p,0)}(\xi)e^{2\pi i(Fs_{p,0}-s_{p,0},\xi)}.$$

Moreover, by (5.14) we have $m_{(p,i)}(\xi) = m_{(p,0)}(E^{(i)*}\xi)$. Similarly, it can be proved that $m_{(p,i)}$ is $\mathcal{H}_{p,i}$-symmetric with respect to the center $s_{p,i}$. Namely, for all $\widetilde{F} \in \mathcal{H}_{p,i}$

$$m_{(p,i)}(\widetilde{F}^*\xi) = m_{(p,i)}(\xi)e^{2\pi i(\widetilde{F}s_{p,i}-s_{p,i},\xi)}.$$

It is easy to check that the same symmetric properties are valid for the dual wavelet masks, since

$$\widetilde{m}_{(p,i)}(\xi) = \frac{1}{m}e^{2\pi i(s_{p,i},\xi)} - \frac{1}{\sqrt{m}}\overline{\mu_{0,p,i}(M^*\xi)}.$$

The vanishing moments of order n for the dual wavelet masks $\widetilde{m}_\nu$ are provided by Lemma 3.4.1.◊

Thus, the matrix extension (5.28) with $U = \widetilde{U} = I_{m-1}$ leads to the wavelet masks that are mutually symmetric, i.e., some wavelet masks are reflected or rotated copies of the others.

Notice that if $\mathcal{H}$ and M are chosen such that $\mathcal{E}_p = \{I_d\}$ for all $p = 0, \dots, \#\Lambda - 1$, then all wavelet masks constructed by Theorem 5.2.3 are $\mathcal{H}$-symmetric. But this is not the only case, when the $\mathcal{H}$-symmetry properties for all wavelet masks can be provided. For an abelian symmetry group $\mathcal{H}$ under certain circumstances, it is possible to extend the polyphase matrices such that all wavelet masks have $\mathcal{H}$-symmetry properties. Let us extend the definition of $\mathcal{H}$-symmetric trigonometric polynomials. A trigonometric polynomial t has the $\mathcal{H}$-symmetry property if for each matrix $E \in \mathcal{H}$

$$t(E^*\xi) = \varepsilon_E e^{2\pi i(r_E,\xi)}t(\xi),$$

where $\varepsilon_E \in \mathbb{C}$, $|\varepsilon_E| = 1$, $r_E \in \mathbb{Z}^d$. This definition is a generalization of (5.2). For example, the new definition includes antisymmetric trigonometric polynomials. Also, trigonometric polynomials, which do not have a symmetry center common to all matrices $E \in \mathcal{H}$, satisfy the new definition.

Next, we show how to symmetrize the row of the polyphase components of an $\mathcal{H}$-symmetric mask, i.e., we want to find a unitary transformation of the row such that each element of the new row has the $\mathcal{H}$-symmetry property. This is possible, if $\mathcal{H}$ is an abelian symmetry group on $\mathbb{Z}^d$.

In this case, $\mathcal{H}$ can be expressed as the direct product of cyclic subgroups due to the fundamental theorem of finite abelian groups. Hence, for any number $p = 0, \dots, \#\Lambda - 1$ the set $\mathcal{E}_p$ is an abelian group. This is a simple corollary from the fundamental theorem of finite abelian groups since $\mathcal{H}_{p,0}$ is a subgroup of $\mathcal{H}$. Thus, group $\mathcal{E}_p$ also can be expressed as the direct product of cyclic subgroups. In order to write it down, we need some additional notations. Let us fix p and let $N_p := \#\mathcal{E}_p$ and γ_p be the number of cyclic subgroups of $\mathcal{E}_p$. Then, there exist unique prime numbers $N_{p,i}$, $i = 1, \dots, \gamma_p$ and matrices $J_1, \dots, J_{\gamma_p} \in \mathcal{E}_p$ such that

$$\mathcal{E}_p = \{I_d, J_1, \dots, J_1^{N_{p,1}-1}\} \times \cdots \times \{I_d, J_{\gamma_p}, \dots, J_{\gamma_p}^{N_{p,\gamma_p}-1}\}.$$

Thus, any element $E \in \mathcal{E}_p$ can be uniquely represented as follows: $E = \prod_{j=1}^{\gamma_p} J_j^{k_j}$, where k_j are some integers from the sets $\{0, \dots, N_{p,j} - 1\}$, respectively. Let $L_{p,1} = 1$, $L_{p,i} = \prod_{j=1}^{i-1} N_{p,j}$, for $i = 2, \dots, \gamma_p$ and $L_{p,\gamma_p+1} = \prod_{j=1}^{\gamma_p} N_{p,j} = N_p$. Thus, each number $k \in \{0, \dots, N_p - 1\}$ can be uniquely represented as $k = \sum_{j=1}^{\gamma_p} k_j L_{p,j}$, where $0 \le k_j \le N_{p,j} - 1$. Or equivalently, there is an one-to-one correspondence between k and $(k_1, \dots, k_{\gamma_p})$. In other words, we get a mixed radix numeral system for all numbers in the set $\{0, \dots, N_p - 1\}$. Let us renumber all matrices in the group $\mathcal{E}_p$ according to this numeral system. Let $E \in \mathcal{E}_p$, then $E = \prod_{j=1}^{\gamma_p} J_j^{k_j}$, so matrix E has number k, where $k = \sum_{j=1}^{\gamma_p} k_j L_{p,j}$. We denote matrix E with number k by $E^{(k)}$.

Let us define addition on the set $\{0, \dots, \#\mathcal{E}_p - 1\}$ as follows:

$$k \oplus l = \sum_{j=1}^{\gamma_p} (k_j \oplus_j l_j) L_{p,j},$$

where $k_j \oplus_j l_j$ is a summation by module $N_{p,j}$, namely

$$k_j \oplus_j l_j = (k_j + l_j) \mod N_{p,j}, \quad \text{where } j \in \{1, \dots, \gamma_p\}.$$

Not hard to see that, $E^{(k)} E^{(l)} = E^{(k \oplus l)}$, where $E^{(k)}, E^{(l)} \in \mathcal{E}_p$.

Using the introduced notations and assumptions, we can rewrite some useful properties. Let us fix p. Then, the matrix $K \in \mathcal{H}$ can be represented as $K = E^{(k)} F$, where $E^{(k)} \in \mathcal{E}_p$, $F \in \mathcal{H}_{p,0}$. In this case, the map $j(p, i, K)$ simply means $i \oplus k$, since $K E^{(i)} = E^{(i \oplus k)} F$ by commutativity. Thus, (5.11) can be written as

$$K s_{p,i} = M r_{p,i}^K + s_{p,i \oplus k} + K c - c, \tag{5.29}$$

where $r_{p,i}^K = M^{-1} E^{(i \oplus k)} M r_{p,0}^F$. Let m_0 be an $\mathcal{H}$-symmetric mask with respect to the origin. Then, condition (5.14) becomes

$$\mu_{0,p,i}((M^{-1} E^{(k)} M)^* \xi) = \mu_{0,p,i \oplus k}(\xi), \quad \forall E^{(k)} \in \mathcal{E}_p, \tag{5.30}$$

for $i = 0, \ldots, \#\mathcal{E}_p - 1$; condition (5.12) for matrix $K \in \mathcal{H}$ such that $K = E^{(k)}F$, $E^{(k)} \in \mathcal{E}_p$, $F \in \mathcal{H}_{p,0}$, becomes

$$\mu_{0,p,i}((M^{-1}KM)^*\xi) = e^{2\pi i(r^K_{p,i},\xi)}\mu_{0,p,i\oplus k}(\xi), \tag{5.31}$$

where $r^K_{p,i} = M^{-1}E^{(i\oplus k)}Mr^F_{p,0}$, $r^F_{p,0}$ is from (5.9).

For the integers $N_{p,i}$ we denote $\varepsilon_{N_{p,i}} := e^{\frac{2\pi i}{N_{p,i}}}$. For any $p \in \{1, \ldots, \#\Lambda - 1\}$, define the matrix of the discrete Fourier transform

$$W_{N_{p,i}} = \frac{1}{\sqrt{N_{p,i}}}\{\varepsilon^{kl}_{N_{p,i}}\}_{k,l=0,N_{p,i}-1}. \tag{5.32}$$

It is known that $W_{N_{p,i}}$ is a unitary and symmetric matrix, i.e., $W_{N_{p,i}}W^*_{N_{p,i}} = I_m$, $W^T_{N_{p,i}} = W_{N_{p,i}}$. Define $\mathcal{W}_p = W_{N_{p,1}} \otimes \cdots \otimes W_{N_{p,\gamma_p}}$, where operation $\otimes$ is the Kronecker product. $\mathcal{W}_p$ is also a unitary matrix and its elements can be expressed as follows:

$$[\mathcal{W}_p]_{k,l} = \left(\prod_{j=1}^{\gamma_p}\frac{1}{\sqrt{N_{p,i}}}\right)\varepsilon^{k_1l_1}_{N_{p,1}}\ldots\varepsilon^{k_{\gamma_p}l_{\gamma_p}}_{N_{p,\gamma_p}},$$

where k, l are corresponded to $(k_1, \ldots, k_{\gamma_p})$ and $(l_1, \ldots, l_{\gamma_p})$, respectively, according to the above mixed radix numeral system.

Some of the properties of the matrix $\mathcal{W}_p$ are

$$[\mathcal{W}_p]_{k,l}[\mathcal{W}_p]_{n,l} = [\mathcal{W}_p]_{k\oplus n,l}, \quad [\mathcal{W}_p]_{k,l}\overline{[\mathcal{W}_p]_{k,l}} = 1, \quad k, l, n = 0, \ldots, \#\mathcal{E}_p - 1.$$

The next Lemma shows that $\mathcal{W}_p$ symmetrizes the part of the row of polyphase components.

Lemma 5.2.4 *Let a dilation matrix M and a symmetry center c be appropriate for an abelian symmetry group $\mathcal{H}$. Suppose m_0 is an $\mathcal{H}$-symmetric with respect to the origin mask and $\mu_{0,p,i}$ are its polyphase components, T_p is defined by $T_p = (\mu_{0,p,0}, \ldots, \mu_{0,p,\#\mathcal{E}_p-1})$. Suppose that $r^F_{p,0} = M^{-1}EMr^F_{p,0}$ for all $F \in \mathcal{H}_{p,0}$ and $E \in \mathcal{E}_p$. Then, each element of the row $T'_p := T_p\mathcal{W}_p$ has the $\mathcal{H}$-symmetry property.*

Proof Denote by $\mu'_{0,p,r}$, $r = 0, \ldots, \#\mathcal{E}_p - 1$, the elements of the new row T'_p. By definition, we have $\mu'_{0,p,r}(\xi) = \sum_{k=0}^{N_p-1}[\mathcal{W}_p]_{k,r}\,\mu_{0,p,k}(\xi)$. Firstly, we show that $\mu'_{0,p,r}(\xi)$ has the $M^{-1}\mathcal{E}_pM$-symmetry property. Indeed, by (5.30) and the properties of $\mathcal{W}_p$ we have for all $E^{(l)} \in \mathcal{E}_p$

$$\mu'_{0,p,r}((M^{-1}E^{(l)}M)^*\xi) = \sum_{k=0}^{\#\mathcal{E}_p-1} [\mathcal{W}_p]_{k,r}\, \mu_{0,p,k}((M^{-1}E^{(l)}M)^*\xi) =$$
$$= \sum_{k=0}^{\#\mathcal{E}_p-1} [\mathcal{W}_p]_{l\oplus k,r}\, \overline{[\mathcal{W}_p]_{l,r}}\mu_{0,p,l\oplus k}(\xi) = \overline{[\mathcal{W}_p]_{l,r}}\mu'_{0,p,r}(\xi).$$

Now show that $\mu'_{0,p,r}(\xi)$, $r = 0, \ldots, \#\mathcal{E}_p - 1$, are $M^{-1}\mathcal{H}_{p,0}M$-symmetric. Indeed, for $F \in \mathcal{H}_{p,0}$, by (5.13), (5.30) we get

$$\mu'_{0,p,r}((M^{-1}FM)^*\xi) = \sum_{k=0}^{\#\mathcal{E}_p-1} [\mathcal{W}_p]_{k,r}\, \mu_{0,p,0}((M^{-1}FE^{(k)}M)^*(\xi)$$
$$= \sum_{k=0}^{\#\mathcal{E}_p-1} [\mathcal{W}_p]_{k,r}\, \mu_{0,p,0}((M^{-1}E^{(k)}M)^*\xi)e^{-2\pi i(M^{-1}E^{(k)}Mr^F_{p,0},\xi)}$$
$$= \mu'_{0,p,r}(\xi)e^{-2\pi i(r^F_{p,0},\xi)}.$$

If $K \in \mathcal{H}$, then K can be represented as $K = E^{(l)}F$, where $F \in \mathcal{H}_{p,0}$ and $E^{(l)} \in \mathcal{E}_p$. Thus, $\mu'_{0,p,r}(\xi)$ has the $\mathcal{H}$-symmetry property

$$\mu'_{0,p,r}((M^{-1}KM)^*\xi) = \overline{[\mathcal{W}_p]_{l,r}}\mu'_{0,p,r}(\xi)e^{-2\pi i(r^F_{p,0},\xi)}.\diamondsuit \tag{5.33}$$

According to this Lemma, we need a *special assumption*

$$r^F_{p,0} = M^{-1}EMr^F_{p,0}, \quad \forall F \in \mathcal{H}_{p,0}, \forall E \in \mathcal{E}_p \tag{5.34}$$

to ensure the $\mathcal{H}$-symmetry property for all components of the row T'_p. Notice that this assumption is used only to provide $M^{-1}\mathcal{H}_{p,0}M$-symmetry for $\mu'_{0,p,r}(\xi)$.

Define a block diagonal unitary matrix $\mathcal{W}$ as follows:

$$\mathcal{W} = \mathtt{diag}(\mathcal{W}_1, \ldots, \mathcal{W}_{\#\Lambda-1}).$$

Let m_0 be an $\mathcal{H}$-symmetric mask with respect to the origin. Suppose the special assumption is valid for all $p = 1, \ldots, \#\Lambda - 1$, i.e., $r^F_{p,0} = M^{-1}EMr^F_{p,0}$ for all $F \in \mathcal{H}_{p,0}$ and $E \in \mathcal{E}_p$. Then, by Lemma 5.2.4 the matrix $\mathcal{W}$ symmetrizes the row $P = (T_1, \ldots, T_{\#\Lambda-1})$ of the polyphase components of m_0, namely all elements of the row

$$P\mathcal{W} = (T_1\mathcal{W}_1, \ldots, T_{\#\Lambda-1}\mathcal{W}_{\#\Lambda-1})$$

have the $\mathcal{H}$-symmetry property. In the following Theorem, it is stated that wavelets with the $\mathcal{H}$-symmetry property can be constructed using matrix $\mathcal{W}$.

Theorem 5.2.5 *Let a dilation matrix M and a center c be appropriate for an abelian symmetry group $\mathcal{H}$, $n \in \mathbb{N}$ and let m_0 and $\widetilde{m}_0$ be $\mathcal{H}$-symmetric with respect to* *the origin masks such that m_0 is interpolatory and has sum rule of order n, $\widetilde{m}_0$ is constructed by (5.27). The special assumption is valid for all $p = 1, \dots, \#\Lambda - 1$, i.e., $r^F_{p,0} = M^{-1}EMr^F_{p,0}$ for all $F \in \mathcal{H}_{p,0}$ and $E \in \mathcal{E}_p$. Then, there exist wavelet masks $m_{(p,i)}$ and $\widetilde{m}_{(p,i)}$, $i = 0, \dots, \#\mathcal{E}_p - 1$, $p = 1, \dots, \#\Lambda - 1$, which have vanishing moments of order n and the $\mathcal{H}$-symmetry property. If refinable functions φ, $\widetilde{\varphi}$ corresponding to refinable masks m_0 and $\widetilde{m}_0$ are in $L_2(\mathbb{R}^d)$, then the corresponding wavelet system is a dual wavelet frame.*

Proof Let $\mu_{0,p,i}$ be the polyphase components of m_0, $i = 0, \dots, \#\mathcal{E}_p - 1$, $p = 1, \dots, \#\Lambda - 1$, and let $T_p = (\mu_{0,p,0}, \dots, \mu_{0,p,N_p-1})$, where $N_p := \#\mathcal{E}_p$, $p = 1, \dots, \#\Lambda - 1$. Thus, we can set $P = (T_1, \dots, T_{\#\Lambda-1})$. Let us consider the matrix extension (5.28) with $U = \widetilde{U} = \mathcal{W}^*$

$$\mathcal{N} = \left(\begin{array}{c|c} \frac{1}{\sqrt{m}} & P \\ \hline -\mathcal{W}^*\widetilde{P}^* & \sqrt{m}(\mathcal{W}^* - \mathcal{W}^*\widetilde{P}^*P) \end{array}\right), \quad \widetilde{\mathcal{N}} = \left(\begin{array}{c|c} \sqrt{m}(1 - \widetilde{P}P^*) & \widetilde{P} \\ \hline -\mathcal{W}^*P^* & \frac{1}{\sqrt{m}}\mathcal{W}^* \end{array}\right) \tag{5.35}$$

Consider the submatrix $\sqrt{m}(\mathcal{W}^* - \mathcal{W}^*\widetilde{P}^*P)$. It is a block matrix and

$$\mathcal{W}^* - \mathcal{W}^*\widetilde{P}^*P =$$
$$\begin{pmatrix} \mathcal{W}_1^* - (\widetilde{T}_1\mathcal{W}_1)^*T_1 & -(\widetilde{T}_1\mathcal{W}_1)^*T_2 & \dots & -(\widetilde{T}_1\mathcal{W}_1)^*T_{\#\Lambda-1} \\ -(\widetilde{T}_2\mathcal{W}_2)^*T_1 & \mathcal{W}_2^* - (\widetilde{T}_2\mathcal{W}_2)^*T_2 & \dots & -(\widetilde{T}_2\mathcal{W}_2)^*T_{\#\Lambda-1} \\ \vdots & & \ddots & \vdots \\ -(\widetilde{T}_{\#\Lambda-1}\mathcal{W}_{\#\Lambda-1})^*T_1 & -(\widetilde{T}_{\#\Lambda-1}\mathcal{W}_{\#\Lambda-1})^*T_2 & \dots & \mathcal{W}_{\#\Lambda-1}^* - (\widetilde{T}_{\#\Lambda-1}\mathcal{W}_{\#\Lambda-1})^*T_{\#\Lambda-1} \end{pmatrix}.$$

Denote the elements of the submatrix $\sqrt{m}(\mathcal{W}^* - \mathcal{W}^*\widetilde{P}^*P)$ as

$$\sqrt{m}(\mathcal{W}^* - \mathcal{W}^*\widetilde{P}^*P) =: \{\mu_{(p,i),(t,j)}\}_{p=1,\dots,\#\Lambda-1,i=0,\dots,N_p-1}^{t=1,\dots,\#\Lambda-1,j=0,\dots,N_t-1}.$$

With this numeration, the element $\mu_{(p,i),(t,j)}$ in the submatrix is in the block (p, t) with position (i, j). Namely, if $p \neq t$, then

$$\mu_{(p,i),(t,j)}(\xi) = \sqrt{m}\,[-(\widetilde{T}_p\mathcal{W}_p)^*T_t]_{i,j} = -\sqrt{m}\,\overline{\widetilde{\mu}'_{0,p,i}(\xi)}\mu_{0,t,j}(\xi);$$

if $p = t$, then

$$\mu_{(p,i),(p,j)}(\xi) = \sqrt{m}\,[\mathcal{W}_p^* - (\widetilde{T}_p\mathcal{W}_p)^*T_p]_{i,j} = \sqrt{m}\left(\overline{[\mathcal{W}_p]_{i,j}} - \overline{\widetilde{\mu}'_{0,p,i}(\xi)}\mu_{0,p,j}(\xi)\right),$$

where $\widetilde{\mu}'_{0,p,i}$, $i = 0, \dots, N_p - 1$, are the elements of the row $\widetilde{T}_p\mathcal{W}_p$, $p = 1, \dots, \#\Lambda - 1$. All $\widetilde{\mu}'_{0,p,i}$ have the $\mathcal{H}$-symmetry property by Lemma 5.2.4. Next, we obtain wavelet masks using the polyphase representation formula. For fixed $p = 1, \dots, \#\Lambda - 1$, $i = 0, \dots, N_p - 1$, we get

$$m_{(p,i)}(\xi) = \sum_{j=0}^{N_p-1} \overline{[W_p]_{i,j}} e^{2\pi i(s_{p,j},\xi)} - \overline{\widetilde{\mu}'_{0,p,i}(M^*\xi)} m_0(\xi).$$

Check that $m_{(p,i)}$ has the $\mathcal{H}$-symmetry property. For a fixed $K \in \mathcal{H}$, with $K = E^{(k)}F$, $E^{(k)} \in \mathcal{E}_p$, $F \in \mathcal{H}_p$, by (5.33), (5.29), the special assumptions and the properties of $\mathcal{W}_p$ we obtain

$$\begin{aligned} m_{(p,i)}(K^*\xi) &= \sum_{j=0}^{N_p-1} \overline{[W_p]_{i,j}} e^{2\pi i(Ks_{p,j},\xi)} - \overline{\widetilde{\mu}'_{0,p,i}(M^*K^*\xi)} m_0(K^*\xi) \\ &= \sum_{j=0}^{N_p-1} \overline{[W_p]_{i,k\oplus j}} [W_p]_{i,k} e^{2\pi i(s_{p,k\oplus j},\xi)} e^{2\pi i(E^{k\oplus j}Mr^F_{p,0},\xi)} - \\ &\qquad - [W_p]_{k,i} \overline{\widetilde{\mu}'_{0,p,i}(M^*\xi)} e^{2\pi i(Mr^F_{p,0},\xi)} m_0(\xi) \\ &= [W_p]_{k,i} e^{2\pi i(Mr^F_{p,0},\xi)} m_{(p,i)}(\xi) \end{aligned}$$

Thus, we get the $\mathcal{H}$-symmetry property for the wavelet mask $m_{(p,i)}$. In our case, the $\mathcal{H}$-symmetry property means that when $K \in \mathcal{H}_{p,0}$, i.e., $K = F = E^{(0)}F$, then $m_{(p,i)}(\xi)$ is $\mathcal{H}_{p,0}$-symmetric with respect to the center $s_{p,0}$. When $K \in \mathcal{E}_p$, i.e., $K = E^{(k)}$, then $m_{(p,i)}(\xi)$ has the $\mathcal{E}_p$-symmetry property, namely $m_{(p,i)}(E^{(k)*}\xi) = [\mathcal{W}_p]_{i,k}\ m_{(p,i)}(\xi)$, since $r^F_{p,0} = r^{I_d}_{p,0} = \mathbf{0}$. It is not hard to see that the dual wavelet masks $\widetilde{m}_{(p,i)}$ have the same $\mathcal{H}$-symmetry property as $m_{(p,i)}$.

The vanishing moments of order n for the wavelet masks are provided by Lemma 3.4.1.◊

Finally, we indicate the symmetric properties of wavelets that correspond to the constructed wavelet masks in Theorems 5.2.3, 5.2.5.

Theorem 5.2.6 *Suppose conditions of Theorem 5.2.3 are valid. Then wavelet functions corresponding to the constructed wavelet masks have the following symmetry properties: for any* $p = 0, \dots, \#\Lambda - 1$

$$\widehat{\psi_{(p,0)}}((M^{-1}FM)^*\xi) = \widehat{\psi_{(p,0)}}(\xi)e^{2\pi i(M^{-1}(F-I_d)s_{p,0},\xi)}, \quad \forall F \in \mathcal{H}_{p,0}.$$

$$\widehat{\psi_{(p,i)}}(\xi) = \widehat{\psi_{(p,0)}}((M^{-1}E^{(i)}M)^*\xi), \quad E^{(i)} \in \mathcal{E}_p, \quad i = 0, \dots \#\mathcal{E}_p - 1,$$

and analogously for dual wavelets $\widetilde{\psi}_{(p,i)}$.

Proof Since φ is $\mathcal{H}$-symmetric with respect to the origin, i.e., $\widehat{\varphi}(K^*\xi) = \widehat{\varphi}(\xi)$ for any $K \in \mathcal{H}$ (by Lemma 5.1.2) and for $p = 0, \dots, \#\Lambda - 1$

$$m_{(p,0)}(F^*\xi) = m_{(p,0)}(\xi)e^{2\pi i(Fs_{p,0}-s_{p,0},\xi)}, \quad \forall F \in \mathcal{H}_{p,0},$$

$$m_{(p,i)}(\xi) = m_{(p,0)}(E^{(i)*}\xi), \quad E^{(i)} \in \mathcal{E}_p, \quad i = 0, \dots \#\mathcal{E}_p - 1,$$

using the definition of wavelets $\widehat{\psi_{(p,i)}}(\xi) = m_{(p,i)}(M^{*-1}\xi)\widehat{\varphi}(M^{*-1}\xi)$, we obtain that $\forall F \in \mathcal{H}_{p,0}$

$$\begin{aligned} \widehat{\psi_{(p,0)}}((M^{-1}FM)^*\xi) &= m_{(p,0)}(F^*M^{*-1}\xi)\widehat{\varphi}(F^*M^{*-1}\xi) \\ &= \widehat{\psi_{(p,0)}}(\xi)e^{2\pi i(Fs_{p,0}-s_{p,0},M^{*-1}\xi)} \end{aligned}$$

and $i = 0, \dots \#\mathcal{E}_p - 1$,

$$\begin{aligned} \widehat{\psi_{(p,i)}}(\xi) &= m_{(p,i)}(M^{*-1}\xi)\widehat{\varphi}(M^{*-1}\xi) \\ &= m_{(p,0)}(M^{*-1}(M^{-1}E^{(i)}M)^*\xi)\widehat{\varphi}(M^{*-1}(M^{-1}E^{(i)}M)^*\xi) \\ &= \widehat{\psi_{(p,0)}}((M^{-1}E^{(i)}M)^*\xi). \end{aligned}$$

Since the symmetry properties for dual masks are the same, we get the analogous relations for dual wavelets.◊

Theorem 5.2.7 *Suppose conditions of Theorem 5.2.5 are valid. Then wavelet functions corresponding to the constructed wavelet masks have the following symmetry properties: for any* $p = 0, \dots, \#\Lambda - 1$, $i = 0, \dots \#\mathcal{E}_p - 1$

$$\widehat{\psi_{(p,i)}}((M^{-1}KM)^*\xi) = [W_p]_{k,i}\widehat{\psi_{(p,0)}}(\xi)e^{2\pi i(M^{-1}(F-I_d)s_{p,0},\xi)}, \quad \forall K \in \mathcal{H},$$

where $K = E^{(k)}F$, $F \in \mathcal{H}_{p,0}$, $E^{(k)} \in \mathcal{E}_p$. *The analogous relation are valid for dual wavelets* $\widetilde{\psi}_{(p,i)}$.

Proof Since φ is $\mathcal{H}$-symmetric with respect to the origin, i.e., $\widehat{\varphi}(K^*\xi) = \widehat{\varphi}(\xi)$ for any $K \in \mathcal{H}$ (by Lemma 5.1.2) and

$$m_{(p,i)}(K^*\xi) = [W_p]_{k,i}e^{2\pi i(Fs_{p,0}-s_{p,0},\xi)}m_{(p,i)}(\xi), \quad \forall K \in \mathcal{H}, \quad K = E^{(k)}F,$$

using the definition of wavelets $\widehat{\psi_{(p,i)}}(\xi) = m_{(p,i)}(M^{*-1}\xi)\widehat{\varphi}(M^{*-1}\xi)$, we obtain that

$$\begin{aligned} \widehat{\psi_{(p,i)}}((M^{-1}KM)^*\xi) &= m_{(p,i)}(KM^{*-1}\xi)\widehat{\varphi}(KM^{*-1}\xi) \\ &= [W_p]_{k,i}\widehat{\psi_{(p,i)}}(\xi)e^{2\pi i(Fs_{p,0}-s_{p,0},M^{*-1}\xi)}. \end{aligned}$$

Since the symmetry properties for dual masks are the same, we get the analogous relations for dual wavelets.◊

5.3 Symmetric Frame-Like Wavelets and Dual Wavelet Frames

In the noninterpolatory case, a problem for the construction of symmetric dual wavelet frames is more difficult since now we do not have an explicit matrix extension formulas. Also these is no simple algorithm for the polyphase matrix extension in the multivariate case. But the rejection of the frame requirements can simplify the construction as it was shown in Chap. 4. The resulting frame-like wavelet systems are also suitable for wavelet analysis. Recall that the basic framework for the construction of frame-like wavelets includes Lemma 4.3.3 and Remark 4.3.4.

We start with an $\mathcal{H}$-symmetric refinable mask that has sum rule of order n. Next, we need to construct an $\mathcal{H}$-symmetric dual refinable mask $\widetilde{m}_0$ such that

$$D^{\beta}\left(1-m_0(\xi)\overline{\widetilde{m}_0(\xi)}\right)\Big|_{\xi=0}=0\ \forall\beta\in\mathbb{Z}_+^d,\ [\beta]<n. \tag{5.36}$$

The method is given in the following Theorem.

Theorem 5.3.1 *Let a dilation matrix M and a center c be appropriate for a symmetry group $\mathcal{H}$, $n\in\mathbb{N}$. Suppose m_0 is an $\mathcal{H}$-symmetric with respect to the center c mask that has sum rule of order n, $m_0(\mathbf{0})=1$. Then, there exists a dual mask $\widetilde{m}_0$ that is $\mathcal{H}$-symmetric with respect to the center c and satisfies condition (5.36).*

Proof Set $(2\pi i)^{[\beta]}\rho_\beta := D^\beta m_0(\mathbf{0})$, for all $\beta\in\mathbb{Z}_+^d$, $[\beta]<n$. And let numbers $\widetilde{\rho}_\beta$ satisfy

$$\sum_{0\le\alpha\le\beta}(-1)^{[\beta-\alpha]}\binom{\beta}{\alpha}\rho_\alpha\overline{\widetilde{\rho}_{\beta-\alpha}}=0,\quad\forall\beta\in\mathbb{Z}_+^d,\ [\beta]<n. \tag{5.37}$$

Numbers $\widetilde{\rho}_\beta$ can be found recursively from (5.37). Define mask $\widetilde{m}_0$ as follows:

$$\widetilde{m}_0(\xi)=\frac{1}{\#\mathcal{H}}\sum_{E\in\mathcal{H}}G(E^*\xi)e^{2\pi i(c-Ec,\xi)}, \tag{5.38}$$

where $G(\xi)$ is a trigonometric polynomial such that $D^\beta G(\mathbf{0})=(2\pi i)^{[\beta]}\widetilde{\rho}_\beta$ for all $\beta\in\mathbb{Z}_+^d[\beta]<n$. It is not hard to check that $\widetilde{m}_0(\xi)$ is $\mathcal{H}$-symmetric with respect to the point c. Let us show that condition (5.36) is also valid. When $\beta=\mathbf{0}$, then $\rho_\mathbf{0}=\widetilde{\rho}_\mathbf{0}=1$, and $m_0(\mathbf{0})=\widetilde{m}_0(\mathbf{0})=1$. For $\beta\ne\mathbf{0}$, condition (5.36) will be valid if

$$D^{\beta}\left(m_0(\xi)\overline{G(E^*\xi)e^{2\pi i(c-Ec,\xi)}}\right)\Big|_{\xi=0}=0,\quad\forall\beta\in\mathbb{Z}_+^d,0<[\beta]<n,\quad\forall E\in\mathcal{H}. \tag{5.39}$$

Since m_0 is $\mathcal{H}$-symmetric, then $m_0(\xi)\overline{G(E^*\xi)e^{2\pi i(c-Ec,\xi)}}=m_0(E^*\xi)\overline{G(E^*\xi)}$. Due to the higher chain rule with the linear change of variables, equalities (5.39) are equivalent to

$$D^{\beta}\left(m_0(\xi)\overline{G(\xi)}\right)\Big|_{\xi=0}=0,\quad \forall\beta\in\mathbb{Z}_+^d, 0<[\beta]<n.$$

These equalities are valid, since $D^{\beta}G(\mathbf{0}) = (2\pi i)^{[\beta]}\widetilde{\rho}_{\beta}$, $\beta \in \mathbb{Z}_+^d$, $[\beta] < n$, and numbers $\widetilde{\rho}_{\beta}$ satisfy (5.37).◊

For applications, it is important to get refinable masks with the minimal spectrum. The general recipe for the construction of an $\mathcal{H}$-symmetric dual refinable mask with the minimal spectrum is based on a fact that for any vector $k \in \mathrm{spec}(G)$, the set of integer points $\{Ek + c - Ec : E \in \mathcal{H}\}$ should be a subset of $\mathrm{spec}(\widetilde{m}_0)$ (see (5.38)). Thus, for some fixed $k_1 \in \mathbb{Z}^d$, one should check whether the system of linear equations $D^{\beta}G(\mathbf{0}) = (2\pi i)^{|\beta|}\widetilde{\rho}_{\beta}$, $\forall\beta \in \mathbb{Z}_+^d$, $[\beta] < n.$, is solvable with $\mathrm{spec}(G) = K_1 := \{Ek_1 + c - Ec : E \in \mathcal{H}\}$. If yes, then one solves the system. If not, one should add another point $k_2 \in \mathbb{Z}^d$ and check the solvability of the system with $\mathrm{spec}(G) = K_1 \cup \{Ek_2 + c - Ec : F \in \mathcal{H}\}$ and so on.

Assume we have two $\mathcal{H}$-symmetric masks m_0 and $\widetilde{m}_0$ such that (5.36) is valid. Let us consider the construction of wavelet systems using the basic matrix extension described in Lemma 4.3.3

$$\mathcal{N}=\left(\begin{array}{c|c} P & 1-P\widetilde{P}^* \\ \hline I_m & -\widetilde{P}^* \end{array}\right),\quad \widetilde{\mathcal{N}}=\left(\begin{array}{c|c} \widetilde{P} & 1 \\ \hline I_m - P^*\widetilde{P} & -P^* \end{array}\right), \tag{5.40}$$

where P and $\widetilde{P}$ are the rows of the polyphase components of masks m_0 and $\widetilde{m}_0$. For convenience, we use the following enumeration of wavelet masks. Let $\mu_{0,p,i}$ and $\widetilde{\mu}_{0,p,i}$, $i = 0, \dots, \#\mathcal{E}_p - 1$, $p = 0, \dots, \#\Lambda - 1$, be the polyphase components of m_0 and $\widetilde{m}_0$,

$$T_p = (\mu_{0,p,0}, \dots, \mu_{0,p,\#\mathcal{E}_p-1}),\qquad \widetilde{T}_p = (\widetilde{\mu}_{0,p,0}, \dots, \widetilde{\mu}_{0,p,\#\mathcal{E}_p-1}).$$

Set $P = (T_0, \dots, T_{\#\Lambda-1})$, $\widetilde{P} = (\widetilde{T}_0, \dots, \widetilde{T}_{\#\Lambda-1})$. The polyphase components for wavelet masks m_ν and $\widetilde{m}_\nu$ are contained in the submatrices I_m and $I_m - P^*\widetilde{P}$ in (5.40). Note that $I_m - P^*\widetilde{P}$ is a block matrix

$$I_m - P^*\widetilde{P} = \begin{pmatrix} I_{N_0-1} - T_0^*\widetilde{T}_0 & -T_0^*\widetilde{T}_1 & \dots & -T_0^*\widetilde{T}_{\#\Lambda-1} \\ -T_1^*\widetilde{T}_0 & I_{N_1-1} - T_1^*\widetilde{T}_1 & \dots & -T_1^*\widetilde{T}_{\#\Lambda-1} \\ \vdots & & \ddots & \vdots \\ -T_{\#\Lambda-1}\widetilde{T}_0 & \dots & \dots & I_{N(\#\Lambda-1)-1} - T_{\#\Lambda-1}\widetilde{T}_{\#\Lambda-1} \end{pmatrix}, \tag{5.41}$$

where $N_p = \#\mathcal{E}_p$. The rows of the submatrix we enumerate by double index (p, i). The (p, i)th row is the ith row in the pth block. A wavelet mask corresponding to the (p, i)th row is denoted by $\widetilde{m}_{(p,i)}$. Analogously wavelet masks $m_{(p,i)}$ are enumerated.

Theorem 5.3.2 *Let a dilation matrix M and a center c be appropriate for a symmetry group $\mathcal{H}$, $n \in \mathbb{N}$. Suppose m_0 and $\widetilde{m}_0$ are $\mathcal{H}$-symmetric with respect to the center c masks such that mask m_0 has sum rule of order n, mask $\widetilde{m}_0$ satisfies condition (5.36).*

Then, wavelet masks $m_{(p,i)}$ and $\widetilde{m}_{(p,i)}$ constructed using matrix extension (5.40) have the following symmetry properties:

1. $m_{(p,0)}, \widetilde{m}_{(p,0)}$ *are* $\mathcal{H}_{p,0}$*-symmetric with respect to the center* $s_{p,0}$,
2. $m_{(p,i)}(\xi) = m_{(p,0)}(E^{(i)*}\xi)e^{2\pi i(c-E^{(i)}c,\xi)}$,
   $\widetilde{m}_{(p,i)}(\xi) = \widetilde{m}_{(p,0)}(E^{(i)*}\xi)e^{2\pi i(c-E^{(i)}c,\xi)}$, $E^{(i)} \in \mathcal{E}_p$,

where $i = 0, \dots, \#\mathcal{E}_p - 1$, $p = 0, \dots, \#\Lambda - 1$. *Wavelet masks* $\widetilde{m}_{(p,i)}$ *have vanishing moments of order* n. *The corresponding MRA-based dual wavelet system is almost frame-like in* S'.

Proof Wavelet masks from submatrix $I_m - P^*\widetilde{P}$ in (5.40) are given by

$$\widetilde{m}_{(p,i)}(\xi) = \frac{1}{\sqrt{m}}e^{2\pi i(s_{p,i},\xi)} - \overline{\mu_{0,p,i}(M^*\xi)}\widetilde{m}_0(\xi),$$

$i = 0, \dots, \#\mathcal{E}_p - 1$, $p = 0, \dots, \#\Lambda - 1$. Then, mask $m_{(p,0)}$ is $\mathcal{H}_{p,0}$-symmetric with respect to the center $s_{p,0}$. Indeed, due to (5.9) and (5.13) for all $F \in \mathcal{H}_{p,0}$ we have

$$\begin{aligned}\widetilde{m}_{(p,0)}(F^*\xi) &= \frac{1}{\sqrt{m}}e^{2\pi i(Fs_{p,0},\xi)} - \overline{\mu_{0,p,0}(M^*F^*\xi)}\widetilde{m}_0(F^*\xi) \\ &= \widetilde{m}_{(p,0)}(\xi)e^{2\pi i(Fs_{p,0}-s_{p,0},\xi)}.\end{aligned}$$

Also, by (5.8) and (5.14) we have

$$\begin{aligned}\widetilde{m}_{(p,0)}(E^{(i)*}\xi) &= \frac{1}{\sqrt{m}}e^{2\pi i(E^{(i)}s_{p,0},\xi)} - \overline{\mu_{0,p,0}(M^*E^{(i)*}\xi)}\widetilde{m}_0(E^{(i)*}\xi) \\ &= \widetilde{m}_{(p,i)}(\xi)e^{2\pi i(E^{(i)}c-c,\xi)}.\end{aligned}$$

Similarly, it can be proved that $\widetilde{m}_{(p,i)}$ is $\mathcal{H}_{p,i}$-symmetric with respect to the center $s_{p,i}$, i.e., for all $\widetilde{F} \in \mathcal{H}_{p,i} : \widetilde{m}_{(p,i)}(\widetilde{F}^*\xi) = \widetilde{m}_{(p,i)}(\xi)e^{2\pi i(\widetilde{F}s_{p,i}-s_{p,i},\xi)}$. It is easy to check that the same symmetric properties are valid for wavelet masks $m_{(p,i)}$, since

$$m_{(p,i)}(\xi) = \frac{1}{\sqrt{m}}e^{2\pi i(s_{p,i},\xi)}, \quad i = 0, \dots, \#\mathcal{E}_p - 1, \, p = 0, \dots, \#\Lambda - 1.$$

Vanishing moments of order n for the dual wavelet masks $\widetilde{m}_{(p,i)}$ are provided by Lemma 4.3.3.◊

Thus, the matrix extension (5.40) leads to wavelet masks that are mutually symmetric, i.e., some wavelet masks are reflected or rotated copies of the others. Notice that if for some $\mathcal{H}$ and M we get that $\mathcal{E}_p = \{I_d\}$ for all $p = 0, \dots, \#\Lambda - 1$, then all wavelet masks constructed by Theorem 5.3.2 are $\mathcal{H}$-symmetric.

Based on Theorem 5.3.2, it is possible to improve frame-like wavelet to dual wavelet frames. The key step here is the so-called lifting scheme that was introduced by Sweldens [1]. It is a tool for designing wavelets and performing the discrete wavelet transform. The lifting scheme has lots of different applications and useful properties. One important feature is that the lifting scheme allows to improve properties of a given wavelet system. In particular, it allows to provide additional vanishing moments for wavelet masks (see, e.g., [2]). Since vanishing moments are necessary and sufficient conditions for wavelet system to be a dual wavelet frame, the lifting scheme can help us to improve frame-like wavelet systems to frames. In this section, we also give a method that helps to preserve symmetry properties of wavelets during the improvement.

Let refinable masks m_0, $\widetilde{m}_0$ and wavelet masks m_ν, $\widetilde{m}_\nu$, $\nu = 1, \dots, r$ be such that the corresponding polyphase matrices $\mathcal{M}$, $\widetilde{\mathcal{M}}$ satisfy $\mathcal{M}^*\widetilde{\mathcal{M}} = I_m$. Let $L_1, \dots, L_r$ be trigonometric polynomials. Define $r \times r$ matrices $\mathcal{L}$, $\widetilde{\mathcal{L}}$ as follows:

$$\mathcal{L} := \begin{pmatrix} 1 & \mathbf{0} \\ L^T & I_{r-1} \end{pmatrix}, \ \widetilde{\mathcal{L}} := \begin{pmatrix} 1 & -\overline{L} \\ \mathbf{0} & I_{r-1} \end{pmatrix},$$

where $L = (L_1, \dots, L_r)$. It is easy to check that $\mathcal{L}^*\widetilde{\mathcal{L}} = I_r$. Define new polyphase matrices $\mathcal{M}_{new} := \mathcal{L}\mathcal{M}$ and $\widetilde{\mathcal{M}}_{new} := \widetilde{\mathcal{L}}\widetilde{\mathcal{M}}$. Note that the equality $\mathcal{M}^*_{new}\widetilde{\mathcal{M}}_{new} = I_m$ is preserved. Denote the elements of the new matrices as follows: $\mathcal{M}_{new} = \{\mu^n_{\nu j}\}_{\nu=0,r}^{j=0,m-1}$, $\widetilde{\mathcal{M}}_{new} = \{\widetilde{\mu}^n_{\nu j}\}_{\nu=0,r}^{j=0,m-1}$. The transformation formulas for the polyphase components are the following:

$$\begin{aligned} &\mu^n_{0,j}(\xi) = \mu_{0,j}(\xi), && \mu^n_{\nu j}(\xi) = \mu_{\nu j}(\xi) + L_\nu(\xi)\mu_{0,j}(\xi), \\ &\widetilde{\mu}^n_{0,j}(\xi) = \widetilde{\mu}_{0,j}(\xi) - \sum_{i=1}^{r} \overline{L_i}(\xi)\widetilde{\mu}_{i,j}(\xi), && \widetilde{\mu}^n_{\nu j}(\xi) = \widetilde{\mu}_{\nu j}(\xi). \end{aligned}$$

Let new masks m^n_ν, $\widetilde{m}^n_\nu$ be constructed from the new polyphase components. Thus, the transformation formulas for the masks are

$$\begin{aligned} &m^n_0(\xi) = m_0(\xi), && m^n_\nu(\xi) = m_\nu(\xi) + L_\nu(M^*\xi)m_0(\xi), \\ &\widetilde{m}^n_0(\xi) = \widetilde{m}_0(\xi) - \sum_{i=1}^{r} \overline{L_i}(M^*\xi)\widetilde{m}_i(\xi), && \widetilde{m}^n_\nu(\xi) = \widetilde{m}_\nu(\xi). \end{aligned} \tag{5.42}$$

The above transformation of the masks is called a *lifting scheme transformation*. Note that the choice of trigonometric polynomials L_ν, $\nu = 1, \dots, r$ is not restricted.

The aim is to find a lifting scheme transformation such that all new masks preserve their symmetry properties and all new wavelet masks have vanishing moments at least of order 1.

Theorem 5.3.3 *Let a dilation matrix M and a center c be appropriate for a symmetry group $\mathcal{H}$, $n \in \mathbb{N}$. Suppose m_0 and $\widetilde{m}_0$ are as in Theorem 5.3.2. Suppose $m_{(p,i)}$, $\widetilde{m}_{(p,i)}$ are wavelet masks constructed using Theorem 5.3.2. Assume that trigonometric polynomials $L_{p,i}$, satisfy $L_{p,i}(\mathbf{0}) = -m_{p,i}(\mathbf{0})$ and*

$$L_{p,0}(M^*F^*M^{*-1}\xi) = L_{p,0}(\xi)e^{2\pi i(r^F_{p,0},\xi)}, \quad \forall F \in \mathcal{H}_{p,0}, \tag{5.43}$$

$$L_{p,0}(M^* E^{(i)*} M^{*-1}\xi) = L_{p,i}(\xi), \quad E^{(i)} \in \mathcal{E}_p, \tag{5.44}$$

for $i = 0, \dots, \#\mathcal{E}_p - 1$, $p = 0, \dots, \#\Lambda - 1$. *New masks* m_0^n, $\widetilde{m}_0^n$, $m_{(p,i)}^n$, $\widetilde{m}_{(p,i)}^n$ *defined by the lifting scheme transformation (5.42) preserve the symmetry properties of masks* m_0, $\widetilde{m}_0$, $m_{(p,i)}$, $\widetilde{m}_{(p,i)}$, *respectively. New wavelet masks* $m_{(p,i)}^n$ *have vanishing moments at least of order 1. If new refinable functions* φ, $\widetilde{\varphi}$ *corresponding to new refinable masks* m_0^n, $\widetilde{m}_0^n$ *are in* $L_2(\mathbb{R}^d)$, *then the resulting wavelet system is a dual wavelet frame.*

Proof Vanishing moments at least of order 1 for wavelet masks $m_{(p,i)}^n$, are provided by conditions $L_{p,i}(\mathbf{0}) = -m_{p,i}(\mathbf{0})$. This follows from (5.42).

Next, we show that the symmetry properties are preserved. For $m_{(p,0)}^n$, Item 1 in Theorem 5.3.2 is preserved, since conditions (5.42) and (5.43) yield

$$\begin{aligned} m_{(p,0)}^n(F^*\xi) &= m_{(p,0)}(F^*\xi) + L_{(p,0)}(M^* F^*\xi) m_0(F^*\xi) \\ &= m_{(p,0)}(\xi) e^{2\pi i (F s_{p,0} - s_{p,0}, \xi)} + L_{(p,0)}(M^* F^*\xi) m_0(\xi) e^{2\pi i (Fc - c, \xi)} \\ &= m_{(p,0)}^n(\xi) e^{2\pi i (F s_{p,0} - s_{p,0}, \xi)}, \end{aligned} \tag{5.45}$$

where $r_{p,0}^F = M^{-1}(c - s_{p,0}) - M^{-1}F(c - s_{p,0})$. Next, Item 2 in Theorem 5.3.2 for $m_{(p,i)}^n$ is preserved, since conditions (5.42) and (5.44) yield

$$\begin{aligned} m_{(p,0)}^n(E^{(i)*}\xi) &= m_{(p,0)}(E^{(i)*}\xi) + L_{(p,0)}(M^* E^{(i)*}\xi) m_0(E^{(i)*}\xi) \\ &= m_{(p,i)}(\xi) e^{2\pi i (E^{(i)}c - c, \xi)} + L_{(p,i)}(M^*\xi) m_0(\xi) e^{2\pi i (E^{(i)}c - c, \xi)} \\ &= m_{(p,i)}^n(\xi) e^{2\pi i (E^{(i)}c - c, \xi)}, \end{aligned} \tag{5.46}$$

where $E^{(i)} \in \mathcal{E}_p$, $i = 0, \dots, \#\mathcal{E}_p - 1$, $p = 0, \dots, \#\Lambda - 1$.

Now, we show that new refinable mask $\widetilde{m}_0^n$ remains $\mathcal{H}$-symmetric with respect to the point c. Let us fix p. Recall that $\mathcal{H}$ can be uniquely represented as follows: $\mathcal{H} = \mathcal{E}_p \times \mathcal{H}_{p,0}$. Namely, for a matrix K and $E^{(i)}$ in $\mathcal{H}$ there exist unique matrices $E^{(j)} \in \mathcal{E}_p$ and $F \in \mathcal{H}_{p,0}$ such that $KE^{(i)} = E^{(j)}F$. In other words, this representation defines a mapping $j(p, \cdot, K)$ from the set of indices $\{0, \dots, \#\mathcal{E}_p - 1\}$ to itself where index j is uniquely defined by index i and matrix $K \in \mathcal{H}$. Let $K \in \mathcal{H}$. Let us consider

$$\widetilde{m}_0^n(K^*\xi) = \widetilde{m}_0(\xi) e^{2\pi i (Kc - c, \xi)} - \sum_{p=0}^{\#\Lambda - 1} \sum_{i=0}^{\#\mathcal{E}_p - 1} \overline{L_{p,i}(M^* K^*\xi)} \widetilde{m}_{p,i}(K^*\xi). \tag{5.47}$$

Let us fix p and i. Then, $KE^{(i)} = E^{(j)}F$, where $F \in \mathcal{H}_{p,0}$, $E^{(j)} \in \mathcal{E}_p$, and $j = j(p, i, K)$. With conditions on $L_{p,i}$, we obtain

$$\begin{aligned}\overline{L_{p,i}(M^*K^*\xi)} = \overline{L_{p,0}(M^*E^{(i)*}K^*\xi)} &= \overline{L_{p,0}(M^*F^*E^{(j)*}\xi)}\\ &= \overline{L_{p,0}(M^*E^{(j)*}\xi)}e^{-2\pi i(Mr^F_{p,0},E^{(j)*}\xi)}\\ &= \overline{L_{p,j}(M^*\xi)}e^{-2\pi i(E^{(j)}Mr^F_{p,0},\xi)},\end{aligned}$$

where $-E^{(j)}Mr^F_{p,0} = E^{(j)}s_{p,0} - E^{(j)}c - KE^{(i)}(s_{p,0} - c)$. Using (5.45) and (5.46) we get

$$\begin{aligned}\widetilde{m}_{(p,i)}(K^*\xi) &= \widetilde{m}_{(p,0)}(E^{(i)*}K^*\xi)e^{2\pi i(c-E^{(i)}c,K^*\xi)}\\ &= \widetilde{m}_{(p,0)}(F^*E^{(j)*}\xi)e^{2\pi i(c-E^{(i)}c,K^*\xi)}\\ &= \widetilde{m}_{(p,0)}(E^{(j)*}\xi)e^{2\pi i(Fs_{p,0}-s_{p,0},E^{(j)*}\xi)}e^{2\pi i(c-E^{(i)}c,K^*\xi)}\\ &= \widetilde{m}_{(p,j)}(\xi)e^{2\pi i(E^{(j)}c-c,\xi)}e^{2\pi i(E^{(j)}(Fs_{p,0}-s_{p,0}),\xi)}e^{2\pi i(Kc-KE^{(i)}c,\xi)}.\end{aligned}$$

So, finally we have

$$\overline{L_{p,i}(M^*K^*\xi)}\widetilde{m}_{(p,i)}(K^*\xi) = \widetilde{m}_{(p,j)}(\xi)\overline{L_{p,j}(M^*\xi)}e^{2\pi i(R,\xi)},$$

where

$$\begin{aligned}R &= -E^{(j)}Mr^F_{p,0} + E^{(j)}c - c + E^{(j)}(Fs_{p,0} - s_{p,0}) + Kc - KE^{(i)}c\\ &= E^{(j)}s_{p,0} - E^{(j)}c - KE^{(i)}(s_{p,0} - c) + E^{(j)}c - c +\\ &\qquad\qquad KE^{(i)}s_{p,0} - E^{(j)}s_{p,0} + Kc - KE^{(i)}c\\ &= Kc - c.\end{aligned}$$

Thus, by (5.47) we obtain that $\widetilde{m}^n_0$ is $\mathcal{H}$-symmetric with respect to the center c.◊

An alternative approach for the construction of dual wavelet frames is based on Algorithm 1 in Sect. 3.4. Here, we modify this algorithm, such that the constructed wavelets have the symmetry properties. Firstly, we give a constructive description of the method. Let a dilation matrix M and a center c be appropriate for $\mathcal{H}$, $n \in \mathbb{N}$.

Algorithm

Step 1. Find a refinable mask m_0 that is $\mathcal{H}$-symmetric with respect to the center c and has sum rule of order n. Let $(2\pi i)^{[\beta]}\rho_\beta := D^\beta m_0(\mathbf{0})$, for all $\beta \in \mathbb{Z}^d_+$, $[\beta] < n$.
Step 2. Next, define numbers $\widetilde{\rho}_\beta$, for all $\beta \in \mathbb{Z}^d_+$, $[\beta] < n$ such that (5.37) is valid and find a dual mask $\widetilde{m}'_0$ such that $\widetilde{m}'_0$ is $\mathcal{H}$-symmetric with respect to the center c and has sum rule of order n and $(2\pi i)^{[\beta]}\widetilde{\rho}_\beta := D^\beta\widetilde{m}'_0(\mathbf{0})$.
Step 3. Let $\widetilde{\mu}'_{00}, \dots, \widetilde{\mu}'_{0,m-1}$ be the polyphase components of mask $\widetilde{m}'_0$. Set

$$\sigma := \sum_{l=0}^{m-1}\overline{\mu_{0l}}\,\widetilde{\mu}'_{0l}, \quad \widetilde{\mu}_{0k} := (2-\sigma)\widetilde{\mu}'_{0k}, k = 0, \dots, m-1, \tag{5.48}$$

$$\mu_{0m} := \overline{(1-\sigma)}, \ \widetilde{\mu}_{0m} := 1 - \sigma. \tag{5.49}$$

Due to Corollary 3.3.10, we have $D^\beta\sigma(\mathbf{0}) = 0$ for all $\beta \in \mathbb{Z}_+^d$, $0 < |\beta| < n$. Therefore, $D^\beta\left(\overline{1-\sigma}\right)(\mathbf{0}) = 0$ for all $\beta \in \mathbb{Z}_+^d$, $0 \le |\beta| < n$. It follows that $D^\beta\widetilde{\mu}_{0k}(\mathbf{0}) = D^\beta\widetilde{\mu}'_{0k}(\mathbf{0})$ for all $\beta \in \mathbb{Z}_+^d$, $|\beta| < n$. Therefore,

$$1 - \sum_{k=0}^{m-1} \overline{\mu_{0k}}\widetilde{\mu}_{0k} = 1 - (2-\sigma)\sum_{k=0}^{m-1} \overline{\mu_{0k}}\widetilde{\mu}'_{0k} = (1-\sigma)^2,$$

which yields $\sum_{k=0}^m \overline{\mu_{0k}}\widetilde{\mu}_{0k} = 1$. Dual refinable mask $\widetilde{m}_0$ is constructed from the polyphase components $\widetilde{\mu}_{0k}$. Note that $\widetilde{m}_0$ has sum rule of order n. Since σ is $\mathcal{H}$-symmetric with respect to the origin (the proof is below in Theorem 5.3.4), then dual refinable mask $\widetilde{m}_0$ is $\mathcal{H}$-symmetric with respect to the center c.

Step 4. Set $r = m$ if $\sigma \equiv 1$; otherwise, we set $r = m+1$, $\mu_{0,m+1} \equiv 0$, $\widetilde{\mu}_{0,m+1} \equiv 0$. Matrix extension $\mathcal{N} := \{\mu_{\nu k}\}_{\nu,k=0}^r$, $\widetilde{\mathcal{N}} := \{\widetilde{\mu}_{\nu k}\}_{\nu,k=0}^r$ can be realized as follows, if $r = m$,

$$\mathcal{N} := \begin{pmatrix} P & 0 \\ I_m - \widetilde{P}^*P & \widetilde{P}^* \end{pmatrix}, \widetilde{\mathcal{N}} := \begin{pmatrix} \widetilde{P} & 0 \\ I_m - P^*\widetilde{P} & P^* \end{pmatrix}, \tag{5.50}$$

or, if $r = m+1$

$$\mathcal{N} := \begin{pmatrix} P & \mu_{0m} & 0 \\ I_m - \widetilde{P}^*P & \mu_{0m}\widetilde{P}^* & \widetilde{P}^* \\ -\overline{\widetilde{\mu}_{0m}}P & 1 - \overline{\widetilde{\mu}_{0m}}\mu_{0m} & \overline{\widetilde{\mu}_{0m}} \end{pmatrix},$$

$$\widetilde{\mathcal{N}} := \begin{pmatrix} \widetilde{P} & \widetilde{\mu}_{0m} & 0 \\ I_m - P^*\widetilde{P} & \widetilde{\mu}_{0m}P^* & P^* \\ -\overline{\mu_{0m}}\widetilde{P} & 1 - \overline{\mu_{0m}}\widetilde{\mu}_{0m} & \overline{\mu_{0m}} \end{pmatrix}, \tag{5.51}$$

where $P = (\mu_{00}, \dots, \mu_{0,m-1})$, $\widetilde{P} = (\widetilde{\mu}_{00}, \dots, \widetilde{\mu}_{0,m-1})$. It is not difficult to see that the matrices satisfy $\mathcal{N}\widetilde{\mathcal{N}}^* = I_r$. This yields that the columns of the polyphase matrices $\mathcal{M}, \widetilde{\mathcal{M}}$ are biorthonormal. Wavelet masks $m_\nu, \widetilde{m}_\nu$, $\nu = 1, \dots, r$, are constructed from the polyphase components.♢♢

The next Theorem states that wavelet masks constructed by Algorithm 1 have symmetry properties. Again, for convenience, we enumerate wavelet masks using double index as in Theorem 5.3.2. For the submatrices $I_m - \widetilde{P}^*P$ and $I_m - P^*\widetilde{P}$, we use notations as in (5.41).

Theorem 5.3.4 *Let a dilation matrix M and a center c be appropriate for a symmetry group $\mathcal{H}$, $n \in \mathbb{N}$. Suppose refinable and wavelet masks are constructed by Algorithm 1. Then, wavelet masks $m_{(p,i)}$ and $\widetilde{m}_{(p,i)}$ have symmetry properties, i.e.,*

- *$m_{(p,0)}$, $\widetilde{m}_{(p,0)}$ are $\mathcal{H}_{p,0}$-symmetric with respect to the center $s_{p,0}$*
- *$m_{(p,i)} = m_{(p,0)}(E^{(i)*}\xi)e^{2\pi i(c-E^{(i)}c,\xi)}$, $\widetilde{m}_{(p,i)} = \widetilde{m}_{(p,0)}(E^{(i)*}\xi)e^{2\pi i(c-E^{(i)}c,\xi)}$, $E^{(i)} \in \mathcal{E}_p$,*

$i = 0, \dots, \#\mathcal{E}_p - 1$, $p = 0, \dots, \#\Lambda - 1$. If $r = m + 1$, then wavelet masks m_r and $\widetilde{m}_r$ defined by the last rows of matrices $\mathcal{N}$ and $\widetilde{\mathcal{N}}$ are $\mathcal{H}$-symmetric with respect to the center c. All wavelet masks have vanishing moments of order n. If refinable functions φ, $\widetilde{\varphi}$ corresponding to refinable masks m_0, $\widetilde{m}_0$ are in $L_2(\mathbb{R}^d)$, then the resulting wavelet system is a dual wavelet frame.

Proof Firstly, we prove that σ defined in (5.48) is $\mathcal{H}$-symmetric with respect to the origin. Let us rewrite σ using another enumeration of the polyphase components and show that $\sigma(M^*K^*M^{*-1}\xi) = \sigma(\xi)$ for all $K \in \mathcal{H}$. Thus,

$$\sigma(M^*K^*M^{*-1}\xi) = \sum_{p=0}^{\#\Lambda-1} \sum_{i=0}^{\#\mathcal{E}_p-1} \overline{\mu_{0,p,i}(M^*K^*M^{*-1})}\, \widetilde{\mu}'_{0,p,i}(M^*K^*M^{*-1}) =$$

$$\sum_{p=0}^{\#\Lambda-1} \sum_{i=0}^{\#\mathcal{E}_p-1} \overline{\mu_{0,p,0}(M^*E^{(i)*}K^*M^{*-1}\xi)}\, \widetilde{\mu}'_{0,p,0}(M^*E^{(i)*}K^*M^{*-1}\xi)$$

Let us fix p. Recall that $\mathcal{H}$ can be uniquely represented as follows: $\mathcal{H} = \mathcal{E}_p \times \mathcal{H}_{p,0}$. Recall that for a matrix K and fixed p there exist unique matrices $E^{(j)} \in \mathcal{E}_p$ and $F \in \mathcal{H}_{p,0}$ such that $KE^{(i)} = E^{(j)}F$, where $j = j(p, i, K)$. Using (5.13), (5.14) and continuing the above equalities, we get

$$\sigma(M^*K^*M^{*-1}\xi) =$$

$$\sum_{p=0}^{\#\Lambda-1} \sum_{i=0}^{\#\mathcal{E}_p-1} \overline{\mu_{0,p,0}(M^*F^*E^{(j(p,i,K))*}M^{*-1}\xi)}\, \widetilde{\mu}'_{0,p,0}(M^*F^*E^{(j(p,i,K))*}M^{*-1}\xi) =$$

$$\sum_{p=0}^{\#\Lambda-1} \sum_{i=0}^{\#\mathcal{E}_p-1} \overline{\mu_{0,p,j(p,i,K)}(\xi)}\, \widetilde{\mu}'_{0,p,j(p,i,K)}(\xi) = \sigma(\xi).$$

Since σ is $\mathcal{H}$-symmetric with respect to the origin, then polyphase components $\widetilde{\mu}'_{0k}$ and $\widetilde{\mu}_{0k}$ defined in (5.49) have the same symmetry properties. Therefore, dual refinable mask $\widetilde{m}_0$ is $\mathcal{H}$-symmetric with respect to the center c. Symmetry properties of wavelet masks $m_{(p,i)}, \widetilde{m}_{(p,i)}$ defined by submatrices $I_m - \widetilde{P}^*P$ and $I_m - P^*\widetilde{P}$ in (5.50) or (5.51) can be proved analogously to the proof of Theorem 5.3.2. If $r = m + 1$, then the last wavelet masks m_r and $\widetilde{m}_r$ defined by the last rows of matrices $\mathcal{N}$ and $\widetilde{\mathcal{N}}$ are $\mathcal{H}$-symmetric with respect to the center c. Indeed, since

$$m_r(\xi) = -\overline{\widetilde{\mu}_{0m}(M^*\xi)} m_0(\xi), \quad \widetilde{m}_r(\xi) = -\overline{\mu_{0m}(M^*\xi)} \widetilde{m}_0(\xi),$$

then for all $K \in \mathcal{H}$

$$\begin{aligned} m_r(K^*\xi) &= -\overline{\widetilde{\mu}_{0m}(M^*K^*M^{*-1}M^*\xi)}m_0(K^*\xi) \\ &= -\overline{\widetilde{\mu}_{0m}(M^*\xi)}m_0(\xi)e^{2\pi i(Kc-c,\xi)} \\ &= m_r(\xi)e^{2\pi i(Kc-c,\xi)} \end{aligned}$$

and analogously with $\widetilde{m}_r$.

Vanishing moments for all wavelet masks are provided by Algorithm 1 in Sect. 3.4.$\Diamond$

Note that again wavelet masks $m_{(p,i)}$, $\widetilde{m}_{(p,i)}$ are $\mathcal{H}_{p,i}$–symmetric with respect to the center $s_{p,i}$.

Next, we show how to achieve the $\mathcal{H}$-symmetry property for all wavelet functions constructed in Theorems 5.3.2, 5.3.3, and 5.3.4. Due to considerations in Sect. 5.2, this can be done for an abelian symmetry group $\mathcal{H}$. The key step is to find a unitary transformation of the row of the polyphase components of an $\mathcal{H}$-symmetric mask such that each element of the new row has the $\mathcal{H}$-symmetry property. According to Lemma 5.2.4, we need a *special assumption*

$$r_{p,0}^F = M^{-1}EMr_{p,0}^F, \quad \forall F \in \mathcal{H}_{p,0}, \forall E \in \mathcal{E}_p$$

to ensure the $\mathcal{H}$-symmetry properties.

To provide the $\mathcal{H}$-symmetry properties for all wavelet functions constructed in Theorem 5.3.2, we extend the polyphase matrices in the following form

$$\mathcal{N} = \begin{pmatrix} P & 1 - P\widetilde{P}^* \\ U & -U\widetilde{P}^* \end{pmatrix}, \quad \widetilde{\mathcal{N}} = \begin{pmatrix} \widetilde{P} & 1 \\ \widetilde{U} - \widetilde{U}P^*\widetilde{P} & -\widetilde{U}P^* \end{pmatrix}. \tag{5.52}$$

where U and $\widetilde{U}$ are matrices such that $U\widetilde{U}^* = I_m$.

Define a block diagonal unitary matrix $\mathcal{W}$ as follows:

$$\mathcal{W} = \mathtt{diag}(\mathcal{W}_0, \dots, \mathcal{W}_{\#\Lambda-1}), \tag{5.53}$$

where $\mathcal{W}_p$ is defined by 5.32, $p = 0, \dots, \#\Lambda - 1$. Let m_0 be an $\mathcal{H}$-symmetric mask. Suppose (5.34) is valid for all $p = 0, \dots, \#\Lambda - 1$. Then, by Lemma 5.2.4 matrix $\mathcal{W}$ symmetrizes the row $P = (T_0, \dots, T_{\#\Lambda-1})$ of the polyphase components of m_0, namely all elements of the row

$$P\mathcal{W} = (T_0\mathcal{W}_0, \dots, T_{\#\Lambda-1}\mathcal{W}_{\#\Lambda-1})$$

have the $\mathcal{H}$-symmetry property. The following theorem states that wavelets with the $\mathcal{H}$-symmetry property can be constructed using matrix $\mathcal{W}$.

Theorem 5.3.5 *Let a dilation matrix M and a center c be appropriate for an abelian symmetry group $\mathcal{H}$, $n \in \mathbb{N}$. Suppose m_0 and $\widetilde{m}_0$ are $\mathcal{H}$-symmetric with respect to the center c refinable masks such that mask m_0 has sum rule of order n, mask $\widetilde{m}_0$ satisfies condition (5.36). Assume that condition (5.34) is valid for all $p =$*

$0, \ldots, \#\Lambda - 1$. *Then, there exist wavelet masks* $m_{(p,i)}$ *and* $\widetilde{m}_{(p,i)}$ *which have the* $\mathcal{H}$*-symmetry property, wavelet masks* $\widetilde{m}_{(p,i)}$ *have vanishing moments of order* n, $i = 0, \ldots, \#\mathcal{E}_p - 1$, $p = 0, \ldots, \#\Lambda - 1$. *The corresponding MRA-based dual wavelet system is almost frame-like in* S'.

Proof Let $\mu_{0,p,i}$ and $\widetilde{\mu}_{0,p,i}$ be the polyphase components of m_0 and $\widetilde{m}_0$, $i = 0, \ldots, \#\mathcal{E}_p - 1$, $p = 1, \ldots, \#\Lambda - 1$, and let $T_p = (\mu_{0,p,0}, \ldots, \mu_{0,p,\#\mathcal{E}_p-1})$, and $\widetilde{T}_p = (\widetilde{\mu}_{0,p,0}, \ldots, \widetilde{\mu}_{0,p,\#\mathcal{E}_p-1})$. Set $P = (T_0, \ldots, T_{\#\Lambda-1})$, $\widetilde{P} = (\widetilde{T}_0, \ldots, \widetilde{T}_{\#\Lambda-1})$. Let us consider matrix extension (5.52) with $U = \widetilde{U} = \mathcal{W}^*$

$$\mathcal{N} = \left(\begin{array}{c|c} P & 1 - P\widetilde{P}^* \\ \hline \mathcal{W}^* & -\mathcal{W}^*\widetilde{P}^* \end{array}\right), \quad \widetilde{\mathcal{N}} = \left(\begin{array}{c|c} \widetilde{P} & 1 \\ \hline \mathcal{W}^* - \mathcal{W}^* P^* \widetilde{P} & -\mathcal{W}^* P^* \end{array}\right). \tag{5.54}$$

Let us consider submatrix $\mathcal{W}^* - \mathcal{W}^* P^* \widetilde{P}$. It is a block matrix

$$\mathcal{W}^* - \mathcal{W}^* P^* \widetilde{P} = \begin{pmatrix} \mathcal{W}_0^* - (T_0\mathcal{W}_0)^*\widetilde{T}_0 & -(T_0\mathcal{W}_0)^*\widetilde{T}_1 & \ldots & -(T_0\mathcal{W}_0)^*\widetilde{T}_{\#\Lambda-1} \\ -(T_1\mathcal{W}_1)^*\widetilde{T}_0 & \mathcal{W}_1^* - (T_1\mathcal{W}_1)^*\widetilde{T}_1 & \ldots & -(T_1\mathcal{W}_1)^*\widetilde{T}_{\#\Lambda-1} \\ \vdots & & \ddots & \vdots \\ -(T_{\#\Lambda-1}\mathcal{W}_{\#\Lambda-1})^*\widetilde{T}_0 & \ldots & \ldots & \mathcal{W}_{\#\Lambda-1}^* - (T_{\#\Lambda-1}\mathcal{W}_{\#\Lambda-1})^*\widetilde{T}_{\#\Lambda-1} \end{pmatrix}.$$

Denote the elements of the submatrix as $\mathcal{W}^* - \mathcal{W}^* P^* \widetilde{P} = \{\widetilde{\mu}_{(p,i),(t,j)}\}_{p=0,\ldots,\#\Lambda-1, i=0,\ldots,\#\mathcal{E}_p-1}^{t=0,\ldots,\#\Lambda-1, j=0,\ldots,\#\mathcal{E}_t-1}$. With this enumeration, the element $\widetilde{\mu}_{(p,i),(t,j)}$ in the submatrix is the element in the block (p, t) with position (i, j) in this block. Namely, if $p \neq t$, then

$$\widetilde{\mu}_{(p,i),(t,j)}(\xi) = [-(T_p\mathcal{W}_p)^*\widetilde{T}_t]_{i,j} = -\overline{\mu'_{0,p,i}(\xi)}\,\widetilde{\mu}_{0,t,j}(\xi);$$

if $p = t$, then

$$\widetilde{\mu}_{(p,i),(p,j)}(\xi) = [\mathcal{W}_p^* - (T_p\mathcal{W}_p)^*\widetilde{T}_p]_{i,j} = \overline{[\mathcal{W}_p]_{i,j}} - \overline{\mu'_{0,p,i}(\xi)}\,\widetilde{\mu}_{0,p,j}(\xi),$$

where $\mu'_{0,p,i}$, $i = 0, \ldots, \#\mathcal{E}_p - 1$, are the elements of the row $T_p\mathcal{W}_p$, $p = 0, \ldots, \#\Lambda - 1$. All $\mu'_{0,p,i}$ have the $\mathcal{H}$-symmetry property by Lemma 5.2.4.

Next, we collect wavelet masks by the polyphase representation formula. For fixed $p = 0, \ldots, \#\Lambda - 1$, $i = 0, \ldots, \#\mathcal{E}_p - 1$, we get

$$\widetilde{m}_{(p,i)}(\xi) = \frac{1}{\sqrt{m}} \sum_{j=0}^{\#\mathcal{E}_p-1} \overline{[W_p]_{i,j}} e^{2\pi i (s_{p,j},\xi)} - \overline{\mu'_{0,p,i}(M^*\xi)}\,\widetilde{m}_0(\xi).$$

Check that $\widetilde{m}_{(p,i)}$ has the $\mathcal{H}$-symmetry property. For a fixed $F \in \mathcal{H}_{p,0}$, by (5.33) and by the properties of $\mathcal{W}_p$ we obtain

$$\begin{aligned}\widetilde{m}_{(p,i)}(F^*\xi) &= \frac{1}{\sqrt{m}} \sum_{j=0}^{\#\mathcal{E}_p-1} \overline{[W_p]_{i,j}} e^{2\pi i(Fs_{p,j},\xi)} - \overline{\mu'_{0,p,i}(M^*F^*\xi)}\widetilde{m}_0(F^*\xi) \\ &= e^{2\pi i(Fs_{p,0}-s_{p,0},\xi)}\widetilde{m}_{(p,i)}(\xi),\end{aligned}$$

since $Fs_{p,j} = F(E^{(j)}s_{p,0} + c - E^{(j)}c) = s_{p,j} + Fs_{p,0} - s_{p,0}$ by (5.34), (5.9), (5.8). For a fixed $E^{(k)} \in \mathcal{E}_p$, by (5.33), and by the properties of $\mathcal{W}_p$ we obtain

$$\begin{aligned}\widetilde{m}_{(p,i)}(E^{(k)}\xi) &= \frac{1}{\sqrt{m}} \sum_{j=0}^{\#\mathcal{E}_p-1} \overline{[W_p]_{i,j}} e^{2\pi i(E^{(k)}s_{p,j},\xi)} - \overline{\mu'_{0,p,i}(M^*E^{(k)}\xi)}\widetilde{m}_0(E^{(k)}\xi) \\ &= [\mathcal{W}_p]_{k,i} e^{2\pi i(E^{(k)}c-c,\xi)}\widetilde{m}_{(p,i)}(\xi),\end{aligned}$$

since $E^{(k)}s_{p,j} = s_{p,j\oplus k} + E^{(k)}c - c$ by (5.8). Therefore, for a fixed $K \in \mathcal{H}$, with $K = E^{(k)}F$, $E^{(k)} \in \mathcal{E}_p$, $F \in \mathcal{H}_{p,0}$,

$$\widetilde{m}_{(p,i)}(K^*\xi) = [W_p]_{k,i} e^{2\pi i(Kc-c+Mr^F_{p,0},\xi)}\widetilde{m}_{(p,i)}(\xi).$$

Thus, we get the $\mathcal{H}$-symmetry property for the wavelet mask $\widetilde{m}_{(p,i)}$. It is not hard to see that the wavelet masks $m_{(p,i)}$ have the same $\mathcal{H}$-symmetry property as $\widetilde{m}_{(p,i)}$.

The vanishing moments of order n for the wavelet masks $\widetilde{m}_{(p,i)}$ are provided by Lemma 4.3.3.◊

Note that if (5.34) is not valid for some p, then we can take the corresponding matrix $\mathcal{W}_p$ equal to the identity matrix. Thus, for this index p the corresponding set of wavelet masks $m_{(p,i)}$, $\widetilde{m}_{(p,i)}$ will remain mutually symmetric.

Next, we introduce a lifting scheme transformation which allows to keep the $\mathcal{H}$-symmetry property of the wavelet masks constructed by Theorem 5.3.5. The corresponding trigonometric polynomials for the transformation we defined are as follows: Let $\mathcal{L}_p = (L_{p,0}, \dots, L_{p,\#\mathcal{E}_p-1})$, where $L_{p,i}$ are defined in Theorem 5.3.3. Let $\mathcal{L}'_p = \mathcal{L}_p\mathcal{W}^*_p$ and denote the elements of the row $\mathcal{L}'_p = (L'_{p,0}, \dots, L'_{p,\#\mathcal{E}_p-1})$.

Theorem 5.3.6 *Let a dilation matrix M and a center c be appropriate for an abelian symmetry group $\mathcal{H}$, $n \in \mathbb{N}$. Suppose m_0 and $\widetilde{m}_0$ are as in Theorem 5.3.5 and condition (5.34) is valid for all $p = 0, \dots, \#\Lambda - 1$. Suppose $m_{(p,i)}$, $\widetilde{m}_{(p,i)}$ are wavelet masks constructed using Theorem 5.3.5. Assume that trigonometric polynomials $L_{p,i}$ are as in Theorem 5.3.3. New masks $m^n_{(p,i)}$, $\widetilde{m}^n_{(p,i)}$ defined by the lifting scheme transformation with trigonometric polynomials $L'_{p,i}$ have the same symmetry properties as masks $m_{(p,i)}$, $\widetilde{m}_{(p,i)}$. New wavelet masks $m^n_{(p,i)}$ have vanishing moments at least of order 1. If new refinable functions φ, $\widetilde{\varphi}$ corresponding to new refinable masks m^n_0, $\widetilde{m}^n_0$ are in $L_2(\mathbb{R}^d)$, then the resulting wavelet system is a dual wavelet frame.*

Proof By direct computations, we can state that

$$L'_{p,i}((M^{-1}E^{(k)}M)^*\xi) = \left[\mathcal{W}_p\right]_{k,i} L'_{p,i}(\xi), \quad \forall E^{(k)} \in \mathcal{E}_p$$

and

$$L'_{p,i}((M^{-1}FM)^*\xi) = L'_{p,i}(\xi)e^{2\pi i(r^F_{p,0},\xi)} \quad \forall F \in \mathcal{H}_{p,0},$$

$i = 0, \ldots, \#\mathcal{E}_p - 1,\ p = 0, \ldots, \#\Lambda - 1$. Therefore, we obtain

$$m^n_{(p,i)}(F^*\xi) = e^{2\pi i(Fs_{p,0}-s_{p,0},\xi)} m^n_{(p,i)}(\xi)$$

and

$$m^n_{(p,i)}(E^{(k)*}\xi) = [\mathcal{W}_p]_{k,i}e^{2\pi i(E^{(k)}c-c,\xi)} m^n_{(p,i)}(\xi),$$

$i = 0, \ldots, \#\mathcal{E}_p - 1,\ p = 0, \ldots, \#\Lambda - 1$. Thus, wavelet mask $m^n_{(p,i)}$ has the same symmetry properties as $m_{(p,i)}$. Also, by direct computations it can be checked that $\widetilde{m}^n_0$ remains $\mathcal{H}$-symmetric with respect to the center c.$\diamondsuit$

Symmetrization step also can be done in Theorem 5.3.4. Again, we assume that $\mathcal{H}$ is an abelian symmetry group and (5.34) is valid for all $p = 0, \ldots, \#\Lambda - 1$. For $r = m$, instead of matrix extension (5.50) we consider

$$\mathcal{Q} := \begin{pmatrix} P & 0 \\ \mathcal{W}^* - \mathcal{W}^*\widetilde{P}^*P & \mathcal{W}^*\widetilde{P}^* \end{pmatrix}, \quad \widetilde{\mathcal{Q}} := \begin{pmatrix} \widetilde{P} & 0 \\ \mathcal{W}^* - \mathcal{W}^*P^*\widetilde{P} & \mathcal{W}^*P^* \end{pmatrix}, \tag{5.55}$$

where matrix $\mathcal{W}$ is a symmetrization matrix defined in (5.53). Note that $\mathcal{Q}^*\widetilde{\mathcal{Q}} = I_r$, since $\mathcal{Q} = \mathtt{diag}(1, \mathcal{W}^*)\mathcal{N}$ and $\widetilde{\mathcal{Q}} = \mathtt{diag}(1, \mathcal{W}^*)\widetilde{\mathcal{N}}$.

For $r = m + 1$, instead of matrix extension (5.51) we consider

$$\mathcal{Q} := \begin{pmatrix} P & \mu_{0m} & 0 \\ \mathcal{W}^* - \mathcal{W}^*\widetilde{P}^*P & \mu_{0m}\mathcal{W}^*P^* & \mathcal{W}^*\widetilde{P}^* \\ -\overline{\widetilde{\mu}_{0m}}P & 1 - \overline{\widetilde{\mu}_{0m}}\mu_{0m} & \overline{\widetilde{\mu}_{0m}} \end{pmatrix},$$

$$\widetilde{\mathcal{Q}} := \begin{pmatrix} \widetilde{P} & \widetilde{\mu}_{0m} & 0 \\ \mathcal{W}^* - \mathcal{W}^*P^*\widetilde{P} & \widetilde{\mu}_{0m}\mathcal{W}^*P^* & \mathcal{W}^*P^* \\ -\overline{\mu_{0m}}\widetilde{P} & 1 - \overline{\mu_{0m}}\widetilde{\mu}_{0m} & \overline{\mu_{0m}} \end{pmatrix}, \tag{5.56}$$

Note that $\mathcal{Q}^*\widetilde{\mathcal{Q}} = I_r$, since $\mathcal{Q} = \mathtt{diag}(1, \mathcal{W}^*, 1)\mathcal{N}$, $\widetilde{\mathcal{Q}} = \mathtt{diag}(1, \mathcal{W}^*, 1)\widetilde{\mathcal{N}}$.

Theorem 5.3.7 *Let a dilation matrix M and a center c be appropriate for an abelian symmetry group $\mathcal{H}$, $n \in \mathbb{N}$. Suppose masks m_ν, $\widetilde{m}_\nu$, $\nu = 0, \ldots, r$ are constructed by Algorithm 1. Then, wavelet masks $m_{(p,i)}$ and $\widetilde{m}_{(p,i)}$ have the $\mathcal{H}$-symmetry properties, i.e., for $K \in \mathcal{H}$, with $K = E^{(k)}F$, $E^{(k)} \in \mathcal{E}_p$, $F \in \mathcal{H}_{p,0}$,*

$$\widetilde{m}_{(p,i)}(K^*\xi) = [W_p]_{k,i}e^{2\pi i(Kc-c+Mr^F_{p,0},\xi)}\widetilde{m}_{(p,i)}(\xi),$$

$i = 0, \ldots, \#\mathcal{E}_p - 1,\ p = 0, \ldots, \#\Lambda - 1$. If $r = m + 1$, then the last wavelet masks m_r and $\widetilde{m}_r$ defined by the last rows of matrices $\mathcal{N}$ and $\widetilde{\mathcal{N}}$ are $\mathcal{H}$-symmetric with respect to the center c. All wavelet masks have vanishing moments of order n. If

refinable functions φ, $\widetilde{\varphi}$ *corresponding to refinable masks* m_0, $\widetilde{m}_0$ *are in* $L_2(\mathbb{R}^d)$, *then the resulting wavelet system is a dual wavelet frame.*

The proof can be done analogously to the proof of Theorem 5.3.5.

Finally, we indicate the symmetric properties of wavelets that correspond to the constructed wavelet masks in Theorems 5.3.2, 5.3.5.

Theorem 5.3.8 *Suppose conditions of Theorem 5.3.2 are valid. Then wavelet functions corresponding to the constructed wavelet masks have the following symmetry properties:*

$$\widehat{\psi_{(p,0)}}((M^{-1}FM)^*\xi) = \widehat{\psi_{(p,0)}}(\xi)e^{2\pi i(M^{-1}(F-I_d)(s_{p,0}-C,\xi)}, \quad \forall F \in \mathcal{H}_{p,0}.$$

$$\widehat{\psi_{(p,i)}}(\xi) = \widehat{\psi_{(p,0)}}((M^{-1}E^{(i)}M)^*\xi)e^{2\pi i(M^{-1}E^{(i)}MC-C,\xi)} \quad \forall E^{(i)} \in \mathcal{E}_{p,0},$$

where $c = (I_d - M)C$.

Proof Since by Lemma 5.1.2 φ is $\mathcal{H}$-symmetric with respect to the center C, i.e., $\widehat{\varphi}(K^*\xi) = e^{2\pi i(C-EC,\xi)}\widehat{\varphi}(\xi)$ for any $K \in \mathcal{H}$, and for $p = 0, \dots, \#\Lambda - 1$

$$m_{(p,0)}(F^*\xi) = m_{(p,0)}(\xi)e^{2\pi i(Fs_{p,0}-s_{p,0},\xi)}, \quad \forall F \in \mathcal{H}_{p,0},$$

$$m_{(p,i)}(\xi) = m_{(p,0)}(E^{(i)*}\xi)e^{2\pi i(c-E^{(i)}c,\xi)}, \quad E^{(i)} \in \mathcal{E}_p, \quad i = 0, \dots \#\mathcal{E}_p - 1,$$

using the definition of wavelets $\widehat{\psi_{(p,i)}}(\xi) = m_{(p,i)}(M^{*-1}\xi)\widehat{\varphi}(M^{*-1}\xi)$, we obtain that $\forall F \in \mathcal{H}_{p,0}$

$$\begin{aligned}\widehat{\psi_{(p,0)}}((M^{-1}FM)^*\xi) &= m_{(p,0)}(F^*M^{*-1}\xi)\widehat{\varphi}(F^*M^{*-1}\xi) \\ &= \widehat{\psi_{(p,0)}}(\xi)e^{2\pi i(Fs_{p,0}-s_{p,0}+C-FC,M^{*-1}\xi)} \\ &= \widehat{\psi_{(p,0)}}(\xi)e^{2\pi i(M^{-1}(F-I_d)(s_{p,0}-C),\xi)}\end{aligned} \tag{5.57}$$

and for $i = 0, \dots \#\mathcal{E}_p - 1$,

$$\begin{aligned}\widehat{\psi_{(p,0)}}((M^{-1}E^{(i)}M)^*\xi) &= m_{(p,0)}(E^{*(i)}M^{*-1}\xi)\widehat{\varphi}(E^{*(i)}M^{*-1}\xi) \\ &= \widehat{\psi_{(p,i)}}(\xi)e^{2\pi i(E^{(i)}c-c+C-E^{(i)}C,M^{*-1}\xi)} \\ &= \widehat{\psi_{(p,i)}}(\xi)e^{2\pi i(C-M^{-1}E^{(i)}MC,\xi)}.\end{aligned} \tag{5.58}$$

Since the symmetry properties for dual masks are the same, we get the analogous relations for dual wavelets.◊

Proof Since φ is $\mathcal{H}$-symmetric with respect to the origin, i.e., $\widehat{\varphi}(K^*\xi) = \widehat{\varphi}(\xi)$ for any $K \in \mathcal{H}$ (by Lemma 5.1.2) and

$$m_{(p,i)}(K^*\xi) = [W_p]_{k,i}e^{2\pi i(Fs_{p,0}-s_{p,0},\xi)}m_{(p,i)}(\xi), \quad \forall K \in \mathcal{H}, \quad K = E^{(k)}F,$$

using the definition of wavelets $\widehat{\psi_{(p,i)}}(\xi) = m_{(p,i)}(M^{*-1}\xi)\widehat{\varphi}(M^{*-1}\xi)$, we obtain that

$$\begin{aligned}\widehat{\psi_{(p,i)}}((M^{-1}KM)^*\xi) &= m_{(p,i)}(KM^{*-1}\xi)\widehat{\varphi}(KM^{*-1}\xi)\\ &= [W_p]_{k,i}\widehat{\psi_{(p,i)}}(\xi)e^{2\pi i(Fs_{p,0}-s_{p,0},M^{*-1}\xi)}.\end{aligned}$$

Since the symmetry properties for dual masks are the same, we get the analogous relations for dual wavelets.◊

Theorem 5.3.9 *Suppose conditions of Theorem 5.3.5 are valid. Then wavelet functions corresponding to the constructed wavelet masks have the following symmetry properties: for any* $p = 0, \dots, \#\Lambda - 1$, $i = 0, \dots \#\mathcal{E}_p - 1$

$$\widehat{\psi_{(p,i)}}((M^{-1}KM)^*\xi) = [W_p]_{k,i}\widehat{\psi_{(p,i)}}(\xi)e^{2\pi i((K-F)c+(F-I_d)s_{p,0}-(K-I_d)C,M^{*-1}\xi)},$$

$\forall K \in \mathcal{H}$ *where* $K = E^{(k)}F$, $F \in \mathcal{H}_{p,0}$, $E^{(k)} \in \mathcal{E}_p$, *where* $c = (I_d - M)C$. *The analogous relations are valid for dual wavelets* $\widetilde{\psi}_{(p,i)}$.

Proof By Lemma 5.1.2 φ is $\mathcal{H}$-symmetric with respect to the center C, i.e., $\widehat{\varphi}(K^*\xi) = e^{2\pi i(C-EC,\xi)}\widehat{\varphi}(\xi)$. Also for any $K \in \mathcal{H}$, and

$$\widetilde{m}_{(p,i)}(K^*\xi) = [W_p]_{k,i}e^{2\pi i(Kc-c+F(s_{p,0}-c)-(s_{p,0}-c),\xi)}\widetilde{m}_{(p,i)}(\xi) \quad \forall K \in \mathcal{H},$$

where $K = E^{(k)}F$, $F \in \mathcal{H}_{p,0}$, $E^{(k)} \in \mathcal{E}_p$. Using the definition of wavelets $\widehat{\psi_{(p,i)}}(\xi) = m_{(p,i)}(M^{*-1}\xi)\widehat{\varphi}(M^{*-1}\xi)$, we obtain that

$$\widehat{\psi_{(p,i)}}((M^{-1}KM)^*\xi) = m_{(p,i)}(K^*M^{*-1}\xi)\widehat{\varphi}(K^*M^{*-1}\xi) =$$

$$[W_p]_{k,i}\widehat{\psi_{(p,i)}}(\xi)e^{2\pi i((K-F)c+(F-I_d)s_{p,0}-(K-I_d)C,M^{*-1}\xi)}$$

for $i = 0, \dots \#\mathcal{E}_p - 1$. Since the symmetry properties for dual masks are the same, we get the analogous relations for dual wavelets.◊

5.4 Examples

In this section, we give several examples which illustrate the results of the chapter. All examples are based on the construction of $\mathcal{H}$-symmetric refinable masks by Theorem 5.1.12 and wavelet masks by Theorems 5.2.3, 5.2.5, 5.3.2, 5.3.3, 5.3.4, 5.3.5, and 5.3.7.

1. Let $\mathcal{H}$ be a hexagonal symmetry group on $\mathbb{Z}^2$, namely

$$\mathcal{H} = \left\{ \pm I_2, \pm \begin{pmatrix} 0 & 1 \\ 1 & 0 \end{pmatrix}, \pm \begin{pmatrix} 1 & 0 \\ 1 & -1 \end{pmatrix}, \pm \begin{pmatrix} 1 & -1 \\ 1 & 0 \end{pmatrix}, \pm \begin{pmatrix} 0 & 1 \\ -1 & 1 \end{pmatrix}, \pm \begin{pmatrix} -1 & 1 \\ 0 & 1 \end{pmatrix} \right\}$$

$c = \mathbf{0}$, $M = \begin{pmatrix} 2 & -1 \\ 1 & 1 \end{pmatrix}$. The set of digits is $D(M) = \{s_0 = (0,0), s_1 = (0,1), s_2 = (0,-1)\}$, $m = 3$. Let us construct an interpolatory refinable mask that is $\mathcal{H}$-symmetric with respect to the origin and has sum rule of order $n = 3$. The digits are renumbered as $s_{0,0} = (0,0)$, $s_{1,0} = (0,1)$, $s_{1,1} = (0,-1)$. And $\mathcal{H}_{0,0} = \mathcal{H}$, $\mathcal{E}_0 = \{I_2\}$,

$$\mathcal{H}_{1,0} = \left\{ I_2, \begin{pmatrix} 0 & 1 \\ 1 & 0 \end{pmatrix}, \begin{pmatrix} -1 & 0 \\ -1 & 1 \end{pmatrix}, \begin{pmatrix} -1 & 1 \\ -1 & 0 \end{pmatrix}, \begin{pmatrix} 0 & -1 \\ 1 & -1 \end{pmatrix}, \begin{pmatrix} 1 & -1 \\ 0 & -1 \end{pmatrix} \right\},$$

$\mathcal{E}_1 = \left\{ I_2, \begin{pmatrix} 1 & 0 \\ 1 & -1 \end{pmatrix} \right\}$. According to Theorem 5.1.12, mask m_0 can be constructed as follows:

$$m_0 : \begin{pmatrix} 0 & 0 & -\frac{1}{27} & 0 & -\frac{1}{27} \\ 0 & 0 & \frac{4}{27} & \frac{4}{27} & 0 \\ -\frac{1}{27} & \frac{4}{27} & \mathbf{\frac{1}{3}} & \frac{4}{27} & -\frac{1}{27} \\ 0 & \frac{4}{27} & \frac{4}{27} & 0 & 0 \\ -\frac{1}{27} & 0 & -\frac{1}{27} & 0 & 0 \end{pmatrix}$$

with support in $[-2, 2]^2 \bigcap \mathbb{Z}^2$. Note that $\mu_{00} = \frac{1}{\sqrt{3}}$. Checking shows (see Sect. 6.8) that the corresponding refinable function φ is in $L_2(\mathbb{R}^2)$. The admissible dual mask $\widetilde{m}_0$ is given by

$$\widetilde{m}_0 : \begin{pmatrix} 0 & 0 & 0 & 0 & 0 & 0 & -\frac{2}{243} & 0 & 0 \\ 0 & 0 & 0 & 0 & \frac{8}{243} & 0 & 0 & \frac{8}{243} & 0 \\ 0 & 0 & -\frac{2}{243} & 0 & -\frac{1}{27} & -\frac{16}{243} & -\frac{1}{27} & 0 & -\frac{2}{243} \\ 0 & 0 & 0 & -\frac{16}{243} & \frac{4}{27} & \frac{4}{27} & -\frac{16}{243} & 0 & 0 \\ 0 & \frac{8}{243} & -\frac{1}{27} & \frac{4}{27} & \mathbf{\frac{47}{81}} & \frac{4}{27} & -\frac{1}{27} & \frac{8}{243} & 0 \\ 0 & 0 & -\frac{16}{243} & \frac{4}{27} & \frac{4}{27} & -\frac{16}{243} & 0 & 0 & 0 \\ -\frac{2}{243} & 0 & -\frac{1}{27} & -\frac{16}{243} & -\frac{1}{27} & 0 & -\frac{2}{243} & 0 & 0 \\ 0 & \frac{8}{243} & 0 & 0 & \frac{8}{243} & 0 & 0 & 0 & 0 \\ 0 & 0 & -\frac{2}{243} & 0 & 0 & 0 & 0 & 0 & 0 \end{pmatrix}$$

with support in $[-4, 4]^2 \bigcap \mathbb{Z}^2$, $\widetilde{m}_0$ has sum rule of order 3. Checking shows (see Sect. 6.8) that the corresponding refinable function $\widetilde{\varphi}$ is in $L_2(\mathbb{R}^2)$.

Since the hexagonal symmetry group $\mathcal{H}$ is not abelian, the wavelet masks are constructed by Theorem 5.2.3:

$$m_{(1,0)}:\begin{pmatrix} 0 & 0 & 0 & 0 & -\frac{1}{243} & 0 & -\frac{1}{243} & 0 & 0 \\ 0 & 0 & 0 & 0 & \frac{4}{243} & \frac{8}{243} & 0 & \frac{4}{243} & 0 \\ 0 & 0 & -\frac{1}{243} & \frac{8}{243} & \frac{1}{27} & -\frac{8}{243} & -\frac{2}{27} & 0 & -\frac{1}{243} \\ 0 & 0 & 0 & -\frac{8}{243} & -\frac{8}{81} & -\frac{4}{27} & -\frac{8}{243} & \frac{8}{243} & 0 \\ 0 & \frac{4}{243} & -\frac{2}{27} & -\frac{4}{27} & \frac{64}{81} & -\frac{8}{81} & \frac{1}{27} & \frac{4}{243} & -\frac{1}{243} \\ 0 & 0 & -\frac{8}{243} & -\frac{8}{81} & -\boldsymbol{\frac{4}{27}} & -\frac{8}{243} & \frac{8}{243} & 0 & 0 \\ -\frac{1}{243} & \frac{8}{243} & \frac{1}{27} & -\frac{8}{243} & -\frac{2}{27} & 0 & -\frac{1}{243} & 0 & 0 \\ 0 & \frac{4}{243} & \frac{8}{243} & 0 & \frac{4}{243} & 0 & 0 & 0 & 0 \\ -\frac{1}{243} & 0 & -\frac{1}{243} & 0 & 0 & 0 & 0 & 0 & 0 \end{pmatrix},$$

$$\widetilde{m}_{(1,0)}:\begin{pmatrix} 0 & 0 & \frac{1}{27} & 0 & 0 \\ 0 & 0 & 0 & -\frac{4}{27} & 0 \\ 0 & -\frac{4}{27} & \frac{1}{3} & 0 & \frac{1}{27} \\ 0 & 0 & -\boldsymbol{\frac{4}{27}} & 0 & 0 \\ \frac{1}{27} & 0 & 0 & 0 & 0 \end{pmatrix},$$

$$m_{(1,1)} = m_{(1,0)}(E^*\xi), \quad \widetilde{m}_{(1,1)} = \widetilde{m}_{(1,0)}(E^*\xi),$$

where $E = \begin{pmatrix} 1 & 0 \\ 1 & -1 \end{pmatrix}$. The bold element in the matrices corresponds to the coefficient $h_{\mathbf{0}}$ of the masks. The wavelet masks $m_{(1,0)}, m_{(1,1)}, \widetilde{m}_{(1,0)}, \widetilde{m}_{(1,1)}$ have vanishing moments of order 3, the masks $m_{(1,0)}, \widetilde{m}_{(1,0)}$ are $\mathcal{H}_{1,0}$-symmetric. The corresponding wavelet system $\{\psi_{jk}^{(\nu)}\}, \{\widetilde{\psi}_{jk}^{(\nu)}\}$ is a dual wavelet frame in $L_2(\mathbb{R}^d)$ providing approximation order 3.

2. Let $\mathcal{H} = \left\{\pm\begin{pmatrix} 1 & 0 \\ 0 & 1 \end{pmatrix}, \pm\begin{pmatrix} 0 & 1 \\ 1 & 0 \end{pmatrix}\right\}$, $M = \begin{pmatrix} 1 & -2 \\ 2 & -1 \end{pmatrix}$. The set of digits is $D(M) = \{s_0 = (0,0), s_1 = (-1,0), s_2 = (0,1)\}$, $m = 3$. Let us construct an interpolatory refinable mask m_0 that is $\mathcal{H}$-symmetric with respect to the origin and has sum rule of order $n = 3$. The digits are renumbered as $s_{0,0} = (0,0)$, $s_{1,0} = (-1,0)$, $s_{1,1} = (0,1)$, $\mathcal{H}_{0,0} = \mathcal{H}$, $\mathcal{E}_0 = \{I_2\}$,

$$\mathcal{H}_{1,0} = \left\{I_2, \begin{pmatrix} 0 & 1 \\ 1 & 0 \end{pmatrix}\right\}, \quad \mathcal{E}_1 = \left\{I_2, \begin{pmatrix} 0 & -1 \\ -1 & 0 \end{pmatrix}\right\}.$$

Refinable mask m_0 constructed by Theorem 5.1.12 and the admissible dual mask $\widetilde{m}_0$ are

$$m_0: \begin{pmatrix} 0 & -\frac{1}{54} & 0 & 0 & -\frac{1}{27} \\ -\frac{1}{54} & 0 & \frac{1}{6} & \frac{2}{27} & 0 \\ 0 & \frac{1}{6} & \frac{1}{3} & \frac{1}{6} & 0 \\ 0 & \frac{2}{27} & \frac{1}{6} & 0 & -\frac{1}{54} \\ -\frac{1}{27} & 0 & 0 & -\frac{1}{54} & 0 \end{pmatrix}, \quad \widetilde{m}_0: \begin{pmatrix} -\frac{1}{216} & 0 & 0 & \frac{7}{648} & 0 & 0 & \frac{1}{81} \\ 0 & \frac{19}{324} & -\frac{1}{24} & 0 & -\frac{47}{648} & 0 & 0 \\ 0 & -\frac{1}{24} & -\frac{61}{648} & \frac{11}{72} & \frac{1}{9} & -\frac{47}{648} & 0 \\ \frac{7}{648} & 0 & \frac{11}{72} & \frac{103}{162} & \frac{11}{72} & 0 & \frac{7}{648} \\ 0 & -\frac{47}{648} & \frac{1}{9} & \frac{11}{72} & -\frac{61}{648} & -\frac{1}{24} & 0 \\ 0 & 0 & -\frac{47}{648} & 0 & -\frac{1}{24} & \frac{19}{324} & 0 \\ \frac{1}{81} & 0 & 0 & \frac{7}{648} & 0 & 0 & -\frac{1}{216} \end{pmatrix},$$

with support in $[-2,2]^2 \bigcap \mathbb{Z}^2$ and in $[-3,3]^2 \bigcap \mathbb{Z}^2$, respectively, $\widetilde{m}_0$ has sum rule of order 2. Checking shows (see Sect. 6.8) that the corresponding refinable functions $\varphi, \widetilde{\varphi}$ are in $L_2(\mathbb{R}^d)$. Since $\mathcal{H}$ is an abelian group, wavelet masks can be constructed by Theorem 5.2.5. Note that the special assumption (5.34) is valid in this case. To avoid roots in the coefficients of wavelet mask, we take matrix $\mathcal{W}$ as

$$\mathcal{W} = \begin{pmatrix} 1 & 0 & 0 \\ 0 & 1 & 1 \\ 0 & 1 & -1 \end{pmatrix}, \quad \widetilde{\mathcal{W}} = \begin{pmatrix} 1 & 0 & 0 \\ 0 & \frac{1}{2} & \frac{1}{2} \\ 0 & \frac{1}{2} & -\frac{1}{2} \end{pmatrix},$$

where $\widetilde{\mathcal{W}}$ is a paraunitary matrix for $\mathcal{W}$, i.e., $\mathcal{W}^*\widetilde{\mathcal{W}} = I_3$. The wavelet masks $m_{(1,0)}, m_{(1,1)}$ are

$$\begin{pmatrix} 0 & -\frac{1}{216} & 0 & 0 & -\frac{1}{324} & 0 & 0 & \frac{1}{81} \\ -\frac{1}{216} & 0 & \frac{19}{324} & \frac{2}{81} & 0 & -\frac{7}{324} & -\frac{2}{81} & 0 \\ 0 & \frac{19}{324} & \frac{1}{12} & -\frac{61}{648} & -\frac{10}{81} & -\frac{1}{9} & -\frac{7}{324} & 0 \\ 0 & \frac{2}{81} & -\frac{61}{648} & -\frac{11}{36} & \frac{109}{162} & -\frac{10}{81} & 0 & -\frac{1}{324} \\ -\frac{1}{324} & 0 & -\frac{10}{81} & \frac{109}{162} & \mathbf{-\frac{11}{36}} & -\frac{61}{648} & \frac{2}{81} & 0 \\ 0 & -\frac{7}{324} & -\frac{1}{9} & -\frac{10}{81} & -\frac{61}{648} & \frac{1}{12} & \frac{19}{324} & 0 \\ 0 & -\frac{2}{81} & -\frac{7}{324} & 0 & \frac{2}{81} & \frac{19}{324} & 0 & -\frac{1}{216} \\ \frac{1}{81} & 0 & 0 & -\frac{1}{324} & 0 & 0 & -\frac{1}{216} & 0 \end{pmatrix}, \begin{pmatrix} 0 & 0 & 0 & 0 & -\frac{1}{162} & 0 & 0 & -\frac{1}{81} \\ 0 & 0 & 0 & -\frac{1}{162} & 0 & \frac{1}{18} & \frac{2}{81} & 0 \\ 0 & 0 & 0 & 0 & \frac{1}{18} & \frac{1}{9} & \frac{1}{18} & 0 \\ 0 & \frac{1}{162} & 0 & 0 & -\frac{26}{27} & \frac{1}{18} & 0 & -\frac{1}{162} \\ \frac{1}{162} & 0 & -\frac{1}{18} & \frac{26}{27} & \mathbf{0} & 0 & -\frac{1}{162} & 0 \\ 0 & -\frac{1}{18} & -\frac{1}{9} & -\frac{1}{18} & 0 & 0 & 0 & 0 \\ 0 & -\frac{2}{81} & -\frac{1}{18} & 0 & \frac{1}{162} & 0 & 0 & 0 \\ \frac{1}{81} & 0 & 0 & \frac{1}{162} & 0 & 0 & 0 & 0 \end{pmatrix}$$

and the dual wavelet masks $\widetilde{m}_{(1,0)}, \widetilde{m}_{(1,1)}$ are

$$\begin{pmatrix} \frac{1}{54} & 0 & 0 & -\frac{1}{54} \\ 0 & -\frac{1}{6} & \frac{1}{6} & 0 \\ 0 & \frac{1}{6} & \mathbf{-\frac{1}{6}} & 0 \\ -\frac{1}{54} & 0 & 0 & \frac{1}{54} \end{pmatrix}, \begin{pmatrix} 0 & 0 & 0 & \frac{1}{18} \\ 0 & 0 & -\frac{1}{6} & 0 \\ 0 & \frac{1}{6} & \mathbf{0} & 0 \\ -\frac{1}{18} & 0 & 0 & 0 \end{pmatrix},$$

where the bold element in the matrices corresponds to the coefficient h_0 of the masks. The wavelet masks $m_{(1,0)}, m_{(1,1)}$ have vanishing moments of order 2, $\widetilde{m}_{(1,0)}, \widetilde{m}_{(1,1)}$ have vanishing moments of order 3. The $\mathcal{H}$-symmetry property is also valid, namely all wavelet masks are $\mathcal{H}_{p,0}$-symmetric with respect to $s_{p,0}$, $m_{(1,0)}(E^*\xi) = m_{(1,0)}(\xi)$, $\widetilde{m}_{(1,0)}(E^*\xi) = \widetilde{m}_{(1,0)}(\xi)$, $E \in \mathcal{E}_1$; $m_{(1,1)}(E^{(1)*}\xi) = -m_{(1,1)}(\xi)$, $\widetilde{m}_{(1,1)}(E^{(1)*}\xi) =$

$-\widetilde{m}_{(1,1)}(\xi)$, where $E^{(1)} = \begin{pmatrix} 0 & -1 \\ -1 & 0 \end{pmatrix}$. The corresponding wavelet system $\{\psi_{jk}^{(\nu)}\}$, $\{\widetilde{\psi}_{jk}^{(\nu)}\}$ is a dual wavelet frame in $L_2(\mathbb{R}^d)$ providing approximation order 3.

3. Let $c = \mathbf{0}$, $M = \begin{pmatrix} 1 & -2 \\ 2 & -1 \end{pmatrix}$, $\mathcal{H} = \{\pm I_2\}$. The set of digits is $D(M) = \{s_0 = (0,0), s_1 = (0,-1), s_2 = (0,1)\}$, $m = 3$. Let us construct a mask m_0 that is symmetric with respect to the origin and satisfies condition (5.17) with $n = 4$ and $\lambda'_\gamma = \delta_{\gamma\mathbf{0}}$, $[\gamma] < 4$. In this case, the digits are renumbered as $s_{0,0} = (0,0)$, $s_{1,0} = (-1,0)$, $s_{1,1} = (0,1)$, $\mathcal{H}_{0,0} = \mathcal{H}$, $\mathcal{E}_0 = \{I_2\}$, $\mathcal{H}_{1,0} = \{I_2\}$, $\mathcal{E}_1 = \mathcal{H}$. According to Theorem 5.1.12, the mask m_0 can be constructed as follows:

$$m_0 : \frac{1}{243}\begin{pmatrix} 0 & 0 & 0 & 0 & -3 & 0 & 0 \\ 0 & 0 & 0 & -5 & 0 & 3 & -4 \\ -1 & 0 & 0 & 24 & 33 & 0 & -5 \\ 0 & 3 & 36 & \mathbf{81} & 36 & 3 & 0 \\ -5 & 0 & 33 & 24 & 0 & 0 & -1 \\ -4 & 3 & 0 & -5 & 0 & 0 & 0 \\ 0 & 0 & -3 & 0 & 0 & 0 & 0 \end{pmatrix}$$

with support in $[-3,3]^2 \bigcap \mathbb{Z}^2$. Note that $\mu_{00} = \frac{1}{\sqrt{3}}$. Checking shows (see Sect. 6.8) that the corresponding refinable function φ is in $L_2(\mathbb{R}^2)$. As for the dual masks, we simply take $\widetilde{m}_0 \equiv 1$. Condition (5.36) is obviously valid and $\widetilde{\varphi}(x)$ is the Dirac delta function. Moreover, we can modify the matrix extension as follows:

$$\mathcal{N} = \begin{pmatrix} \frac{1}{\sqrt{3}} & \mu_{01} & \mu_{02} \\ 0 & \sqrt{3} & 0 \\ 0 & 0 & \sqrt{3} \end{pmatrix}, \quad \widetilde{\mathcal{N}} = \begin{pmatrix} \sqrt{3} & 0 & 0 \\ -\overline{\mu_{01}} & \frac{1}{\sqrt{3}} & 0 \\ -\overline{\mu_{02}} & 0 & \frac{1}{\sqrt{3}} \end{pmatrix}.$$

This allows us to reduce the number of the wavelet functions. By Theorem 5.3.5, we obtain the wavelet masks

$$m_1(\xi) = e^{2\pi i(s_1,\xi)} + e^{2\pi i(s_2,\xi)}, \quad m_2(\xi) = e^{2\pi i(s_1,\xi)} - e^{2\pi i(s_2,\xi)};$$

$$\widetilde{m}_1 : \frac{1}{486}\begin{pmatrix} 0 & 0 & 0 & 5 & 0 & 0 & 4 \\ 0 & 0 & 0 & 0 & -30 & 0 & 0 \\ 0 & 0 & -36 & 81 & 0 & -6 & 0 \\ 6 & 0 & 0 & \mathbf{-48} & 0 & 0 & 6 \\ 0 & -6 & 0 & 81 & -36 & 0 & 0 \\ 0 & 0 & -30 & 0 & 0 & 0 & 0 \\ 4 & 0 & 0 & 5 & 0 & 0 & 0 \end{pmatrix}, \quad \widetilde{m}_2 : \frac{1}{486}\begin{pmatrix} 0 & 0 & 0 & 5 & 0 & 0 & 4 \\ 0 & 0 & 0 & 0 & -36 & 0 & 0 \\ 0 & 0 & -36 & 81 & 0 & 0 & 0 \\ 4 & 0 & 0 & \mathbf{0} & 0 & 0 & -4 \\ 0 & 0 & 0 & -81 & 36 & 0 & 0 \\ 0 & 0 & 36 & 0 & 0 & 0 & 0 \\ -4 & 0 & 0 & -5 & 0 & 0 & 0 \end{pmatrix}$$

with supports in $[-3,3]^2 \bigcap \mathbb{Z}^2$. The wavelet masks $\widetilde{m}_1$, $\widetilde{m}_2$ have vanishing moments of order 4. Each wavelet function $\widetilde{\psi}^{(1)}$, $\widetilde{\psi}^{(2)}$ is the linear combination of translations of the Dirac delta function. Thus, we are in the conditions of Theorem 4.3.6. The corresponding symmetric almost frame-like wavelet system provides approximation order 4 according to (4.27).

4. Let $c = (1/2, 1/2)$, $M = \begin{pmatrix} 2 & 0 \\ 0 & 2 \end{pmatrix}$. The set of digits is $D(M) = \{s_0 = (0, 0), s_1 = (0, 1), s_2 = (1, 0), s_3 = (1, 1)\}$, $m = 4$. Let us construct a mask m_0 that is $\mathcal{H}^{axis}$-symmetric with respect to the center c and satisfies condition (5.17) with $n = 2$ and $\lambda'_\gamma = \delta_{\gamma\mathbf{0}}$, $[\gamma] < 2$. In this case, the digits are renumbered as $s_{0,0} = (0, 0)$, $s_{0,1} = (0, 1)$, $s_{0,2} = (1, 0)$, $s_{0,3} = (1, 1)$. $\mathcal{H}_{0,0} = \{I_2\}$, $\mathcal{E}_0 = \mathcal{H}$. According to Theorem 5.1.12, the mask m_0 can be constructed as follows:

$$m_0 : \begin{pmatrix} 0 & 1/16 & 1/16 & 0 \\ 1/16 & 1/8 & 1/8 & 1/16 \\ 1/16 & \mathbf{1/8} & 1/8 & 1/16 \\ 0 & 1/16 & 1/16 & 0 \end{pmatrix}$$

with support in $[-1, 2]^2 \bigcap \mathbb{Z}^2$. Checking shows (see Sect. 6.8) that the corresponding refinable function φ is in $L_2(\mathbb{R}^2)$. As for the dual masks, we take

$$\widetilde{m}_0 : \begin{pmatrix} 1/4 & 1/4 \\ \mathbf{1/4} & 1/4 \end{pmatrix}$$

with support in $[0, 1]^2 \bigcap \mathbb{Z}^2$. Condition (5.36) is valid. Checking shows (see Sect. 6.8) that the corresponding dual refinable function $\widetilde{\varphi}$ is in $L_2(\mathbb{R}^2)$.

The symmetrization matrix $\mathcal{W}$ is given by

$$\mathcal{W} = \frac{1}{2} \begin{pmatrix} 1 & 1 & 1 & 1 \\ 1 & -1 & 1 & -1 \\ 1 & 1 & -1 & -1 \\ 1 & -1 & -1 & 1 \end{pmatrix}.$$

By Theorem 5.3.5, we obtain the wavelet masks

$$m_{(0,0)} : \frac{1}{2} \begin{pmatrix} 1 & 1 \\ \mathbf{1} & 1 \end{pmatrix} \quad m_{(0,1)} : \frac{1}{2} \begin{pmatrix} -1 & 1 \\ \mathbf{1} & -1 \end{pmatrix} \quad m_{(1,0)} : \frac{1}{2} \begin{pmatrix} -1 & -1 \\ \mathbf{1} & 1 \end{pmatrix} \quad m_{(1,1)} : \frac{1}{2} \begin{pmatrix} 1 & -1 \\ \mathbf{1} & -1 \end{pmatrix}$$

with supports in $[0, 1]^2 \bigcap \mathbb{Z}^2$.

Dual wavelet masks are

$$\widetilde{m}_{(0,0)} : \frac{1}{8} \begin{pmatrix} 0 & 0 & -1/8 & -1/8 & 0 & 0 \\ 0 & 0 & -1/8 & -1/8 & 0 & 0 \\ -1/8 & -1/8 & 1/2 & 1/2 & -1/8 & -1/8 \\ -1/8 & -1/8 & \mathbf{1/2} & 1/2 & -1/8 & -1/8 \\ 0 & 0 & -1/8 & -1/8 & 0 & 0 \\ 0 & 0 & -1/8 & -1/8 & 0 & 0 \end{pmatrix} \quad \widetilde{m}_{(0,1)} : \frac{1}{8} \begin{pmatrix} -1 & 1 \\ \mathbf{1} & -1 \end{pmatrix}$$

with supports in $[-2,3]^2 \bigcap \mathbb{Z}^2$ and $[0,1]^2 \bigcap \mathbb{Z}^2$ accordingly and

$$\widetilde{m}_{(0,2)} : \frac{1}{8}\begin{pmatrix} 1/8 & 1/8 \\ 1/8 & 1/8 \\ -1 & -1 \\ \mathbf{1} & 1 \\ -1/8 & -1/8 \\ -1/8 & -1/8 \end{pmatrix} \quad \widetilde{m}_{(0,3)} : \frac{1}{8}\begin{pmatrix} -1/8 & -1/8 & 1 & -1 & 1/8 & 1/8 \\ -1/8 & -1/8 & \mathbf{1} & -1 & 1/8 & 1/8 \end{pmatrix}.$$

with supports in $[0,1] \times [-2,3] \bigcap \mathbb{Z}^2$ and $[-2,3] \times [0,1] \bigcap \mathbb{Z}^2$ accordingly. The wavelet masks $\widetilde{m}_{(0,i)}$, $i = 1,2,3,4$, have vanishing moments of order 2. The corresponding axial symmetric/antisymmetric frame-like wavelet system provides approximation order 2 according to Theorem 5.3.5.

5. Let us illustrate now how Theorem 5.1.15 about the general form of all symmetric masks can be used for the construction of orthogonal masks. Let $\mathcal{H}$ be a hexagonal abelian symmetry group on $\mathbb{Z}^2$, namely

$$\mathcal{H} = \left\{ I_2, \begin{pmatrix} 0 & -1 \\ 1 & -1 \end{pmatrix}, \begin{pmatrix} -1 & 1 \\ -1 & 0 \end{pmatrix} \right\},$$

$c = \mathbf{0}$, $M = \begin{pmatrix} 2 & 0 \\ 0 & 2 \end{pmatrix}$. The set of digits is $D(M) = \{s_0 = (0,0), s_1 = (0,1), s_2 = (1,0), s_3 = (1,1)\}$, $m = 4$. Let us construct a refinable mask that is $\mathcal{H}$-symmetric with respect to the origin and has sum rule of order $n = 3$. In our case, the digits are renumbered as $s_{0,0} = (0,0)$, $s_{1,0} = (1,0)$. And $\mathcal{H}_{0,0} = \mathcal{H}$, $\mathcal{E}_0 = \{I_2\}$, $\mathcal{H}_{1,0} = \{I_2\}$, $\mathcal{E}_1 = \mathcal{H}$. According to Theorem 5.1.12, mask m_0 can be constructed as follows:

$$m_0 : \begin{pmatrix} 0 & 0 & 0 & -\frac{1}{32} & 0 & 0 & 0 \\ 0 & 0 & 0 & 0 & 0 & 0 & 0 \\ 0 & 0 & 0 & \frac{3}{16} & \frac{3}{32} & 0 & 0 \\ 0 & 0 & \frac{3}{32} & \mathbf{\frac{1}{4}} & \frac{3}{16} & 0 & -\frac{1}{32} \\ 0 & 0 & \frac{3}{16} & \frac{3}{32} & 0 & 0 & 0 \\ 0 & 0 & 0 & 0 & 0 & 0 & 0 \\ -\frac{1}{32} & 0 & 0 & 0 & 0 & 0 & 0 \end{pmatrix}$$

with spectrum in $[-3,3]^2 \bigcap \mathbb{Z}^2$. Checking shows (see Sect. 6.8) that the corresponding refinable function φ is in $L_2(\mathbb{R}^2)$.

Next, we illustrate how we can use the general form of all masks (Theorem 5.1.15) in order to construct $\mathcal{H}$-symmetric orthogonal masks. Suppose now $n = 2$. $\mathcal{H}$, M are as above. In this case, the simplest mask m_0^s is

$$\begin{pmatrix} 0 & \frac{1}{8} & \frac{1}{8} \\ \frac{1}{8} & \mathbf{\frac{1}{4}} & \frac{1}{8} \\ \frac{1}{8} & \frac{1}{8} & 0 \end{pmatrix}$$

with spectrum in $[-1, 1]^2 \bigcap \mathbb{Z}^2$. The corresponding refinable function φ is in $L_2(\mathbb{R}^2)$ and $\nu_2(\varphi^s) \geq 2$. If we introduce 4 additional parameters for trigonometric function G_1 during the construction by Theorem 5.1.12, we get the following mask

$$\begin{pmatrix} 0 & 0 & 0 & \frac{f}{4} & 0 & \frac{a}{4} & 0 \\ 0 & 0 & 0 & 0 & \frac{b}{4} & 0 & \frac{1}{4}(a-b+c) \\ 0 & \frac{1}{4}(a-b+c) & \frac{c}{4} & \frac{1}{4}\left(-3a+b-2c-2f+\frac{1}{2}\right) & \frac{1}{4}\left(a-b+f+\frac{1}{2}\right) & \frac{c}{4} & 0 \\ 0 & 0 & \frac{1}{4}\left(a-b+f+\frac{1}{2}\right) & \mathbf{\frac{1}{4}} & \frac{1}{4}\left(-3a+b-2c-2f+\frac{1}{2}\right) & 0 & \frac{f}{4} \\ \frac{a}{4} & \frac{b}{4} & \frac{1}{4}\left(-3a+b-2c-2f+\frac{1}{2}\right) & \frac{1}{4}\left(a-b+f+\frac{1}{2}\right) & \frac{b}{4} & 0 & 0 \\ 0 & 0 & \frac{c}{4} & 0 & \frac{a}{4} & 0 & 0 \\ \frac{f}{4} & 0 & \frac{1}{4}(a-b+c) & 0 & 0 & 0 & 0 \end{pmatrix}.$$

Note that mask m_0^s corresponds to parameters $a = b = c = f = 0$. By the conditions for orthogonal mask 2.27, we can found several sets of parameters when mask is orthogonal. For example, $a = 0.289381, b = 0.155585, c = -0.281659, f = -0.19962$ and mask is

$$\begin{pmatrix} 0. & 0. & 0. & -0.049905 & 0. & 0.0723452 & 0. \\ 0. & 0. & 0. & 0. & 0.0388962 & 0. & -0.0369657 \\ 0. & -0.0369657 & -0.0704147 & 0.1875 & 0.108544 & -0.0704147 & 0. \\ 0. & 0. & 0.108544 & \mathbf{0.25} & 0.1875 & 0. & -0.049905 \\ 0.0723452 & 0.0388962 & 0.1875 & 0.108544 & 0.0388962 & 0. & 0. \\ 0. & 0. & -0.0704147 & 0. & 0.0723452 & 0. & 0. \\ -0.049905 & 0. & -0.0369657 & 0. & 0. & 0. & 0. \end{pmatrix}$$

with spectrum in $[-3, 3]^2 \bigcap \mathbb{Z}^2$. Checking shows (see Sect. 6.8) that the corresponding refinable function φ is in $L_2(\mathbb{R}^2)$.

References

1. Sweldens, W.: The Lifting scheme: A custom-design construction of biorthogonal wavelets. Appl. Comput. Harmon. Anal. **2**, 186–200 (1996)
2. Bhatt, G.: Construction of wavelet frames using lifting like schemes. Int. J. Math. Anal. **5**(48), 1583–1593 (2011)

Chapter 6
Smoothness of Wavelets

Abstract The regularity of multivariate wavelet frames with an arbitrary dilation is studied by the matrix approach. The formulas for the Hölder exponents in spaces C and L_p are obtained in terms of the joint spectral radius of the corresponding transition matrices. Some results on higher order regularity, on the local regularity, and on the asymptotics of the moduli of continuity in various spaces are presented.

In this chapter, we address the problem of regularity of compactly supported wavelets. How to decide whether the wavelet function belongs to a given functional space and how to compute its exponent of regularity? Since the wavelet function is compactly supported, it is generated by refinement Eq. (2.16) with finitely many nonzero terms. It will be convenient to change the notation to the form used in most of literature: We denote coefficients $h_i = c_i$ and change the summation index from k to $-k$. Thus, we consider refinement equation of the form

$$\varphi(x) = \sum_{k \in \mathcal{I}} c_k \, \varphi(Mx \; - \; k) \, . \tag{6.1}$$

where $\boldsymbol{c} = \{c_k\}_{k \in \mathcal{I}}$ is a finite set of coefficients, $\mathcal{I} \subset \mathbb{Z}^d$ is a finite index set. The corresponding transition operator (2.49) now gets the form:

$$[T\varphi](x) = \sum_{k \in \mathcal{I}} c_k \, \varphi(Mx \; - \; k) \, . \tag{6.2}$$

Thus, the solution of refinement Eq. (6.1) is the eigenvector of T with the eigenvalue 1: $T\varphi = \varphi$. Clearly, the problem of regularity of compactly supported wavelets is reduced to the regularity of solutions of Eq. (6.1). For the univariate refinement equations, there are several efficient methods to compute or to estimate the exponents of regularity of solutions: the brute force method, the method of estimating invariant cycles, the method of computing the Sobolev regularity, and the matrix method. The first and the second ones approximate the exponents of regularity, while the latter two compute the exact values. For the multivariate equations with an arbitrary dilation matrix, only the methods of computing of the Sobolev regularity can be efficiently extended (see, for instance, [1–4], and references therein). The other methods are

A. Krivoshein et al., *Multivariate Wavelet Frames*,
Industrial and Applied Mathematics, DOI 10.1007/978-981-10-3205-9_6

generalized only in some special cases. For instance, when the dilation matrix M is isotropic, i.e., it is orthogonal in some basis [4–7]. Another approach is to consider regularity in special Besov spaces corresponding to the matrix M [8]. We focus on the matrix method that computes the Hölder regularity in the spaces C and L_p by means of the joint spectral radius of *transition matrices* associated with refinement equations. Let us recall that the *Hölder exponent* of a function $\varphi \in C(\mathbb{R}^d)$ is

$$\alpha_\varphi = \sup\left\{\alpha \geq 0 \,\middle|\, \|\varphi(\cdot + h) - \varphi(\cdot)\|_\infty \leq C\,\|h\|^\alpha\right\}.$$

The results of this chapter are based on recent work [9], where it was shown that the matrix method can be generalized to multivariate refinement equations with arbitrary dilation matrices, not necessarily isotropic. First, we are going to establish several facts on multivariate refinement equations.

6.1 Tiles, Self-similar Tilings, and Supports of Refinable Functions

The first step of the matrix method is to present a refinement equation as an equation for a vector function. That vector function is defined on a special set called a tile. In the univariate case, a tile is a segment. In the multivariate case, if the dilation matrix M is diagonal with positive diagonal elements, then the tile is a parallelotope. For general dilation matrices, the tiles may have a more complicated structure.

We fix a set of digits $D(M) = \{s_0, \ldots, s_{m-1}\} \in \mathbb{Z}^d$, $m = |\det M|$. For every integer point $p \in \mathbb{Z}^d$, we denote by M_p, the affine operator $M_p x = Mx - p$, $x \in \mathbb{R}^d$. We use the notation $0.s_1 s_2 \ldots = \sum_{k=1}^{+\infty} M^{-k} s_k$, where $\{s_k\}_{k\in\mathbb{N}}$ is a sequence of digits from $D(M)$. Consider the following set

$$G = \left\{0.s_1 s_2 \ldots \;\middle|\; s_k \in D(M), k \in \mathbb{N}\right\}. \tag{6.3}$$

We need some basic facts on this set.

Proposition 6.1.1 *For every dilation matrix M and for an arbitrary set of digits $D(M)$, the set G is compact and possesses the following properties:*

(a) *the Lebesgue measure of G is equal to some natural number q;*

(b) $G = \cup_{s\in D(M)} M_s^{-1} G$ *and the sets $M_s^{-1} G$ have pairwise intersections of zero measure;*

(c) *the indicator function $\chi = \chi_G(x)$ of G satisfies the refinement equation* $\chi(x) = \sum_i \chi(Mx - s_i)$;

(d) $\sum_{k\in\mathbb{Z}^d} \chi(x + k) \equiv q$, *i.e., integer shifts of G cover $\mathbb{R}^d$ with q layers.*

Proof Since $\rho(M) < 1$, it follows that $\sum_k \|M^{-k}\| \leq C$ for some constant C. Hence, $\|0.s_1 s_2 \ldots\| \leq C \max_k \|s_k\|$, and therefore the set G is bounded. Furthermore, let a sequence of points $\{\mathbf{s}^{(j)}\}_{j\in\mathbb{N}}$, where $\mathbf{s}^{(j)} = 0.s_1^{(j)} s_2^{(j)} \ldots \in G$, converges to some point $\mathbf{s} \in \mathbb{R}^d$. Passing to a subsequence, it can be assumed that the first digits $s_1^{(j)}$ coincide for all j with some digit s_1. Then pass to the next subsequence with the second digit s_2, etc. Then, $0.s_1 s_2 \ldots = \mathbf{s}$ and hence $\mathbf{s} \in G$. This proves that G is closed and therefore compact.

Since every point from $\mathbb{R}^d$ can be presented in the form $n + 0.s_1 s_2 \ldots$ with a suitable $n \in \mathbb{Z}^d$, it follows that the integer shifts of G cover the whole $\mathbb{R}^d$. Therefore, G has a positive measure. Since $M_s^{-1}(0.s_1 \ldots) = 0.ss_1 \ldots$, we see that $G = \cup_{s \in D(M)} M_s^{-1} G$. On the other hand, $|\det M^{-1}| = 1/m$ and hence $\mu(M_s^{-1} G) = \frac{1}{m} \mu(G)$ for all $s = 0, \ldots, m-1$. This implies that all those sets have pairwise intersections of zero measure. This proves property c. Hence $\chi(x)$ satisfies the Strang-Fix condition (Theorem 4.2.11), and so $\sum_k \chi(x + k) \equiv \mu(G)$ a.e. This means that the integer shifts of G cover the whole $\mathbb{R}^d$ with $\mu(G)$ layers, which yields that the measure of G is integer.◊

Corollary 6.1.2 *The following properties of the set G are equivalent:*

(a) $\mu(G) = 1$*;*

(b) *the integer shifts of G cover* $\mathbb{R}^d$ *and their pairwise intersections are of measure zero;*

(c) *the system of functions* $\{\chi_G(\cdot + k)\}_{k \in \mathbb{Z}^d}$ *is orthonormal.*

If the set G possesses those properties, then it is called a *tile*. The system of all integer shifts of a tile is called *tiling*. Thus, tiling can be defined as follows

Definition 6.1.3 A tiling generated by a dilation matrix M and by a set of digits $D(M)$ is a collection of sets $\mathcal{G} = \{G + k\}_{k \in \mathbb{Z}^d}$, where G is a compact set such that

(a) those sets cover $\mathbb{R}^d$ and the intersection of each two sets has Lebesgue measure zero;

(b) $G = \cup_{s \in D(M)} M_s^{-1} G$.

Corollary 6.1.2 implies that the set (6.3) is a tile if and only if $\mu(G) = 1$. Thus, tile is a fractal set whose self-similarity is defined by several affine operators with the same integer-matrix linear part and with different integer shifts. It is known (see [7]) that in cases $d = 2, 3$ and in case $|\det M| > d$, for every dilation matrix M, there is a digit set $D(M)$ such that G is a tile. In what follows, we assume that we chose that set of digits and that G is a tile.

Example 6.1.4 In the univariate case, let M be a natural number and $D(M) = \{0, \ldots, M-1\}$. Then, the tile G is a segment $[0, M-1]$.

In the multivariate case, if $M = \mathrm{diag}\,(m_1, \ldots, m_d)$ with natural $m_1, \ldots, m_d$ and $D(M) = M([0,1)^d) \cap \mathbb{Z}^d = \{(z_1, \ldots, z_d) \in \mathbb{Z}^d | z_i \in [0, m_i - 1]\}$, then G is a parallelotope $[0, m_1 - 1] \times \cdots \times [0, m_d - 1]\}$.

The set constructed in Example 2.3.7 is a tile. In that case, it is a parallelogram.

If $M = \begin{pmatrix} 1 & 1 \\ -1 & 1 \end{pmatrix}$ and $D(M) = \{(0,0), (1,0)\}$, then G is the *Dragon set* (see, for instance, [10]).

We denote $\mathcal{G}^n = M^{-n}\mathcal{G} = \{M^{-n}(k + G)\}_{k \in \mathbb{Z}^d}$. For a sequence of indices $k_1, \ldots, k_n \in \mathbb{Z}^d$, we often use a short multiindex notation $\boldsymbol{k} = (k_1, \ldots, k_n)$, n is said to be the length of $\boldsymbol{k}$. We denote $G_{\boldsymbol{k}} = G_{k_1 \dots k_n} = M_{k_1}^{-1} \cdots M_{k_n}^{-1} G$. Thus, $\mathcal{G}_n = \{G_{k_1 \dots k_n} | k_1, \dots, k_n \in D(M)\}$.

Consider now refinement Eq. (6.1) with coefficients $\boldsymbol{c} = \{c_k\}_{k \in \mathcal{I}}$.

Proposition 6.1.5 *The support of φ is a subset of the set*

$$S = \{x \in \mathbb{R}^d : x = \sum_{j=1}^{\infty} M^{-j} a_j, a_j \in \mathcal{I}\}. \tag{6.4}$$

Proof First of all, $\mathcal{I}$ is compact, which is proved in the same way as for the set G in Proposition 6.1.1. If $x = \sum_{j=1}^{\infty} M^{-j} a_j$, then for every $a_0 \in \mathcal{I}$, we have $M^{-1}(x + a_0) = \sum_{j=1}^{\infty} M^{-j} a_{j-1}$. Thus, $M^{-1}(x + a_0)S \subset S$ for each $a_0 \in \mathcal{I}$. Therefore, the space of distributions supported on the compact set S is invariant under the transition operator $[T\varphi](x) = \sum_{k \in \mathcal{I}} c_k \varphi(Mx - k)$. Taking an arbitrary function $f \in L_1(\mathbb{R}^d)$ supported on S such that $\int_{\mathbb{R}^d} f_0(x)dx = 1$ we see that all the functions $T^k f$, $k \in \mathbb{N}$, are also supported on S. On the other hand, by Theorem 2.6.4, we have $T^k f \to \varphi$, where φ is the solution of the refinement equation $T\varphi = \varphi$. Consequently, supp $\varphi \subset S$.$\Diamond$

We take a finite subset $\mathcal{I}_c \subset \mathbb{Z}^d$ such that $S \subset G + \mathcal{I}_c$. We define $\mathcal{I}_c$ as the minimal set of integer points that possesses this property, although an arbitrary set of integers with this property will suffice.

6.2 The Matrix Method of Computing the Hölder Regularity

For the sake of simplicity, we assume further that the mask satisfies sum rule of order one:

$$\sum_{k \in \Lambda} c_{Mk-s} = 1, \qquad s \in D(M). \tag{6.5}$$

This conditions arise naturally in the context of subdivision and of compactly supported wavelets. It is necessary for the convergence of a subdivision scheme associated with this refinement equation. Also it is necessary for the existence of compactly supported refinable functions with stable integer translates, i.e., translates that possess the Riesz basis property [5].

The main idea of the matrix approach is to pass from a function $f : \mathbb{R}^d \to \mathbb{R}$ supported on S to the vector function $v(x) = v_f(x) = \big(f(x + k)\big)_{k \in \mathcal{I}_c} \in \mathbb{R}^N$. Then, transition operator (6.2) restricted to the space $\{f \in L_1(\mathbb{R}^d) | \text{supp} f \subset S\}$ becomes the following *self-similarity operator* $\boldsymbol{A}$ on the space $L_1(G)$:

$$[Av](x) = T_s v(Mx - s), \qquad x \in M^{-1}(G + s), \quad s \in D(M), \tag{6.6}$$

where $v : G \to \mathbb{R}^N$ is a vector function, T_s is the $N \times N$ *transition matrix* defined for $s \in D(M)$ as follows

$$(T_s)_{ab} = c_{Ma-b-s}, \qquad a, b \in \mathcal{I}_c\,. \tag{6.7}$$

Here, we numerate the rows and columns of matrices by elements from the set $\mathcal{I}_c \subset \mathbb{Z}^d$. If one associates a usual number to any element of that set, then we obtain a "usual" enumeration of vector and matrix entries. The refinement equation becomes the self-similarity equation $\mathbf{A}v = v$ on the vector function $v(x)$:

$$v(x) = T_s v(Mx - s), \qquad x \in M^{-1}(G + s), \quad s \in D(M)\,, \tag{6.8}$$

We denote $\mathcal{T} = \{T_s, s \in D(M)\}$ and $\mathcal{T}^k$ is the set of products of length k of operators from $\mathcal{T}$ (the products with repetitions permitted). For a multiindex $\boldsymbol{s} = (s_1 \ldots s_k) \in D^k(M)$, let $\Pi_{\boldsymbol{s}} = T_{s_1} \cdots T_{s_k}$ be the corresponding product from $\mathcal{T}^k$.

Sum rule of order one (6.5) implies that every matrix T_s has the sum of elements in each column equal to one. Therefore, all matrices T_s have a common invariant affine subspace $V = \{x \in \mathbb{R}^d \mid \sum_k x_k = 1\} \subset \mathbb{R}^N$ and a common invariant linear subspace $W = \{x \in \mathbb{R}^d \mid \sum_k x_k = 0\} \subset \mathbb{R}^n$.

For every function f supported on S, we consider the extension $\tilde{v}$ of the function $v(x) = v_f(x)$ onto the whole $\mathbb{R}^d$. This function is defined on $\mathbb{R}^d$ by the same formula

$$\tilde{v}(x) \quad = \quad \big(f(x+k)\big)_{k \in \mathcal{I}_c}\,. \tag{6.9}$$

Clearly, $v = \tilde{v}|_G$. Take an arbitrary $k \in \mathbb{Z}^d$. For every $x \in G - k$, we have

$$\big(\tilde{v}(x)\big)_j \quad = \quad \begin{cases} \big(v(x+k)\big)_{j-k}\,, & j + k \in \mathcal{I}_c\,; \\ 0 & , \text{otherwise}\,. \end{cases} \tag{6.10}$$

Thus, on each set $G - k$, $k \in \mathbb{Z}^d$ of the tiling $\mathcal{G}$, the vector $\tilde{v}(x)$ has its jth component equal to the $(j+k)$th component of the vector $v(x+k)$, provided $j + k \in \mathcal{I}_c$ and equal to zero otherwise. Observe that for every $k \neq 0$, the vector $v(x)$ necessarily has at least one zero component at all $x \in G - k$.

Let $U = \operatorname{span}\big\{v(x) - v(y) \big| x - y \in \mathbb{R}^d\big\}$ be the space of differences of the values of the function v. We denote $\dim U = n$. Since $U \subset W$, we always have $n \le N - 1$. Note that U is a common invariant linear space for the operators T_s, $s \in D(M)$ (Lemma 6.2.3). Hence, one can define restrictions $A_s = T_s|_U$, $s \in D(M)$ of those operators to the subspace U. We assume a basis of U to be fixed, so A_s is associated with an $n \times n$ matrix. If the family $\mathcal{T}$ is irreducible on U, then $A_s = T_s|_W$. We denote $\mathcal{A} = \{A_s, s \in D(M)\}$ and $\mathcal{A}^k = \{A_{s_1} \cdots A_{s_k} \,|\boldsymbol{s} \in D^k(M)\}$ is the set of products of length k of operators from $\mathcal{A}$.

We are going to establish two results. The first one, Theorem 6.2.4 gives a criterion of continuity of the refinable function and the formula for its Hölder regularity. The criterion is formulated in terms of the joint spectral radius of linear operators.

Definition 6.2.1 The *joint spectral radius* of a compact family of finite-dimensional linear operators $\mathcal{A}$ is the following limit:

$$\rho(\mathcal{A}) = \lim_{k\to\infty} \max_{A_{s_i}\in\mathcal{A},\, i=1,\dots,k} \|A_{s_1}\dots A_{s_k}\|^{1/k}$$

This limit always exists and does not depend on the operator norm [11]. The joint spectral radius measures the simultaneous contractibility of operators of the family $\mathcal{A}$. We have $\rho(\mathcal{A}) < 1$ if and only if there is a norm in $\mathbb{R}^n$ in which all $A \in \mathcal{A}$ are contractions. In general, $\rho(\mathcal{A})$ is equal to the infimum of numbers β such that there is a norm in which $\|A\| < \beta$ for all $A \in \mathcal{A}$.

In the criterion one uses the operators A_s that are restrictions of the transition operators T_s to the space U. However, the space U is well defined only for a continuous refinable function φ and hence cannot be used in the criterion of its continuity. Nevertheless, this subspace can be defined without involving the values of φ.

Proposition 6.2.2 *If the function φ is continuous, then the subspace U coincides with the smallest by inclusion common invariant subspace of matrices T_s, $s \in D(M)$, that contains d vectors $T_s v_0 - v_0$, $s \in D(M)$, where v_0 is an eigenvector of T_0 corresponding to the eigenvalue one.*

Let us recall that $0 = s_0 \in D(M)$, which justifies the notation T_0. Proposition 6.2.2 will be proved in the next section. It actually gives an alternative definition of the subspace U. Let us recall that due to the sum rule each matrix T_s has an eigenvalue one. In rare cases, however, this eigenvalue can be multiple and the eigenvector v_0 is not unique up to normalization. However, there is at most one eigenvector v_0 such that the restrictions of the family $\mathcal{T}$ to its common invariant subspace spanned by vectors $T_s v_0 - v_0$, $s \in D(M)$ have the joint spectral radius less than one (Proposition 6.4.3). Thus,

The space U is the minimal common invariant subspace T_s, $s \in D(M)$, that contains vectors $T_s v_0 - v_0$, $s \in D(M)$, and such that $\rho(\mathcal{T}|_U) < 1$, where $v_0 \neq 0$ is an eigenvector of T_0 corresponding to the eigenvalue 1.

In Theorem 6.2.4 below, we see that the condition $\rho(\mathcal{T}|_U) < 1$ is necessary and sufficient for the continuity of the refinable function. For the definition of U, we need this condition only in the rare cases, when there are several noncollinear eigenvectors v_0. Otherwise, U is well defined for any refinable equations satisfying the sum rule. In the next section, we consider in detail the algorithm of constructing U and of computing the refinable function φ on an everywhere dense set of points.

We denote by λ_i the eigenvalues of M counting multiplicity, $|\lambda_1| \le \dots \le |\lambda_d|$, let $r_1 < \dots < r_q$ be all different values of modules of those eigenvalues, exactly n_i eigenvalues have modulus r_i. We always have $|\lambda_1| = r_1 > 1$. If M is isotropic, then $q = 1$. For $i = 1, \dots, q$, let $J_i \subset \mathbb{R}^d$ be the span of root subspaces of M corresponding to the eigenvalues of modulus r_i. Thus, J_i is a subspace of dimension n_i and the operator $M|_{J_i}$ has all its eigenvalues equal to r_i by modulus. The whole space $\mathbb{R}^d$ is a direct sum of the subspaces $J_1, \ldots, J_q$. The corresponding factorization of the matrix M has the following block diagonal form:

$$M = \begin{pmatrix} M^{(1)} & 0 & \cdots & 0 \\ 0 & M^{(2)} & 0 & \cdots \\ \vdots & \cdots & \cdots & 0 \\ 0 & \cdots & 0 & M^{(q)} \end{pmatrix} \qquad (6.11)$$

Let $U_i = \operatorname{span}\{v(x) - v(y) \mid y - x \in J_i\}$. Thus, $U_i \subset U$ is a subspace spanned by differences of the function v corresponding to the shifts of the argument from the space J_i. Since J_i is an invariant subspace for M, it follows that U_i is a common invariant subspace for the family $\mathcal{A} = \{A_s = T_s|_U\}_{s \in D(M)}$, as guaranteed by the following simple lemma.

Lemma 6.2.3 *If J is an invariant subspace for the matrix M, then $L = \operatorname{span}\{v(y) - v(x) | y - x \in J\}$ is a common invariant subspace for $\mathcal{A}$.*

Proof If $u \in L$, it is spanned by several vectors of the form $v(y) - v(x)$ with $y - x \in J$. For every $s \in D(M)$, we define $x' = M^{-1}(x + s)$, $y' = M^{-1}(y + s)$ and have

$$v(y') - v(x') = A_s\big(v(My' - s) - v(Mx' - s)\big) = A_s\big(v(y) - v(x)\big).$$

Hence, $A_s\big(v(y) - v(x)\big) \in L$ for each pair (x, y), and therefore $A_s u \in L$.$\Diamond$

Thus, $U_i = \operatorname{span}\{v(y) - v(x) | y - x \in J_i\}$ is a common invariant subspace of the family $\mathcal{A}$. It is seen easily that the spaces $\{U_i\}_{i=1}^q$ span the whole space U, but their sum may not be direct. Those subspaces, unlike the subspaces $\{J_i\}_{i=1}^q$, may have nontrivial intersections. For example, they can all coincide with U. We denote $\rho_i = \rho(\mathcal{A}|_{U_i})$.

Theorem 6.2.4 *A refinable function is continuous if and only if $\rho(\mathcal{A}) < 1$. In this case,*

$$\alpha_\varphi = \min_{i=1,\ldots,q} \frac{\log \rho_i}{\log(1/r_i)} \qquad (6.12)$$

This follows directly from the next result, which will be proved in Sect. 6.5. For an arbitrary subspace $J \subset \mathbb{R}^d$, let $\alpha_{\varphi,J} = \sup\{\alpha \ge 0 \,|\, \|\varphi(y) - \varphi(x)\| \le C\|y - x\|^{\alpha}, y - x \in J\}$ is the Hölder exponent of φ along the subspace J.

Theorem 6.2.5 *For a continuous refinable function φ, we have*

$$\alpha_{v,J_i} = \frac{\log \rho_i}{\log(1/r_i)}, \qquad i = 1, \ldots, q. \qquad (6.13)$$

Corollary 6.2.6 *If $\rho(\mathcal{T}|_W) < 1$, then φ is continuous and $\alpha_\varphi \ge -\log_{\rho(M)} \rho(\mathcal{T}_W)$.*

Proof Since $U \subset W$, it follows that $\rho(\mathcal{A}) = \rho(\mathcal{T}|_U) \le \rho(\mathcal{T}|_W)$. Hence, by Theorem 6.2.4, if $\rho(\mathcal{T}|_W) < 1$, then φ is continuous. Moreover, since $r_i \le \rho(M)$ for all i, equality (6.13) implies that $\alpha_\varphi \ge -\log_{\rho(M)} \rho(\mathcal{T}_W)$.$\Diamond$

6.3 Special Cases and Examples

The univariate case (d = 1). In this case, M is a number, $|M| = m$, Theorem 6.2.4 becomes a well-known statement that $\alpha_\varphi = -\log_m \rho(\mathcal{T}|_U)$. If φ is stable, i.e., its integer translates are linearly independent, then we always have $\rho(\mathcal{T}|_U) = \rho(\mathcal{T}|_W)$, although, in general, $U \ne W$ (see [5]). The space U was completely characterized in [12], and it was shown that every refinement equation can be factorized to the case $U = W$. In the multivariate case, however, there is no factorization procedure and some equations, even with stable solutions, cannot be reduced to the case $U = W$ (Example 6.3.3 below).

The case of an isotropic dilation matrix. In the multivariate case ($\mathbf{d} \ge \mathbf{2}$), the simplest situation is when the matrix M is isotropic. Since $\mathbf{q} = \mathbf{1}$, it follows that there is only one subspace $U_1 = U$. Theorem 6.2.4 implies the following fact well known in the literature.

Corollary 6.3.1 *If M is isotropic, then $\alpha_\varphi = -\log_{\rho(M)} \rho(\mathcal{A})$.*

The irreducible case. Another favorable case is not already that special. This is the case when the set of matrices $\mathcal{A} = \mathcal{T}|_U$ is irreducible, i.e., they do not share common invariant subspaces. Applying Theorem 6.2.4, we obtain

Corollary 6.3.2 *If the family $\mathcal{A}$ is irreducible, then $\alpha_\varphi = -\log_{\rho(M)} \rho(\mathcal{A})$.*

The irreducibility assumption, however, may fail in many important cases. For instance, if φ is a tensor product of two refinable functions of smaller number of variables, then $\mathcal{A}$ is always reducible.

Example 6.3.3 Let $\varphi_1 \in C^1(\mathbb{R})$ be a univariate refinable function with $m = 2$ and $\varphi_2 \in C^1(\mathbb{R})$ be a univariate refinable function with $m = 3$. Then, the function $\varphi = \varphi_1 \otimes \varphi_2$ satisfies the refinement equation with the symbol $\boldsymbol{m}(z_1, z_2) = \boldsymbol{m}_1(z_1)\boldsymbol{m}_2(z_2)$ and with the diagonal matrix $M = \mathrm{diag}\{2, 3\}$. We have $\rho_1 = \rho(\mathcal{A}|_{U_1}) = \frac{1}{2}$, $\rho_2 = \rho(\mathcal{A}|_{U_2}) = \frac{1}{3}$. By Theorem 6.2.4, $\alpha_\varphi = \min\{-\log_2 \rho_1, -\log_3 \rho_2\} = 1$, which is natural, because $\varphi \in C^1(\mathbb{R}^2)$. On the other hand, $\rho = \rho(\mathcal{A}) = \max\left\{\frac{1}{2}, \frac{1}{3}\right\} = \frac{1}{2}$. Hence, $-\log_{\rho(M)} \rho(\mathcal{A}) = -\log_3 \frac{1}{2} < 1$. Thus, $\alpha_\varphi > -\log_{\rho(M)} \rho(\mathcal{A})$ in this case. Note that if the functions φ_1, φ_2 are both stable, then so is φ. Nevertheless, unlike the univariate case, the regularity of φ is not determined by the value $\log_{\rho(M)} \rho(\mathcal{A})$.

The following result shows that the case of irreducible family $\mathcal{A}$ is not generic.

Corollary 6.3.4 *If the matrix M is not isotropic and the refinable function φ is Lipschitz continuous, then the family $\mathcal{A}$ is reducible.*

Proof We show that in this case $\alpha_\varphi > -\log_{\rho(M)} \rho(\mathcal{A})$. Indeed, $\alpha_\varphi = 1$, hence this inequality is equivalent to $\rho(\mathcal{A}) > 1/\rho(M)$. Assume the contrary: $\rho(\mathcal{A}) \le 1/\rho(M)$. Since M is not isotropic, factorization (6.11) contains $q \ge 2$ blocks, and hence $r_i < \rho$ for some i. By Theorem 6.2.5, we have $\alpha_{\varphi, J_i} = -\log_{r_i} \rho(\mathcal{A}) > -\log_{\rho(M)} \rho(\mathcal{A}) \ge 1$. Therefore, φ is an identical constant on every affine subspace $v + U_i$. Hence, this function is an identical zero, because it is compactly supported. The contradiction completes the proof.◊

The case of a dominant invariant subspace. According to many practical observations, this case is much more generic than the irreducible case.

A subspace $U' \subset U$ is dominant for a family of operators $\mathcal{A}$ if
(1) U' is a common invariant subspace for $\mathcal{A}$;
(2) U' is contained in all common invariant subspaces of the family $\mathcal{A}$;
(3) $\rho(\mathcal{A}|_{U'}) = \rho(\mathcal{A})$.

If we take a basis of a dominant subspace U' and complement it to a basis of the whole space U, then all the matrices $A \in \mathcal{A}$ get a block lower triangular form:

$$A = \begin{pmatrix} A^{(1)} & 0 \\ * & A^{(2)} \end{pmatrix}, \qquad (6.14)$$

where the block $A^{(2)}$ corresponds to the subspace U'. If $\rho(\mathcal{A}^{(2)}) \ge \rho(\mathcal{A}^{(1)})$, then

$$\rho(\mathcal{A}|_{U'}) = \rho(\mathcal{A}^{(2)}) = \max\left\{ \rho(\mathcal{A}^{(1)}), \rho(\mathcal{A}^{(2)})\right\} = \rho(\mathcal{A}).$$

Thus, the case of dominant subspace satisfies the form (6.14) of matrices $A \in \mathcal{A}$ with $\rho(\mathcal{A}^{(2)}) \ge \rho(\mathcal{A}^{(1)})$. Since any common invariant subspace of $\mathcal{A}$ contains U', it follows that the joint spectral radius of $\mathcal{A}$ restricted to any common invariant subspace is equal to $\rho(\mathcal{A})$. Therefore, we have proved

Corollary 6.3.5 *If the family $\mathcal{A}$ possesses a dominant subspace, then $\alpha_\varphi = -\log_{\rho(M)} \rho(\mathcal{A})$.*

6.4 Construction of the Continuous Refinable Function

The continuity of the refinable function depends on the joint spectral radius of the matrices T_s on the space U. The main question is how to evaluate this space and construct the solution φ. For many equations, U coincides with W, but this is not always the case. For the univariate equations, the method of evaluation of the space U

was elaborated in [13]. In multivariate case, the method is actually similar, with several significant modifications, especially in the proofs.

Construction of the space U. For each $i \in D(M)$, we take a point $z_i = (M - I)^{-1}i$. This is a fixed point of the contraction affine map M_i^{-1}. Let us recall that this map is defined as $M_i^{-1}x = M^{-1}(x + i)$ and that it maps G to G_i. Hence, $z_i \in G_i$. The value of the function φ at the point z_i can be computed. Indeed, since $Mz_i - i = z_i$, from Eq. (6.8), it follows that $T_i v(z_i) = v(z_i)$. Thus, for each $i \in D(M)$, the vector $v(z_i)$ is an eigenvector of the matrix T_i with the eigenvalue one, normalized by the condition $\bigl(e, v(z_i)\bigr) = 1$, where $e = (1, \dots, 1) \in \mathbb{R}^N$ is the vector of ones. Then for an arbitrary multiindex $\boldsymbol{d} = d_1 \dots d_k$, we denote $z_{i,\boldsymbol{d}} = M_{d_1}^{-1} \cdots M_{d_k}^{-1} z_i$. For $\boldsymbol{d} = \emptyset$, we set $z_{i,\boldsymbol{d}} = z_i$. Observe that $z_{i,\boldsymbol{d}} \in G_{i\,\boldsymbol{d}}$ and that $z_{i,\boldsymbol{d}\,i^r} = z_{i,\boldsymbol{d}}$ for every $r \geq 1$.

Let $Q_k = \bigl\{z_{i,\boldsymbol{d}} \mid \boldsymbol{d} \in D^k(M),\ \ i \in D(M)\bigr\}$. This is a set of m^{k+1} points, one point in each set $G_{i\,\boldsymbol{d}}$, $(i\,\boldsymbol{d}) \in D^{k+1}(M)$. In particular, $Q_0 = \{z_i\}_{i \in D(M)}$. Since $z_{i,\boldsymbol{d}} = z_{i,\boldsymbol{d}\,i} \in Q_{k+1}$, it follows that $Q_k \subset Q_{k+1}$ for all $k \geq 0$. Thus, $\{Q_k\}_{k \geq 0}$ is an embedded system of sets. Each set Q_k is an ε_k-net for the tile G with $\varepsilon_k = \operatorname{diam}(G_{i\,\boldsymbol{d}}) < C\,(\rho(M^{-1}) + \varepsilon)^k \to 0$ as $k \to \infty$. Therefore, the set $Q = \cup_{k \geq 0} Q_k$ is everywhere dense in G.

From self-similarity Eq. (6.8), it follows that

$$v(z_{i,\boldsymbol{d}}) = T_{d_1 \cdots d_k}\, v(z_i)\,. \tag{6.15}$$

Hence, we come to the following algorithm of step-by-step construction of the function φ or, equivalently, the function $v = v_{\varphi}$. First we find all $v(z_i)$, $i \in D(M)$, as the eigenvectors of T_i with the eigenvalue one, normalized by the condition $\bigl(e\,,\ v(z_i)\bigr) = 1$. Then, we find v at points of the sets Q_k, sequentially for $k = 1, 2, \dots$. For every $z_{i\,\boldsymbol{d}} \in Q_k$ and for each $j \in D(M)$, we compute $v(z_{i\;j\,\boldsymbol{d}}) = T_j v(z_{i\,\boldsymbol{d}})$, thus evaluating v on the set Q_{k+1}. As a result, v is computed on an everywhere dense set of points Q. Hence, if v is continuous, this algorithm determines v in a unique way. In practice, of course, we stop after finitely many steps k having computed $v(x)$, $x \in Q_k$. Then, the piecewise-constant function

$$v_k(\cdot) \equiv v(z_{i,\boldsymbol{d}}), \qquad x \in G_{i\boldsymbol{d}}\,, \quad \boldsymbol{d} \in D^k(M) \tag{6.16}$$

is an approximation for v, and the difference $\|v - v_k\|_{C(G)}$ can be efficiently estimated by the joint spectral radius of the family $\boldsymbol{A} = T|_U$. Before making this method rigorous, we need to define U and prove the continuity of the function v.

Let us recall that for an arbitrary refinement equation, the space U is defined as the smallest by inclusion common invariant subspace of matrices T_s, $s \in D(M)$ that contains vectors $T_s v(z_0) - v(z_0)$, $d \in D(M)$ (Proposition 6.2.2). We are going to prove that if $\varphi \in C(\mathbb{R}^d)$, then this definition coincides with the definition $U = \operatorname{span}\{v(y) - v(x) | x, y \in G\}$ given in the Introduction. The proof of Proposition 6.2.2 immediately follows from a more general fact given below.

Proposition 6.4.1 *For an arbitrary refinement equation, the space U contains all the vectors $v(y) - v(x)$, $x, y \in Q$.*

Since Q is everywhere dense in G, it follows that if v is continuous, then U contains all the vectors $v(y) - v(x)$, $x, y \in Q$, which proves Proposition 6.2.2.

Proof of Proposition 6.4.1. Let $x \in Q_k$ and $y \in Q_m$ with $k \geq m$. By the embeddedness property, $Q_m \subset Q_k$ and hence $x, y \in Q_k$. Now we apply induction in k. For $k = 0$, $x = z_i$, and $y = z_j$ for some $i, j \in D(M)$, hence $v(y) - v(x) \in U$ by the definition of U. Suppose now $v(b) - v(a) \in U$ for all $a, b \in Q_{k-1}$. Consider an arbitrary pair $x, y \in Q_k$. Let $x = z_{i,\boldsymbol{d}}$ with some $\boldsymbol{d} = d_1 \ldots d_{k-1} d_k$. Denote $a = z_{i,d_1 \ldots d_{k-1} i} = z_{i,d_1 \ldots d_{k-1}} \in Q_{k-1}$. We have

$$v(x) \; - \; v(a) = v\big(d_1 \ldots d_{k-1} d_k\big) - \; v\big(d_1 \ldots d_{k-1} i\big) = T_{d_1} \cdots T_{d_{k-1}} \left(z_{d_k} - z_i\right).$$

Since $z_{d_k} - z_i \in U$ and U is invariant with respect to all operators T_s, we have $v(x) - v(a) \in U$. Similarly we define the point $b \in Q_{k-1}$ for y and prove that $v(y) - v(b) \in U$. Since a and b are both from Q_{k-1}, by the inductive assumption, it follows that $v(b) - v(a) \in U$. Consequently,

$$v(y) \; - \; v(x) = \big(v(y) \; - \; v(b)\big) + \; \big(v(b) \; - \; v(a)\big) + \big(v(a) \; - \; v(x)\big) \; \in U \,.\Diamond$$

Denote $U^{(k)} = \operatorname{span}\Big\{v(y) - v(x) \;\Big|\; x, y \in Q_k\Big\}$. Clearly, $U^{(k)} \subset U^{(k+1)}$ for all $k \geq 0$.

Proposition 6.4.2 *There is $k \leq N - 1$ such that $U = U^{(k)}$.*

Proof Let k be the smallest number such that $U^{(k)} = U^{(k+1)}$. Since

$$U^{(k+1)} = \operatorname{span}\Big\{\, T_s\, U^{(k)} \;\;\Big|\;\; s \in D(M) \Big\}\,,$$

it follows that $U^{(k)}$ is a common invariant subspace for the family $\mathcal{T}$. Since it contains the vectors $v(y) - v(x)$, $x, y = Q_0$, and $U^{(j)} \neq U^{(j+1)}$ for all $j < k$, we see that $U^{(k)}$ is indeed the smallest common invariant subspace containing those vectors, i.e., $U^{(k)} = U$.$\Diamond$

The proof of Proposition 6.4.2 actually suggests an efficient algorithm to compute the space U.

Algorithm of computing U. *Zero iteration.* Find the eigenvector $v_0 \in V$, for which $T_0 v_0 = v_0$ and compute all $N - 1$ vectors $T_s v_0 - v_0$, $s \in D(M) \setminus \{0\}$. Remove redundant vectors, i.e., those spanned by others and leave basis vectors, call them $g_1, \ldots, g_t$.

kth iteration. We have $t + k - 1 \leq N - 1$ linearly independent vectors $g_1, \ldots, g_{t+k-1}$. For each $j = 1, \ldots, t + k - 1$, we compute the m vectors $T_s g_j$. If for some s, the rank of the matrix formed by column vectors $g_1, \ldots, g_{t+k-1}, T_s e_j$ exceeds $t + k - 1$, then we set $g_{t+k} = T_s g_j$, and go to the $(k + 1)$st iteration. Otherwise, if this rank is equal to $k + k - 1$ for all $s \in D(M)$, then we set $U = \operatorname{span}\{g_1, \ldots, g_{t+k-1}\}$ and the algorithm terminates. The end.

Since each iteration increases the number of the linearly independent vectors by one, the algorithm makes at most $N - t - 1$ iterations.

Once U is found, we define the operators $A_s = T_s|_U$, $s \in D(M)$, then we compute $\rho(\mathcal{A})$. If $\rho(\mathcal{A}) < 1$, then φ is continuous. In this case, we step-by-step find the values of v_φ at the points of the set Q_k for sufficiently big k and approximate v by the piecewise-constant function v_k defined in (6.16). Otherwise, if $\rho(\mathcal{A}) \geq 1$, then $\varphi \notin C(\mathbb{R}^d)$. The criterion of the continuity is proved in Theorem 6.2.4 in Sect. 6.2. If $\varphi \notin C(\mathbb{R}^d)$, then we can verify whether $\varphi \in L_p(\mathbb{R}^d)$ by the criterion of Theorem 6.8.4 (Sect. 6.8).

We complete this section with Proposition 6.4.3 which ensures that U is well defined even if there are many eigenvectors $v_0 \in V$ of the matrix T_0 with the eigenvalue one.

Proposition 6.4.3 *For an arbitrary refinement equation, the matrix T_0 has at most one, up to normalization, eigenvector $v_0 \in V$ with the eigenvalue one such that the subspace U generated by this eigenvector possesses the property $\rho(\mathcal{T}|_U) < 1$. If such an eigenvector exists, then φ is continuous and $v_0 = v_\varphi(z_0)$.*

Proof If such an eigenvector v_0 exists, then by Theorem 6.2.4, the refinable function is continuous and span $\{v(y) - v(x) | x, y \in G\} = U$. Applying our procedure of constructing the refinable function, we see that $v_0 = v(z_0)$. If there is another eigenvector v_0' with this property, it generates another refinable function for which $v_0' = v(z_0)$. By the uniqueness of solution of a refinement equation, those two solutions must be proportional; hence the vectors v_0 and v_0' are collinear.$\Diamond$

6.5 Proofs of the Main Theorems

We begin with several facts on the joint spectral radius. Then, we establish the key auxiliary results, after which we will be able to prove Theorems 6.2.4 and 6.2.5.

Several facts on the joint spectral radius

The following characteristic property was established in the work [11], where the joint spectral radius was introduced.

Theorem 6.5.1 *For an arbitrary family of operators $\mathcal{A}$ acting in $\mathbb{R}^n$ and for any $\varepsilon > 0$, there exists a norm $\|\cdot\|_\varepsilon$ in $\mathbb{R}^n$ such that $\|A\|_\varepsilon < \rho + \varepsilon$ for all $A \in \mathcal{A}$.*

Proof Let $\tilde{A} = \frac{1}{\rho+\varepsilon} A$ for all $A \in \mathcal{A}$. Then, $\rho(\tilde{\mathcal{A}}) = \frac{\rho}{\rho+\varepsilon} < 1$. Hence, $\sup_{A_{s_i} \in \tilde{\mathcal{A}},\, i=1,\dots,k,\, k \in \mathbb{N}} \|\tilde{A}_{s_1} \cdots \tilde{A}_{s_k}\| < \infty$. Then the function

$$f(x) = \sup_{\tilde{A}_{s_i} \in \tilde{\mathcal{A}},\, i=1,\dots,k,\, k \in \mathbb{N}} \|\tilde{A}_{s_1} \cdots \tilde{A}_{s_k} x\|$$

is a norm and satisfies $f(Ax) \leq (\rho + \varepsilon) f(x)$, $A \in \mathcal{A}$, $x \in \mathbb{R}^n$.$\Diamond$

The proof of the following result can be found in [14].

Theorem 6.5.2 *For an arbitrary family of operators $\mathcal{A}$ acting in $\mathbb{R}^n$, there exists a point $u \in \mathbb{R}^n$ and a constant $C(u) > 0$ such that*

$$\max_{A_{s_i} \in \mathcal{A}} \|A_{s_1} \cdots A_{s_k} u\| \geq C(u)\, \rho^k, \, k \in \mathbb{N}.$$

If $\mathcal{A}$ is irreducible, then

$$\max_{A_{s_i} \in \mathcal{A}} \|A_{s_1} \cdots A_{s_k}\| \leq C\, \rho^k, \, k \in \mathbb{N},$$

where C is a constant.

Auxiliary results. Proofs of Theorems 6.2.4 and 6.2.5 are based on the key auxiliary fact formulated in Proposition 6.5.6. We will arrive to it after several technical lemmas.

Lemma 6.5.3 *Assume the segment $[0, 1]$ is covered with ℓ closed sets. Then, there are $\ell + 1$ points $0 = a_0 \leq \ldots \leq a_\ell = 1$, such that for each $i = 0, \ldots, \ell - 1$, the points a_i, a_{i+1} belong to one set.*

Proof Let the first set contain the point $0 = a_0$. Put a_1 to be the maximal (in the natural ordering of the real line) point of the first set. If $a_1 \neq 1$, then a_1 must belong to another set. We call this set second and put a_3 to be the maximal point of that set. We keep doing so until $a_{\ell_0} = 1$ for some ℓ_0. We have $\ell_0 \leq \ell$ since there is no repetition of the sets. If $\ell_0 < \ell$, we complement the sequence by points $a_{\ell_0+1}, \ldots, a_\ell = 1$.$\Diamond$

Next, we show that a segment of a given length cannot intersect too many sets of the tiling.

Lemma 6.5.4 *For every tiling $\mathcal{G}$, there is a constant $\eta = \eta(\mathcal{G})$ such that every line segment $[x, y] \in \mathbb{R}^d$ intersects at most $\eta \max\{1\,, \, \|y - x\|\,\}$ sets of that tiling.*

Proof It suffices to prove that the number of tiles intersected by a segment of length one is bounded above by some constant. It will imply that the number of tiles intersected by any segment of length $\leq k$, where k is an integer, bounded by this constant times k, from which the proposition follows. Thus, let a segment $[x, y]$ be of length one. Each tile intersecting this segment is contained in the Minkowski sum of the segment $[x, y]$ with the Euclidean ball of radius diam (G). If we denote by V the volume of this body, we conclude that the total number of tiles intersecting $[x, y]$ does not exceed $\frac{V}{\mu(G)} \leq V$, which completes the proof.$\Diamond$

Let us recall that for an arbitrary function $f : \mathbb{R}^d \to \mathbb{R}$ supported on S, we consider the vector function $v(x) = v_f(x) = \big(f(x + k)\big)_{k \in \mathcal{I}_c}$ defined on the set G and its extension $\tilde{v}$ defined by the same formula on the whole $\mathbb{R}^d$. Our next observation is that for arbitrary pair of points from the tile set G, the norm of difference of the function $\tilde{v}$ at those points becomes smaller after each integer shift of the variable.

Lemma 6.5.5 *For arbitrary points* x, y *from one set* $G' = G - n$ *of the partition* $\mathcal{G}$, $n \in \mathbb{Z}^d$, *we have* $\|\tilde{v}(y) - \tilde{v}(x)\| \le \|v(y+n) - v(x+n)\|$.

Proof By formula (6.10), each component of the vector $\tilde{v}(y) - \tilde{v}(x)$ is either equal to the component of the vector $v(y+n) - v(x+n)$ with the index shifted by n (if the shifted index still belongs to $\mathcal{I}_c$) or vanish (otherwise). Thus, when we pass from the vector $v(y+n) - v(x+n)$ to the vector $\tilde{v}(y) - \tilde{v}(x)$, each component either change its position (and keeps its value) or vanish. Hence, in the Euclidean norm, we have $\|v(y+n) - v(x+n)\| \ge \|\tilde{v}(y) - \tilde{v}(x)\|$.◊

Now we are ready to formulate the main result used in the proofs of theorems.

Proposition 6.5.6 *Let* φ *be a refinable function and* $v = v_\varphi$. *Then for arbitrary points* $x, y \in \mathbb{R}^d$ *and for every* $k \in \mathbb{N}$, *there is* $\ell \le \max \eta \{ 1, \|M^k(x-y)\| \}$, *products* $\Pi_0, \dots, \Pi_{\ell-1} \in \mathcal{T}^k$, *nonnegative numbers* $\{\alpha_j\}_{j=0}^{\ell-1}$ *with the sum equal to one, and collections of points* $\{x_j\}_{j=0}^{\ell-1}, \{y_j\}_{j=0}^{\ell-1}$ *from* G *such that* $y_j - x_j = \alpha_j M^k(y-x)$ *for all* $j = 1, \dots, \ell-1$, *and*

$$\|\tilde{v}(y) - \tilde{v}(x)\| \le \sum_{j=0}^{\ell-1} \left\| \Pi_j \Big(v(y_j) - v(x_j) \Big) \right\|. \tag{6.17}$$

Proof The tiling $\mathcal{G}^k$ covers the segment $[x, y]$. Applying Lemma 6.5.3, we get $\ell + 1$ points $\{a_i\}_{i=0}^{\ell}$ on the segment $[x, y]$ such that $a_0 = x, a_\ell = y$, and each pair of successive points a_j, a_{j+1} belongs to one set of the tiling $\mathcal{G}^k$. First, we estimate the number ℓ. Once ℓ elements of the tiling $\mathcal{G}^k = M^{-k}\mathcal{G}$ cover a segment of length $\|y - x\|$, it follows that the same number of elements of the tiling $\mathcal{G}$ cover a segment of length $\|M^k(y-x)\|$. Therefore, Lemma 6.5.3 yields $\ell \le \eta \max\{1, \|M^k(y - x)\|\}$.

Furthermore, the element of the tiling $\mathcal{G}^k$ containing both a_j and a_{j+1} lies in some set $G' = G - n$ of the tiling $\mathcal{G}$, where $n \in \mathbb{Z}^n$. This means that there is a multiindex $s \in D^k(M)$ such that $a_j, a_{j+1} \in G_s - n$. Note that both s and n depend on j. Thus, for each $j = 0, \dots, \ell-1$, applying Eq. (6.8) to the points $x = a_j + n \in G_s$ and $x = a_{j+1} + n \in G_s$, we obtain

$$\begin{gathered} \Big\| v(a_{j+1}+n) - v(a_j+n) \Big\| = \\ \Big\| \Pi_s \Big(v\big(M_{s_k} \cdots M_{s_1}(a_{j+1}+n)\big) - v\big(M_{s_k} \cdots M_{s_1}(a_j+n)\big) \Big) \Big\|, \end{gathered} \tag{6.18}$$

where $\Pi_s = T_{s_1} \cdots T_{s_k}$. We set

$$x_j = M_{s_k} \cdots M_{s_1} \big(a_j + n\big); \qquad y_j = M_{s_k} \cdots M_{s_1} \big(a_{j+1} + n\big).$$

For each j, we define the number α_j from the equality $\|a_{j+1} - a_j\| = \alpha_j \|y - x\|$. We see that $\sum_{j=1}^{\ell-1} \alpha_j = 1$ and that $y_j - x_j = M^k(a_{j+1} - a_j) = \alpha_j M^k(y-x)$. Equality (6.18) reads

$$\left\| v(a_{j+1}+n) - v(a_j+n) \right\| = \left\| \Pi_s\big(v(y_j) - v(x_j)\big) \right\|. \tag{6.19}$$

On the other hand, the triangle inequality yields

$$\left\|\tilde v(y) - \tilde v(x)\right\| \le \sum_{j=0}^{\ell-1}\left\|\tilde v(a_{j+1}) - \tilde v(a_j)\right\| \le \sum_{j=0}^{\ell-1}\left\| v\left(a_{j+1}+n(j)\right) - v\left(a_j+n(j)\right)\right\|.$$

The latter inequality follows from Lemma 6.5.5: $\left\|\tilde v(a_{j+1}) - \tilde v(a_j)\right\| \le \left\|v(a_{j+1}+n) - v(a_j+n)\right\|$, where $n = n(j)$ is defined above. On the other hand, the value $\left\|v(a_{j+1}+n) - v(a_j+n)\right\|$ is, in view of (6.19), smaller than or equal to $\left\|\Pi_s\left(v(y_j) - v(x_j)\right)\right\|$. Writing $\Pi_s = \Pi_j$, we obtain $\left\|\tilde v(a_{j+1}) - \tilde v(a_j)\right\| \le \left\|\Pi_j\left(v(y_j) - v(x_j)\right)\right\|$. Taking the sum over all $j = 0, \ldots, \ell-1$, we arrive at (6.17).$\Diamond$

We start with proving Theorem 6.2.5, then we give a proof of Theorem 6.2.4.

Proof of Theorem 6.2.5. We choose a small $\varepsilon \in (0, 1-\rho(\mathcal{A}))$.

Let us fix $i = 1, \ldots, q$ and show that $\alpha_{\varphi, J_i} \ge \frac{\log \rho_i}{\log(1/r_i)}$. For arbitrary points $x, y \in G$ such that $y - x \in J_i$ and $\|y-x\| < 1$, define k as the smallest number such that $\|M^k(y-x)\| \ge 1$. Since $y - x \in J_i$, it follows that

$$\|M^k(y-x)\| \quad \le \quad C\,(r_i+\varepsilon)^k\,\|y-x\|, \tag{6.20}$$

where C depends only on M. Consequently,

$$C\,\|y-x\|\,(r_i+\varepsilon)^k \ge 1. \tag{6.21}$$

Now, we apply Proposition 6.5.6 to the points x, y, and to this number k. We see that the norm $\|v(y) - v(y)\|$ is bounded above by the sum (6.17), where

$$\ell \le \eta\,\max\left\{1,\ \|M^k(y-x)\|\right\} = \eta\|M^k(y-x)\|$$

and each $\|v(y_j) - v(x_j)\|$ is bounded above by $2\,\|v(\cdot)\|_{C(G)}$. Hence, inequality (6.17) implies

$$\left\|v(y) - v(x)\right\| \quad\le\quad \sum_{j=0}^{\ell-1}\|\Pi_j\|\cdot 2\,\|v\|_{C(G)} \quad\le$$

$$\le\quad 2\ell\,C\,\|v\|_{C(G)}\,(\rho_i+\varepsilon)^k \quad\le\quad 2\eta\|M^k(y-x)\|\,C\,\|v\|_{C(G)}\,(\rho_i+\varepsilon)^k$$

(we invoked Theorem 6.5.2 to estimate the norm of $\Pi_j = T_{s_1}\cdots T_{s_k}$). By the definition of k, we have $\|M^{k-1}(y-x)\| < 1$ and hence $\|M^k(y-x)\| \le \|M\|\cdot\|M^{k-1}(y-x)\| < \|M\|$. Thus,

$$\left\|v(y) - v(x)\right\| \le \sum_{j=0}^{\ell-1}\|\Pi_j\|\cdot 2\,\|v\|_{C(G)} \le 2\ell\,C\,\|v\|_{C(G)}\,(\rho_i+\varepsilon)^k \le$$

$$2\,\eta\,\|M\|\,C\,\|v\|_{C(G)}\,(\rho_i+\varepsilon)^k$$

Combining this with (6.21), we get

$$\big\|v(y)\ -\ v(x)\big\|\ \le C\,\|x-y\|^{\alpha(\varepsilon)}$$

with $\alpha(\varepsilon)\ =\ -\frac{\log(\rho_i+\varepsilon)}{\log(r_i+\varepsilon)}$ and with some constant C depending on ε. Taking $\varepsilon\to 0$, we conclude the proof of the lower bound.

Now let us establish the inverse inequality: $\alpha_{\varphi,J_i}\le\frac{\log\rho_i}{\log(1/r_i)}$. We take $\varepsilon\in(0,r_i)$. Applying Theorem A2, we find $u\in U_i$ such that for every k, there are indices $s_1,\ldots,s_k\in D(M)$ with the property $\|\Pi_{\mathbf{s}}\,u\|\ \ge\ C(u)\ >\ 0$ for all $\Pi_{\mathbf{s}}\ \in\mathcal{A}^k, k\in\mathbb{N}$.

Since the subspace U_i is spanned by the differences $v(y)-v(x)$, $y-x\in J_i$, there are $n=n_i$ pairs (x_j,y_j) of points from G, $j=1,\ldots,n_i$, such that $u=\sum_{j=1}^{n_i}\gamma_j\big(v(y_j)-v(x_j)\big)$. Denote $x_j^{(k)}=M_{s_1}^{-1}\cdots M_{s_k}^{-1}x_j$ and $y_j^{(k)}=M_{s_1}^{-1}\cdots M_{s_k}^{-1}y_j$. Thus, $x_j^{(k)},y_j^{(k)}\in G_{\mathbf{s}}$ and $\|y_j^{(k)}-x_j^{(k)}\|\ \le\ C_\varepsilon\,(r_i-\varepsilon)^{-k}\|y_j-x_j\|$. We have

$$\sum_{j=1}^{n_i}|\gamma_j|\cdot\big\|v(y_j^{(k)})\ -\ v(x_j^{(k)})\big\|=$$

$$\sum_{j=1}^{n_i}|\gamma_j|\cdot\big\|\Pi_{\mathbf{s}}\big(v\,\big(M_{s_k}\cdots M_{s_1}y_j^{(k)}\big)-\ v\big(M_{s_k}\cdots M_{s_1}x_j^{(k)}\big)\ \big\|=$$

$$\sum_{j=1}^{n_i}|\gamma_j|\cdot\big\|\,\Pi_{\mathbf{s}}\,\big(v(y_j)\ -\ v(x_j)\big)\,\big\|\ge\left\|\sum_{j=1}^{n_j}\gamma_j\,\Pi_{\mathbf{s}}\,\big(v(y_j)\ -\ v(x_j)\big)\right\|=$$

$$\left\|\Pi_{\mathbf{s}}\,\left(\sum_{j=1}^{n}\gamma_j\,\big(v(y_j)\ -\ v(x_j)\big)\right)\right\|\ =\ \big\|\,\Pi_{\mathbf{s}}\,h\,\big\|\ge C(u)\,\rho_i^k\,.$$

Thus,

$$\sum_{j=1}^{n_i}|\gamma_j|\cdot\big\|v(y_j^{(k)})\ -\ v(x_j^{(k)})\big\|\ge C(u)\,\rho^k\,,\qquad k\in\mathbb{N}\,.$$

Consequently, at least one of the n_i numbers $\|v(y_j^{(k)})-v(x_j^{(k)})\|$, $j=1,\ldots,n_i$, is bigger than or equal to $\frac{C(u)}{\sum_j|\gamma_j|}\,\rho^k$. Combining this with the inequality $\|y_j^{(k)}-x_j^{(k)}\|\ \le C_\varepsilon\,(r_i-\varepsilon)^{-k}\|y_j-x_j\|$, we obtain

$$\|v(y_j^{(k)})\ -\ v(x_j^{(k)})\|\ge C\,\big\|\,y_j^{(k)}\ -\ x_j^{(k)}\,\big\|^{\,\alpha}\,,$$

where $\alpha = \frac{\log \rho_i}{\log(1/(r_i - \varepsilon)}$, and the constant $C > 0$ does not depend on k. Since $\|y_j^{(k)} - x_j^{(k)}\| \to 0$ as $k \to \infty$, we wee that there are arbitrary small segments $[x_j^{(k)}, y_j^{(k)}]$ on which the variation of the function v is at least a constant times the length of that segment to the power of $\frac{\log \rho_i}{\log(1/(r_i - \varepsilon)}$. Therefore, $\alpha_{\varphi, J_i} \le \frac{\log \rho_i}{\log(1/(r_i - \varepsilon)}$. Since ε is arbitrary, the proof is completed. $\Diamond$

Proof of Theorem 6.2.4. First, we show that the condition $\rho(\mathcal{A}) < 1$ is sufficient for continuity of the refinable function. We denote $\rho(\mathcal{A}) = \rho$ and choose $\varepsilon \in (0, 1 - \rho)$. Find the values of the function v on the everywhere dense set Q as described in Sect. 6.4. This defines the function φ on the set $\tilde{Q} = \cup_{n \in \mathbb{Z}^d}(Q - n)$. This function is supported on the set $S \cap \tilde{Q}$. Finally, we define the extension $\tilde{v} : \tilde{Q} \to \mathbb{R}$ of the function v as $\tilde{v}(x) = \bigl(\varphi(x + k)\bigr)_{k \in \mathcal{I}_c}$. We are going to prove that $\tilde{v}$ is uniformly continuous on $\tilde{Q}$, from which it will follow that its extension to the entire $\mathbb{R}^d$ is continuous.

Let us first establish that the function v is uniformly bounded on the set Q. Denote $C_0 = \max\{\|v(x) - v(y)\| | x, y \in Q_1\}$. It follows that for every $i \in D(M)$ and $\boldsymbol{d} \in D^k(M)$, we have

$$\max\{\|v(x) - v(y)\| | x, y \in G_{i, \boldsymbol{d}}\} \le C_0 \|A_{d_1} \cdots A_{d_{k-1}}\| \le C_1(\rho + \varepsilon)^k,$$

where C_1 does not depend on $i, \boldsymbol{d}$, nor on k. For every j and $\boldsymbol{d} \in D^j(M)$, we have

$$\|v(z_{i\boldsymbol{d}}) - v(z_i)\| \le \|v(z_{i, d_1}) - v(z_i)\| + \sum_{k=2}^{j} \|v(z_{i, d_1 \dots d_k}) - v(z_{i, d_1 \dots d_{k-1}})\| \le$$

$$\le C_0 \sum_{k=2}^{j} (\rho + \varepsilon)^{k-1} \le \frac{C_0}{1 - \rho - \varepsilon}.$$

Hence $\|v(z)\| \le \max_{i \in D(M)} \|v(z_i)\| + \frac{C_0}{1 - \rho - \varepsilon}$ for each $z \in Q$, which implies the uniform boundedness of v.

Now we take arbitrary points $x, y \in \tilde{Q}$ and estimate the norm $\|\tilde{v}(y) - \tilde{v}(x)\|$ in the same way as in the proof of Theorem 6.2.5. We apply Proposition 6.5.6 and obtain

$$\|\tilde{v}(y) - \tilde{v}(x)\| \le 2\eta \|M\| C \|v\|_{C(Q)} (\rho + \varepsilon)^k,$$

where k is the smallest number such that $\|M^k(y - x)\| \ge 1$. Note that the value $\|v\|_{C(Q)}$ is finite because v is bounded Q. Since $k \to \infty$ as $\|y - x\| \to 0$, we see that $\tilde{v}$ is uniformly continuous on $\tilde{Q}$, which completes the proof of continuity.

Thus, if $\rho < 1$, then $\varphi \in C(\mathbb{R}^d)$. By Theorem 6.2.5, the Holder exponent of φ on shifts along the subspace J_i is equal to $\alpha_i = \frac{\log \rho_i}{\log(1/r_i)}$. We pass to a basis in the space $\mathbb{R}^d$, in which all the subspaces J_i are orthogonal to each other. Using a natural expansion $h = h_1 + \ldots + h_q$, $h_i \in J_i$ we obtain for arbitrary $\varepsilon > 0$

$$\|\varphi(\cdot + h) - \varphi(\cdot)\| \le \sum_{i=1}^{q} \|\varphi(\cdot + h_i) - \varphi(\cdot)\| \le \sum_{i=1}^{q} C\,\|h_i\|^{\alpha_i - \varepsilon} \le C\,d^{\alpha-\varepsilon}\|h\|^{\alpha_i-\varepsilon},$$

where $\alpha = \min_{i=1,\dots,q} \alpha_i$. Consequently, $\alpha_\varphi = \min_{i=1,\dots,q} \alpha_i$. ◊

6.6 Derivatives of Refinable Functions

In the univariate case, the factorization of refinement equations was elaborated in [15–17]. If the solution φ of refinement equation belongs to C^1, then φ is a convolution of a piecewise-constant function and of a continuous solution of a refinement equation of a smaller order. This resolves the question of differentiability of refinable functions and classify all smooth refinable functions. In particular, every C^ℓ-refinable function is a convolution of a refinable spline of order $\ell - 1$ and of a continuous refinable function. This factorization technique, however, cannot be extended to multivariate case. Some results on this direction can be found in [8]. In this section, we are going to see that some factorization of multivariate smooth refinable functions is possible in the following sense: The derivative of a refinable function φ can be found by solving several refinement and generalized refinement equations. The continuous differentiability of φ is equivalent to continuity of solutions of all those equations. The main idea is that the derivatives of φ along vectors of the Jordan basis of M satisfy refinement and generalized refinement equations.

Definition 6.6.1 A generalized refinement equation is an equation of the form $\varphi = T\varphi + g$, where T is a transition operator and g is a known function.

Let $\{e_1\}_{i=1}^d$ be the Jordan basis of the matrix M in $\mathbb{R}^d$, i.e., the basis in which M has Jordan form. The Jordan basis consists of eigenvectors, for which $Me_i = \lambda e_i$ and, and generalized eigenvectors, for which $Me_i = \lambda e_i + e_{i-1}$. Take one Jordan block of dimension p corresponding to an eigenvalue λ. After possible renumbering, it can be assumed that $e_1, \dots, e_p$ is the corresponding vectors of the Jordan basis. Thus, $Me_1 = \lambda e_1$ and $Me_i = \lambda e_i + e_{i-1}, i = 2, \dots, p$. For a vector $a \in \mathbb{R}^d$, we denote by $\mathcal{S}'_a$ the space of compactly supported distributions, whose mean along every straight line parallel to a is equal to zero. Clearly, if φ is compactly supported, then $\frac{\partial\varphi}{\partial a} = (a, \varphi') \in \mathcal{S}'_a$. We use the short notation $\mathcal{S}'_{e_i} = \mathcal{S}'_i$.

Proposition 6.6.2 *Suppose $e_1, \dots, e_p$ are vectors of the Jordan basis of the dilation matrix M corresponding to one Jordan block (e_1 is an eigenvector, the others are generalized eigenvectors), then the function $\varphi_1 = (e_1, \varphi')$ belongs to $\mathcal{S}'_1$ and satisfies the refinement equation*

$$\varphi_1 = \lambda T\varphi_1\,; \tag{6.22}$$

for each $i = 2, \dots, p$, the function $\varphi_i = (e_i, \varphi')$ belongs to $\mathcal{S}'_i$ and satisfies the generalized refinement equation

$$\varphi_i = \lambda T \varphi_i + \sum_{t=1}^{i-1} (-1)^{t-1} \lambda^{-t} \varphi_{i-t} . \tag{6.23}$$

Conversely, the system of Eqs. (6.22), (6.23) possesses a unique solution $\varphi_i \in \mathcal{S}_i'$, $i = 1, \dots, p$, which coincides with the corresponding directional derivatives of the solution φ of the original equation $T\varphi = \varphi$.

Proof If $Me_1 = \lambda e_1$, then

$$\varphi_1(x) = \bigl(e_1, \varphi'(x)\bigr) = \left(e_1, \sum_k c_k \Bigl[\varphi(Mx-k)\Bigr]'\right) = \left(e_1, \sum_k c_k M^* \varphi'(Mx-k)\right) =$$

$$\left(M e_1, \sum_k c_k \varphi'(Mx-k)\right) = \left(\lambda e_1, \sum_k c_k \varphi'(Mx-k)\right) =$$

$$\lambda \sum_k c_k \Bigl(e_1, \varphi'(Mx-k)\Bigr) = \lambda \sum_k c_k \varphi_1(Mx-k) .$$

For the vector e_2, we have $Me_2 = \lambda e_2 + e_1$, and an extra term appears: $\sum_k c_k \varphi_1 (Mx-k) = \lambda^{-1}\varphi_1(x)$. Hence, $\varphi_2 = \lambda T \varphi_2 + \lambda^{-1}\varphi_1$. It remains to apply induction in $i \geq 2$.

Conversely, if Eq. (6.22) possess a solution $\varphi_1 \in \mathcal{S}_1'$, then its primitive along the line e_1 is compactly supported and satisfies the equation $bT\varphi = \varphi$. Hence, by the uniqueness of the solution, that primitive coincides with φ and $\frac{\partial \varphi}{\partial_{e1}} = \varphi_1$. Then, we prove the same for φ_2, etc.$\Diamond$

Theorem 6.6.3 *Let $\{e_i\}_{i=1}^d$ be a Jordan basis of the dilation matrix M. If for each Jordan block, the Eqs. (6.22), (6.23) possess continuous solutions $\varphi_i \in \mathcal{S}_i'$, $i = 1, \dots, p$, then $\varphi \in C^1(\mathbb{R}^d)$ and $\frac{\partial \varphi}{\partial e_i} = \varphi_i$.*

Thus, $\varphi \in C^1(\mathbb{R}^d)$ if and only if d (generalized) refinement equations corresponding to the Jordan basis have continuous solutions in the spaces $\mathcal{S}_i'$. Those solutions are directional derivatives of φ along the vectors of the basis. Clearly, the gradient φ' is easily found from those derivatives.

Proof If there are continuous solutions $\varphi_i \in \mathcal{S}_i'$, $i = 1, \dots, d$, then, for some initial point $a = (a_1, \dots, a_d) \notin S$, we define

$$\varphi(x) = \int_{a_1}^{x_1} \varphi_1(\tau, a_2, \dots, a_d)\, d\tau + \dots + \int_{a_d}^{x_d} \varphi_d(x_1, \dots, x_{d-1}, \tau)\, d\tau .$$

This function is compactly supported, since $\varphi_i \in \mathcal{S}_i'$ belongs to $C^1(\mathbb{R}^d)$ and satisfies the refinement equation $T\varphi = \varphi$.$\Diamond$

Thus, the continuous differentiability of a refinable function is equivalent to the existence of continuous solutions of d corresponding generalized refinement equations. If the dilation matrix M has a basis of eigenvectors, then everything is reduced to d usual refinement equations.

Corollary 6.6.4 *Suppose the dilation matrix has a basis of eigenvectors $e_1, \dots, e_d$. If $\varphi \in C^1(\mathbb{R}^d)$, then each directional derivative $\varphi_i = \frac{\partial \varphi}{\partial e_i} = (e_i, \varphi')$ belongs to $\mathcal{S}'_{e_i}$ and satisfies the refinement equation $\varphi_i = \lambda_i T \varphi_i$.*

Conversely, if all those d equations have solutions in the corresponding spaces $\mathcal{S}'_{e_i}$, then the original equation $T\varphi = \varphi$ has a C^1-solution, whose directional derivatives along the vectors e_i coincide with φ.

6.7 Modulus of Continuity of a Refinable Function

Apart from thc computing of cxact valucs of thc Höldcr cxponcnts in thc paccs C and L_p, the matrix approach can be useful in at least two additional problems: a refine analysis of the modulus of continuity and the local regularity of solutions. In this section, we consider the first problem. Since the results in C and in L_p are formulated similarly, we focus on the space C. Thus, we study the *modulus of continuity* in the space $C(\mathbb{R}^d)$:

$$\omega(\varphi, t) = \sup_{\|h\| \le t} \|\varphi(\cdot + h) - \varphi(\cdot)\|_{C(\mathbb{R}^d)}\,. \tag{6.24}$$

As usual, we are interested in the asymptotics of $\omega(\varphi, t)$ as $t \to +0$. Clearly, the Hölder exponent is the supremum of $\alpha \ge 0$ such that $\omega(\varphi, t) \le C t^{\alpha}$. However, in many cases, the Hölder exponent gives too rough information about the function. For example, if $\alpha_\varphi = 1$, then the function φ may be continuously differentiable or may not even be Lipschitz continuous. The Lipschitz continuity takes place if and only if the exponent $\alpha_\varphi = 1$ is *sharp*.

Definition 6.7.1 The Hölder exponent α of a function φ is sharp if there is a constant such that $\omega(\varphi, t) \le C t^{\alpha}, t \in (0, 1)$.

In the univariate case, it was noted long ago that the Hölder exponent of a refinable function is not always sharp. For example, the derivative of the Dubuc function of the four-point interpolatory scheme with the parameter $w = \frac{1}{16}$ is "almost Lipschitz" with the power one, i.e., $\omega(\varphi', t) \asymp t\,|\log t|$ as $t \to 0$. See [18]. This is the univariate case. In the bivariate case with the matrix $M = 2I$, it was recently shown that the derivative of the function of the Butterfly subdivision scheme in the same case $w = \frac{1}{16}$ is "almost Lipschitz" with the power two, i.e., $\omega(\varphi', t) \asymp t\,|\log t|^2$ as $t \to 0$, see [19]. The matrix approach allows us to make the same analysis of the asymptotic behavior of the modulus of continuity for the multivariate refinable functions with arbitrary dilation matrix. To formulate the main result of this section, we need to introduce some notation.

The *resonance degree* of a compact set of $n \times n$ matrices $\mathcal{A}$ is the smallest natural number $\nu = \nu(\mathcal{A})$ such that $\max_{X_i \in \mathcal{A}, i=1,\dots,k} \|X_1 \cdots X_k\| \le C\,\rho^k\, k^{\nu}$ for all $k \in \mathbb{N}$. By Theorem A2, all irreducible families have the resonance degree one. In general, $\nu \le n-1$. Moreover, ν does not exceed the *valency* of the matrix family [17], i.e., the total number of diagonal blocks $\mathcal{A}^{(j)}$ in the lower triangular Frobenius factorization of the family $\mathcal{A}$ such that $\rho(\mathcal{A}^{(j)}) = \rho(\mathcal{A})$. Every family of matrices has the Frobenius factorization in a suitable basis:

$$A = \begin{pmatrix} A^{(1)} & 0 & \cdots & 0 \\ * & A^{(2)} & 0 & \cdots \\ \vdots & \cdots & \cdots & 0 \\ * & \cdots & * & A^{(r)} \end{pmatrix}, \qquad A \in \mathcal{A}, \tag{6.25}$$

where the family in each diagonal block $\mathcal{A}^{(j)} = \{A^{(j)} | A \in \mathcal{A}\}$ is irreducible, $j = 1, \dots, r$. In particular, if the family is irreducible, then the Frobenius factorization is trivial with $r = 1$ and, of course, the valency is equal to one, we again arrive at Theorem A2. Thus, the resonance degree does not exceed the valency. This estimate was improved in more complicated terms in [20]. Note that for one matrix A, the resonance degree $\nu(A)$ is the largest size of Jordan block corresponding to the biggest by modulus eigenvalue. So, the resonance degree of one matrix can be efficiently found.

Remark 6.7.2 The resonance degree is always integer, by definition. Let us note that there are examples of finite matrix families (even pairs of matrices $\mathcal{A} = \{A_0, A_1\}$) for which $\max_{A_i \in \mathcal{A}} \|X_1 \cdots X_k\| \asymp \rho^k k^{\beta}$ with a non-integer β (see [21]).

Theorem 6.7.3 *For a refinable function* φ*, we have*

$$\omega(\varphi, t) \quad \le \quad C\,t^{\alpha}\,|\log t|^{\alpha\nu(M_i) + \nu(\mathcal{A}|_{U_i})}, \qquad t \in (0\,,\,1/2)\,, \tag{6.26}$$

where $\alpha = \alpha_{\varphi}$ *is the Hölder exponent,* $i \in \{1, \dots, q\}$ *is the index for which the number* $\alpha\nu(M_i) + \nu(\mathcal{A}|_{U_i})$ *is maximal among all indices with the property* $\log_{1/r_i} \rho_i = \alpha_{\varphi}$.

Corollary 6.7.4 *If for all indices* i *with the property* $\log_{1/r_i} \rho_i = \alpha_{\varphi}$*, the matrix* M_i *does not have nontrivial Jordan blocks and* $\nu(\mathcal{A}|_{U_i}) = 1$ *(in particular, if* $\mathcal{A}|_{U_i}$ *is irreducible), then* $\omega(\varphi, t) \le C\,t^{\alpha}$.

Corollary 6.7.5 *Under the assumptions of Corollary 6.7.4, if* $\alpha_{\varphi} = 1$*, then* φ *is Lipschitz continuous.*

Remark 6.7.6 If $\omega(\varphi, t, J) = \sup\{\|\varphi(x+h) - \varphi(x)\| \,|\, \|h\| \le t\ h \in J\}$ denotes the modulus of continuity along a subspace J, then the value $\omega(\varphi, t, J_i)$ for each $i = 1, \dots, q$ is estimated above by inequality (6.26).

Remark 6.7.7 A careful analysis of the proof of Theorem 6.7.3 below makes it possible to construct examples for which the upper bound (6.26) is attained. Thus, inequality (6.26) cannot be improved in that terms. In particular, it is shown easily that if M has the largest Jordan block of a given size $\nu \geq 1$ corresponding to the biggest by modulus eigenvalue and the family $\mathcal{A}$ is irreducible, then $\omega(\varphi, t_k) \geq C t_k^{\alpha} \, |\log t|^{\alpha\nu}$ for a sequence $t_k \to 0$.

Proof of Theorem 6.7.3. In the proof of Theorem 6.2.5, we replace estimate (6.20) by $\|M^k(y-x)\| \leq C\, r_i^k\, k^{\nu(M_i)-1}\, \|y-x\|$, after which inequality (6.21) becomes $C\, \|y-x\|\, r_i^k\, k^{\nu(M_i)-1} \geq 1$. Then, we estimate $\|\Pi_j\|$ by $C\, \rho_i^k\, k^{\nu(\mathcal{A}|_{U_i})-1}$ instead of $C\, (\rho_i + \varepsilon)^k$. Combining those two assertions, we obtain $\|v(y) - v(x)\| \leq C\, \|y - x\|^{\alpha} \, \bigl|\log \|y-x\|\bigr|^{\alpha\nu(M_i)+\nu(\mathcal{A}|_{U_i})}$, where $\alpha = \log_{1/r_i} \rho_i$. Since this holds for each $i = 1, \ldots, q$, the theorem follows. ◊

6.8 Refinement Equations in L_p.

In this section, we obtain a criterion of solvability of a refinement equation in the space $L_p(\mathbb{R}^d)$ for $p \in [1, +\infty)$. Once the continuity of a refinable function depends on the joint spectral radius of matrices $\{A_s | s \in D(M)\}$, the L_p-solvability depends on the so-called p-radius:

Definition 6.8.1 For $p \in [1, +\infty)$, the *L_p-spectral radius* or a *p-radius* of a finite family of finite-dimensional linear operators $\mathcal{A} = \{A_1, \ldots, A_m\}$ is the following limit:

$$\rho(\mathcal{A}) = \lim_{k\to\infty} \left[m^{-k} \sum_{A_{s_i} \in \mathcal{A},\, i=1,\ldots,k} \|A_{s_1} \ldots A_{s_k}\|^p \right]^{1/pk}$$

This limit always exists and does not depend on the operator norm.

Theorem 6.8.2 *For a family of m operators $\mathcal{A}$ acting in $\mathbb{R}^n$, for every $p \geq 1$, and for any $\varepsilon > 0$, there exists a norm $\|\cdot\|$ in $\mathbb{R}^n$ such that*

$$\left[\frac{1}{m} \sum_{j=1}^{m} \| Ax \|^p \right]^{1/p} < (\rho_p + \varepsilon)\, \|x\| .$$

Proof As in the proof of Theorem 6.5.1, we consider the family $\tilde{A} = \frac{1}{\rho_p + \varepsilon} A,\ A \in \mathcal{A}$ and define the function

$$f(x) = \sup_{k \in \mathbb{N}} \left[m^{-k} \sum_{\tilde{A}_{s_i} \in \tilde{\mathcal{A}},\, i=1,\ldots,k} \|\tilde{A}_{s_1} \cdots \tilde{A}_{s_k} x\|^p \right]^{1/p} .$$

Then, we have $f(Ax) \leq (\rho_p + \varepsilon) f(x),\ A \in \mathcal{A},\ x \in \mathbb{R}^n$.◊

Let us fix a family of m operators $\mathcal{A}$ acting in $\mathbb{R}^n$. For $p \in [1, +\infty)$ and for an arbitrary $u \in \mathbb{R}^n$, we denote

$$\mathcal{F}_k(p, u) \quad = \quad \Bigl(m^{-k} \sum_{s_j \in \{1,\dots,m\}} \|A_{s_1} \cdots A_{s_k} u\|^p\Bigr)^{1/p}, \qquad k \in \mathbb{N}.$$

The following result proved in [22] states that the p-radius is the exponent of growth of the value $\mathcal{F}_k(p, u)$ as $k \to \infty$.

Theorem 6.8.3 *[22]. For every $u \in \mathbb{R}^n$ that does not belong to a common invariant subspace of operators from $\mathcal{A}$, there is a constant $C(u) > 0$ such that*

$$\mathcal{F}(p, u) \quad \geq \quad C(u)\, (\rho_p)^k\,, \qquad k \in \mathbb{N}.$$

For $\varphi \in L_p$, the space U is defined in the same way as for continuous φ, with only difference that $v(y) - v(x) \in U$ for almost all pairs $x, y \in \mathbb{R}^d$. As usual, $A_s = T_s|_U$, $s \in D(M)$.

Theorem 6.8.4 *A refinable function belongs to $L_p(\mathbb{R}^d)$ if and only if $\rho_p(\mathcal{A}) < 1$.*

In the proof, we use the following well-known fact.

Lemma 6.8.5 *Let $G \subset \mathbb{R}^n$ be a compact set, $\mu(G) = 1$. For an arbitrary $f \in L_p(G)$, $p < \infty$ and for a partition $\Delta = \{\Delta_j\}_{j=1}^K$ of the set G to measurable sets, we denote $S(f, \Delta)$ a step (piecewise constant) function that on each set Δ_j equals to $\frac{1}{\mu(\Delta_j)} \int_{\Delta_j} f(x)\, dx$. Then $\bigl\| f - S(f, \Delta)\bigr\|_p \to 0$ as* diam $(\Delta) \to 0$, *where* diam $(\Delta) = \max_j$ diam (Δ_j).

Proof We fix $\varepsilon > 0$ and approximate f by a continuous function $\tilde f$ so that $\|f - \tilde f\|_p < \varepsilon$. By the uniform continuity, there exists $\delta > 0$ such that $\|\tilde f(x) - \tilde f(y)\| < \varepsilon$, whenever $\|x - y\| < \delta$. So, if diam $(\Delta_j) < \delta$, then $\|\tilde f(y) - \frac{1}{\mu(\Delta_j)} \int_{\Delta_j} f(x)\, dx\| < \varepsilon$ for every $y \in \Delta_j$. Therefore, $\|\tilde f - S(\tilde f, \Delta)\|_\infty < \varepsilon$, and since $\mu(G) = 1$, we see that $\|\tilde f - S(\tilde f, \Delta)\|_p < \varepsilon$, whenever diam $(\Delta) < \delta$. Finally,

$$\|S(f, \Delta) - S(\tilde f, \Delta)\|_p \le \|f - \tilde f\|_p < \varepsilon\,.$$

Applying the triangle inequality, we obtain $\|f - S(f, \Delta)\|_p \le$

$$\|f - \tilde f\|_p + \|\tilde f - S(\tilde f, \Delta)\|_p + \|S(\tilde f, \Delta) - S(f, \Delta)\|_p < \varepsilon + \varepsilon + \varepsilon = 3\varepsilon\,.$$

This concludes the proof.◊

Proof of Theorem 6.8.4. First, we show that the condition $\rho_p(\mathcal{A}) < 1$ is sufficient for the existence of an L_p-solution of refinement equation. Choose $\varepsilon \in (0, 1 - \rho_p)$ and take the special norm in the space U from Theorem 6.8.2. In that norm,

$\left(m^{-1} \sum_{s\in D(M)} \|A_s x\|^p\right)^{1/p} \le (\rho_p + \varepsilon)\|x\|$ for all $x \in U$. Now we consider the space of functions

$$V_{U,p} = \left\{ f \in L_p(S) \middle| v_f(x) \in V,\, v_f(x) - v_f(y) \in U \text{ a.e.,} x,\, y \in G \right\}$$

with the norm $\|f\| = \left(\int_G \|v_f(x)\|^p \, dx\right)^{1/p}$. This space is nonempty because a piecewise-constant function f for which $v_f(\cdot) \equiv z$ a.e., where $z \in V$ is the eigenvector of the operator $\frac{1}{m}\sum_{s\in D(M)} T_s$ with the eigenvalue one, belongs to it. For every f_1, f_2 from that space, we have

$$\left\|T(f_1 - f_2)\right\| = \left\|A(v_{f_1} - v_{f_2})\right\| \le \rho_p + \varepsilon < 1.$$

Therefore, T is a contraction on that space, and hence it has a unique fixed point $\varphi V_{U,p}$. This is an L_p-solution of the refinement equation $T\varphi = \varphi$.

Now we prove the necessity: if $\varphi \in L_p(\mathbb{R}^d)$, then $\rho_p(\mathcal{A}) < 1$. Thus, let self-similarity Eq. (6.8) possess a solution $v = v_\varphi \in L_p$. Denote $a = \int_G v(x)\, dx$. By the same symbol, we denote the identical function $a(x) \equiv a$ on G. For arbitrary $k \ge 1$ and an arbitrary set G_s of the partition $\mathcal{G}^k$, we have

$$\mu(G_s)^{-1} \int_{G_s} v(x)\, dx = \int_G A_{s_{i_1}} \cdots A_{s_{i_k}} v(x)\, dx =$$

$$A_{s_{i_1}} \cdots A_{s_{i_k}} \int_G v(x)\, dx = A_{s_{i_1}} \cdots A_{s_{i_k}} a.$$

Therefore, the step function $f_k = A^k a$ is equal to the average $\mu(G_s)^{-1} \int_{G_s} v(x)\, dx$ on each set G_s of the partition $\mathcal{G}^k$. The diameter of this partition tends to zero, hence by Lemma 6.8.5, one has $\|f_k - v\|_p \to 0$ as $k \to \infty$.

Assume there is a common linear subspace $L \subset U$ of the operators A_j that contains the vectors $v(x) - a$ for almost all $x \in G$. This subspace can be defined by several equations $(l_q, u) = 0$, $q = 1, \dots, h$, where l_q are some linear functionals in $\mathbb{R}^n$. For each q, we have $\left(l_q, v(x)\right) = \left(l_q, a\right)$ for almost all x. Therefore, for any $j \in \{1, \dots, m\}$ Eq. (6.6) implies

$$\left(l_q\,,\, A_j a\right) = \left(l_q\,,\, \int_G A_j v(x)\, dx\right) = \left(l_q\,,\, \mu(G_j)^{-1} \int_{G_j} v(x)\, dx\right) = \left(l_q\,,\, a\right).$$

This yields that the affine plane $\tilde{L} = a + L$ contains all the points $A_j\, a$, $j = 1, \dots, m$. Hence, for any $u \in L$, we have $A_j(a + u) = A_j\, a + A_j\, u \in \tilde{L}$, because $A_j\, a \in \tilde{L}$ and $A_j\, u \in L$. Thus, $\tilde{L}$ is a common invariant affine plane for the family $\mathcal{A}$. Therefore, the vectors $v(x)$ belong to $\tilde{L}$ for almost all $x \in G$. This means that the differences $v(x) - v(y)$ belong to L for almost all $x, y \in G$. Thus, however, contradicts the definition of U. Hence $L = U$, and therefore is a subset $\Omega \subset G$ of positive Lebesgue measure such that $v(x) - a$ does not belong to any common invariant subspace of the family $\mathcal{A}$, whenever $x \in \Omega$. Now we apply Theorem 6.8.3 for $u = v(x) - a$. We see that the function $C(x) = C(v(x) - a)$ is positive on Ω, consequently there is an $\varepsilon > 0$ and a set of positive measure $\Omega_\varepsilon \subset \Omega$ such that $C(x) \ge \varepsilon$ for all $x \in \Omega_\varepsilon$. Thus, $\mathcal{F}_k(p, v(t) - a) \ge \varepsilon\,(\rho_p)^k$ for all $x \in \Omega_\varepsilon$. On the other hand, for arbitrary L_p-functions f_1, f_2, we have

$$\big\|A^k(f_1 - f_2)\big\|_p = \big\|\mathcal{F}_k\big(p, f_1(\cdot) - f_2(\cdot)\big)\big\|_p, \qquad k \in \mathbb{N}. \tag{6.27}$$

Substituting $f_1 = v$, $f_2 = a$, we obtain

$$\big\|A^k v - A^k a\big\|_p = \big\|\mathcal{F}_k\big(p, v(\cdot) - a\big)\big\|_p .$$

Thus,

$$\big\|v - A^k a\big\|_p \ge \mu(\Omega_\varepsilon)^{1/p}\,\varepsilon\,(\rho_p)^k, \qquad k \in \mathbb{N}. \tag{6.28}$$

The left-hand side tends to zero as $k \to \infty$, and so $\mu(\Omega_\varepsilon)^{1/p}\,\varepsilon\,\rho_p^k \to 0$ as $k \to \infty$, which implies $\rho_p < 1$. $\Diamond$

Remark 6.8.6 We see that the existence of an L_p-solution of a refinement equation under the condition $\rho_p(\mathcal{A}) < 1$ is proved much simpler than a similar result for continuous solution (Theorem 6.2.4). In fact, an elegant argument with a contraction operator T on the affine subspace $V(U, p)$ cannot be directly extended to prove continuity because of one reason: the piecewise-constant function f for which $v_f \equiv z$ is not continuous. That is why it becomes a problem to show that the space of *continuous* functions f such that $v_f(x) - v_f(y) \in U$, $x, y \in G$ is nonempty. We are not aware of any simple proof of this fact for multivariate equations.

Corollary 6.8.7 *If* $\rho_p(\mathcal{T}|_W) < 1$, *then* $\varphi \in L_p(\mathrm{Re}^d)$.

Proof Since $U \subset W$, it follows that $\rho(\mathcal{A}) = \rho(\mathcal{T}|_U) \le \rho(\mathcal{T}|_W)$. Hence, by Theorem 6.8.4, if $\rho(\mathcal{T}|_W) < 1$, then $\varphi \in L_p$.$\Diamond$

Construction of the space U and of L_p-refinable function φ.

The construction of a continuous refinable function described in Sect. 6.4. is realized pointwise and hence is not applicable in the space L_p. Moreover, the vectors $v(z_s)$ are not well defined if $v \in L_p$ since one point z_p is a set of measure zero. That is why the constructions of the space U and of the function φ need to be modified for the L_p case. This can be done in the following way.

From the Eq. (6.6), it follows that the vector $z = \int_G v(x)\,dx$ is an eigenvector of the operator $T = \frac{1}{m}\sum_{s \in D(M)} T_s$ with the eigenvalue 1. Then, the subspace U is the minimal common invariant subspace of the matrices T_s, $s \in D(M)$ that contains m vectors $T_s z - z$, $s \in D(M)$. If $\rho_p(\mathcal{T}|_U) < 1$, then the solution φ belongs to $L_p(\mathbb{R}^d)$. Numerically it can be computed as follows: for every $\boldsymbol{s} \in D^k(M)$, the value $\frac{1}{\mathrm{Vol(G_s)}}\int_{G_s} v(x)\,dx$, which is simply the mean of the function $v = v_\varphi$ on the set of the tile G_s (let us recall that $\mathrm{Vol}(G_s) = m^{-k}$), is equal to $A_{s_1} \cdots A_{s_k} z$. So, we can compute the mean values of the solution φ on all sets of the tiling $\mathcal{G}^k$. This approximates the solution in the space L_p. For instance, if $\chi = \chi_G$ is the characteristic function of the tile G, then the function $\varphi_k = T^k \chi$ is a piecewise-constant approximation of the solution φ. On each set G_s, $\varphi_k(x)$ is equal to the identical constant $A_{s_1} \cdots A_{s_k} z$. The function φ_k converges to φ with a linear rate: $\|\varphi_k - \varphi\|_{L_p} \le C(\rho_p + \varepsilon)^k, k \to \infty$.

Criteria for L_2. It is quite expected that the case $p = 2$ is very special. Indeed, in this case, one can verify that $\varphi \in L_2(\mathbb{R}^d)$ without approximating the p-radius. First of all, the 2-radius can be efficiently computed as an eigenvalue of a special matrix of a bigger dimension. Second, there is another criterion that does not involve the p-radius. We begin with the first strategy.

Consider the operator W_2 acting in the space of symmetric $n \times n$-matrices (i.e., in $\mathbb{R}^{(n^2+n)/2}$) by the formula

$$W_2(X) = \frac{1}{m}\left(A_1 X A_1^* + \cdots + A_m X A_m^*\right). \tag{6.29}$$

Theorem 6.8.8 *The 2-radius of the operators $A_1, \ldots, A_m$ is equal to $\sqrt{\rho(W_2)}$, where $\rho(W_2)$ is the spectral radius of the operator W_2.*

For the proof, see [23]. The operator W_2 can be written in the matrix form with the Kronecker products. For the sake of simplicity, we consider real matrices. The Kronecker product of two matrices $A = (a_{ij})$ and $B = (b_{ij})$, where A is a $m \times n$ matrix and B is a $k \times l$ matrix, is the $mk \times nl$ matrix

$$A \otimes B = \begin{pmatrix} a_{11}B & \cdots & a_{1n}B \\ \vdots & * & \vdots \\ a_{m1}B & \cdots & a_{mn}B \end{pmatrix}$$

Set $A^{\otimes k} = A \otimes \cdots \otimes A$ (k multipliers). If $p = 2r$ is an even integer, then

$$\rho_p(A_1, \cdots, A_m) = \left(\rho(W_p)\right)^{1/p}, \tag{6.30}$$

where $\rho(W_p)$ is the spectral radius of the $n^p \times n^p$ matrix

$$W_p = \frac{1}{m}\left(A_1^{\otimes p} + \cdots + A_m^{\otimes p}\right),$$

In particular, for $p = 2$, the operator W_2 has the following matrix form:

$$W_2 = \frac{1}{m}\left(A_1^{\otimes 2} + \cdots + A_m^{\otimes 2}\right) \tag{6.31}$$

The spectral radius of this operator is equal to $(\rho_2)^2$.

The second approach reduces the L_2-solvability of a refinement equation to estimating the spectral radius of the transition matrix corresponding to the mask $|m_0(\xi)|^2$. The details can be found in [4, 24].

6.9 Computation of the Joint Spectral Radius

To compute the joint spectral radius, we apply Invariant polytope algorithm from [25]. It usually finds the exact value of the joint spectral radius for matrices of size at most 20. We give a brief description of this algorithm here, and the details and the theoretical base can be found in [19, 25]. For other efficient algorithms, see [26, 27].

For a given set $\Omega \subset \mathbb{R}^n$, we denote by $\mathrm{Conv}(\Omega)$ the convex hull of Ω and by $\mathrm{Conv}_0(\Omega) = \mathrm{Conv}\{\Omega, -\Omega\}$ the symmetrized convex hull. The sign $\asymp$ denotes as usual the asymptopic equivalence of two values (i.e., equivalence up to multiplication by a constant).

Let $\mathcal{A} = \{A_1, \dots, A_m\}$ be a set of $n \times n$ matrices. A product $\Pi = A_{s_\ell} \cdots A_{s_1}$ is called a *spectral maximizing product* (s.m.p) if $\rho(\mathcal{A}) = [\rho(\Pi)]^{1/\ell}$. So, if one finds a spectrum maximizing product, then he finds the joint spectral radius. The idea of the algorithm is to prove that a chosen product Π is spectrum maximizing by constructing a convex polytope $P \subset \mathbb{R}^n$ such that $A_i P \subset \rho_\ell P$, $i = 1, \dots, m$, where $\rho_\ell = [\rho(\Pi)]^{1/\ell}$. This means that in the Minkowski norm $\|\cdot\|_P$ defined by the polytope P, we have $\|A_i\|_P \le \rho_\ell$, and therefore $\rho(\mathcal{A}) = \rho_\ell$.

Algorithm 1 (the Invariant polytope algorithm).

`Initialization.` First, we fix some number ℓ_0 and find a product $\Pi = A_{s_\ell} \dots A_{d_1}$ with the maximal value $[\rho(\Pi)]^{1/\ell}$ among all products of lengths $\ell \le \ell_0$. We call this product a *candidate s.m.p.* and try to prove that it is actually an s.m.p. Denote $\rho_\ell = [\rho(\Pi)]^{1/\ell}$ and normalize all the matrices A_i as $\tilde{A}_i = \rho_\ell^{-1} A_i$. Thus, we obtain the family $\tilde{A}$ and the product $\tilde{\Pi} = \tilde{A}_{s_\ell} \dots \tilde{A}_{s_1}$ such that $\rho(\tilde{\Pi}) = 1$. For the sake of simplicity, we assume that the largest by modulo eigenvalue of $\tilde{\Pi}$ is real, in which case it is ± 1. We assume it is 1, the case of -1 is considered in the same way. The eigenvector $v^{(1)}$ corresponding to this eigenvalue is called *leading eigenvector*. The vectors, $v^{(j)} = \tilde{A}_{s_{j-1}} \cdots \tilde{A}_{s_1} v^{(1)}$, $j = 2, \dots, \ell$, are leading eigenvectors of cyclic permutations of $\tilde{\Pi}$. Then, we construct a sequence of finite sets $\mathcal{V}_i \subset \mathbb{R}^n$ and their subsets $\mathcal{R}_i \subset \mathcal{V}_i$ as follows:

`Zero iteration.` We set $\mathcal{V}_0 = \mathcal{R}_0 = \{v^{(1)}, \dots, v^{(\ell)}\}$.

k`th iteration,` $k \ge 1$. We have finite set $\mathcal{V}_{k-1}$ and its subset $\mathcal{R}_{k-1}$. We set $\mathcal{V}_k = \mathcal{V}_{k-1}$, $\mathcal{R}_k = \emptyset$ and for every $v \in \mathcal{R}_{k-1}$, $\tilde{A} \in \tilde{\mathcal{A}}$, check whether $\tilde{A}v$ is an *interior point* of $\mathrm{absco}(\mathcal{V}_k)$ (this is an LP problem). If so, we omit this point and take the next pair $(v, \tilde{\mathcal{A}}) \in \mathcal{R}_{k-1} \times \tilde{\mathcal{A}}$, otherwise we add $\tilde{A}v$ to $\mathcal{V}_k$ and to $\mathcal{R}_k$. When all pairs $(v, \tilde{A})$ are exhausted, both $\mathcal{V}_k$ and $\mathcal{R}_k$ are constructed. Let $P_k = \mathrm{Conv}_0(\mathcal{V}_k)$. We have

$$\mathcal{V}_k = \mathcal{V}_{k-1} \cup \mathcal{R}_k , \quad P_k = \mathrm{co}\,\{\tilde{A}_1 P_{k-1}, \dots, \tilde{A}_m P_{k-1}\} .$$

`Termination`. The algorithm halts when $\mathcal{V}_k = \mathcal{V}_{k-1}$, i.e., $\mathcal{R}_k = \emptyset$ (no new vertices are added in the kth iteration). In this case, $P_{k-1} = P_k$, and hence P_{k-1} is an invariant polytope, Π is an s.m.p., and $\rho(\mathcal{A}) = [\rho(\Pi)]^{1/\ell}$. **End of the algorithm.**

Actually, the algorithm works with the sets $\mathcal{V}_k$ only, the polytopes P_k are needed to illustrate the idea. Thus, in each iteration of the algorithm, we construct a polytope $P_k \subset \mathbb{R}^d$, store all its vertices in the set $\mathcal{V}_k$ and spot the set $\mathcal{R}_k \subset \mathcal{V}_k$ of newly appeared (after the previous iteration) vertices. Every time we check whether $\tilde{\mathcal{A}} P_k \subset P_k$. If so, then P_k is an invariant polytope, $\|\tilde{A}_i\|_{P_k} \le 1$ for all i, and Π is an s.m.p. Otherwise, we update the sets $\mathcal{V}_k$ and $\mathcal{R}_k$ and continue.

If the algorithm terminates within finite time, then it proves that the chosen candidate is indeed an s.m.p. and gives the corresponding polytope norm. Although there are simple examples of matrix families for which the algorithm does not terminate, we believe that such cases are rare in practice. In fact, in all numerical experiments made with randomly generated matrices and with matrices from applications, the algorithm did terminate in finite time providing an invariant polytope.

Remark 6.9.1 We can introduce an arbitrary number of extra initial vertices to the vertices $v^{(1)}, \dots, v^{(n)}$, and sometimes it helps to speed up the algorithm. We use extra vertices in the next section when computing the joint spectral radius of Daubechies matrices.

Remark 6.9.2 If the algorithm does not halt within finite time, then we interrupt it after some Nth iteration and compute the joint spectral radius by means of the double inequality:

$$[\rho(\Pi)]^{1/N} \quad \le \quad \rho \quad \le \quad [\rho(\Pi)]^{1/N} \mu_N ,$$

where $\mu_N = \inf\,\{\mu > 0 | P_N \subset \mu P_{N-1}\}$ (this number is found by solving the corresponding linear programming problem).

References

1. Jia, R.Q.: Characterization of smoothness of multivariate refinable functions in Sobolev spaces. Trans. Amer. Math. Soc. **351**, 4089–4112 (1999)
2. Han, B.: Computing the smoothness exponent of a symmetric multivariate refinable function. SIAM J. Matrix Anal. Appl. **24**(3), 693–714 (2003)
3. Han, B.: Vector cascade algorithms and refinable function vectors in Sobolev spaces. J. Approx. Theory **124**(1), 44–88 (2003)
4. Han, B.: Solutions in Sobolev spaces of vector refinement equations with a general dilation matrix. Adv. Comput. Math. **24**(1–4), 375–403 (2006)
5. Cavaretta, A.S. Dahmen, W., Micchelli, C.A.: Stationary subdivision. Mem. Am. Math. Soc. **93**(453), (1991)
6. Cabrelli, C.A., Heil, C., Molter, U.M.: Accuracy of lattice translates of several refinable multidimensional refinable functions. J. Approx. Theory **95**, 5–52 (1998)

7. Cabrelli, C.A., Heil, C., Molter, U.M., Self-similarity and multiwavelets in higher dimensions. Memoirs. Amer. Math. Soc. **170**(807), (2004)
8. Cohen, A., Gröchenig, K., Villemoes, L.: Regularity of multivariate refinable functions. Constr. Approx. **15**, 241–255 (1999)
9. Charina, M., Protasov, V.Y.: Matrix approach to analyse smoothness of multivariate wavelets, preprint
10. Novikov, I.Y., Protasov, V.Y. Skopina, M.A.: Wavelet Theory. AMS, Providence, RI, Translations Mathematical Monographs, V. 239 (2011)
11. Rota, G.C., Strang, G.: A note on the joint spectral radius. Kon. Nederl. Acad. Wet. Proc. **63**, 379–381 (1960)
12. Protasov V.Y.: Spectral decomposition of 2-block Toeplitz matrices and refinement equations, St. Petersburg Math. J. **18**(4), 607–646 (2007)
13. Collela, D., Heil, C.: Characterization of scaling functions. I. Continuous solutions. SIAM J. Matrix Anal. Appl. **15**, 496–518 (1994)
14. Barabanov, N.E.: Lyapunov indicator for discrete inclusions, I-III. Autom. Remote Control **49**(2), 152–157 (1988)
15. Daubechies, I., Lagarias, J.: Two-scale difference equations. II. Local regularity, infinite products of matrices and fractals. SIAM J. Math. Anal. **23**, 1031–1079 (1992)
16. Villemoes, L.: Wavelet analysis of refinement equations. SIAM J. Math. Anal. **25**(5), 1433–1460 (1992)
17. Protasov, VYu.: Fractal curves and wavelets. Izv. Math. **70**(5), 123–162 (2006)
18. Dubuc, S.: Interpolation through an iterative scheme. J. Math. Anal. Appl. **114**, 185–204 (1986)
19. Guglielmi, N., Protasov, V.Y.: Invariant polytopes of sets of matrices with applications to regularity of wavelets and subdivisions. SIAM J. Matrix Anal. Appl. **37**(1), 18–52 (2016)
20. Chitour, Y., Mason, P., Sigalotti, M.: On the marginal instability of linear switched systems. Syst. Cont. Lett. **61**, 747–757 (2012)
21. Protasov, V.Y., Jungers, R.: Resonance and marginal instability of switching systems. Nonlinear Anal.: Hybrid Syst. **17**, 81–93 (2015)
22. Protasov, V.Y.: Extremal L_p-norms and self-similar functions. Linear Alg. Appl. **428**(10), 2339–2357 (2008)
23. Protasov, V.Y.: The generalized spectral radius. A geometric approach, Izvestiya Math. **61**, 995–1030 (1997)
24. Lawton, W., Lee, S.N., Shen, Z.: Convergence of multidimensional cascade algorithm. Numer. Math. **78**(3), 427–438 (1998)
25. Guglielmi, N., Protasov, VY.: Exact computation of joint spectral characteristics of matrices. Found. Comput. Math. **13**(1), 37–97 (2013)
26. Gripenberg, G.: A necessary and sufficient condition for the existence of father wavelet. Stud. Math. **114**(3), 207–226 (1995)
27. Möller, C., Reif, U.: A tree-based approach to joint spectral radius determination. Linear Alg. Appl. **563**, 154–170 (2014)

Bibliography

1. Averbuch, A.Z., Neitaanmäki, P., Zheludev, V.A.: Spline and spline wavelet methods with applications to signal and image processing, Springer International Publishing, 2016
2. Bratelli, O., Jorgensen, P.: Wavelets Through a Looking Glass: The World of the Spectrum. Birkhauser, 2002
3. Cameron, P.J.: Permutation Groups, Cambridge University Press, 1999
4. Christensen, O.: An introduction to frames and Riesz bases. Birkhäuser, Boston (2003)
5. Christensen, O.: Frames and bases: an introductory course. Applied and Numerical Harmonic Analysis. Birkhäuser Boston Inc, Boston, MA (2008)
6. Christensen, O.: Frames and bases: An introductory course, Birkhauser, 2008
7. Chui, C.K.: An introduction to wavelets. Academic Press, New York (1992)
8. Daubechies, I.: Ten lectures on wavelets. SIAM, CBMS-NSR Series in Appl. Math (1992)
9. Dong, B., Shen, Z.: MRA-based wavelet frames and applications, Ser. 19, AMS, Providence, RI, 2013
10. Hernandes, E., Weis, G.A.: A first cours of wavalets. CRC Press, Boca Raton, FL (1996)
11. Kashin, B.S., Saakyan, A.A.: Orthogonal series. AMS, Providence, RI, Translations Mathematical Monographs **75**, (1999)
12. Mallat, S.: A wavelet tour of signal processing, 3rd edn. Academic Press, New York (2009)
13. Meyer, J.: Wavelets and operators, Lect. Notes Math. 1989. Vol.137
14. Meyer, Y.: Wavelets and operators, Cambridge University Press. Cambridge. 1992. (English translation of [15])
15. Wojtaszczyk, P.: A mathematical introduction to wavelets, London Math. Soc. Student texts **37**. 1997
16. Allen, J. D.: Perfect reconstruction filter banks for the hexagonal grid, in Fifth International Conference on Information, Communications and Signal Processing, 73–76 (2005)
17. Andaloro, G., Cotronei, M., Puccio, L.: A new class of non-separable symmetric wavelets for image processing. Communications to simai congress **3**, 324–336 (2009)
18. Ando, T., Shih, M.-H.: Simultaneous contractibility. SIAM J. Matrix Anal. Appl. **19**(2), 487–498 (1998)
19. Averbuch, A.Z., Zheludev, V.A., Cohen, T.: Multiwavelet frames in signal space originated from Hermite splines. IEEE Transactions on Signal Processing **55**(3), 797–808 (2007)
20. Battle, G.: A block spin construction of ondelettes. Part 1: Lemarier functions, Comm. Math Phys. **110**, 601–615 (1987)
21. Belogay, E., Wang, Y.: Arbitrarily smooth orthogonal non-separable wavelets in R^2. SIAM J Math Anal. **30**(3), 678–697 (1999)
22. Berger, M.A., Wang, Y.: Bounded semigroups of matrices. Linear Alg. Appl. **166**, 21–27 (1992)

A. Krivoshein et al., *Multivariate Wavelet Frames*,
Industrial and Applied Mathematics, DOI 10.1007/978-981-10-3205-9

23. Blondel, V.D., Tsitsiklis, J.N.: The boundedness of all products of a pair of matrices is undecidable. Syst. Control Lett. **41**(2), 135–140 (2000)
24. Blondel, V.D., Tsitsiklis, J.N.: Approximating the spectral radius of sets of matrices in the max-algebra is NP-hard. IEEE Trans. Autom. Control **45**(9), 1762–1765 (2000)
25. Blu, T., Unser, M., van de Ville, D.: On the multidimensional extension of the quincunx subsampling matrix. IEEE Signal Processing Letters **12**(2), 112–115 (2005)
26. de Boor, C., DeVore, R., Ron, A.: On construction of multivariate (pre) wavelets, Constr. Approx., 123–166 (1993)
27. de Boor, C., DeVore, R., Ron, A.: The structure of finitely generated shift-invariant spaces in $L_2(\mathbb{R}^d)$. J. Funct. Anal. **119**(1), 37–78 (1994)
28. de Boor, C., DeVore, R., Ron, A.: Approximation from shift-invariant subspaces of $L_2(\mathbb{R}^d)$. Trans. Am. Math. Soc. **341**(2), 787–806 (1994)
29. de Boor, C., DeVore, R., Ron, A.: Approximation orders of FSI spaces in $L_2(\mathbb{R}^d)$. Constructive Approximation **14**(3), 411–427 (1998)
30. Bownik, M.: Tight frames of multidimensional wavelets. J. Fourier Anal. Appl. **3**, 525–542 (1997)
31. Bownik, M.: A Characterization of Affine Dual Frames in $L_2(\mathbb{R}^n)$. Appl. Comput. Harmon. Anal. **8**(2), 203–221 (2000)
32. Bownik, M., Weber, E.: Affine frames. GMRA's, and the canonical dual, Studia Mathematica **159**(3), 453–479 (2003)
33. Bownik, M., Rzeszotnik, Z.: On the existence of multiresolution analysis for frameletes. Math. Ann. **332**, 705–720 (2005)
34. Bownik, M., Speegle, D.: The Feichtinger Conjecture for Wavelet Frames, Gabor Frames and Frames of Translates. Canad. J. Math. **58**(6), 1121–1143 (2006)
35. Bownik, M., Lemvig, J.: The canonical and alternate duals of a wavelet frame. Appl. Comput. Harmon. Anal. **23**(2), 263–272 (2007)
36. Charina, M., Chui, C.K.: Tight frames of compactly supported multivariate multi-wavelets. Journal of Computational and Applied Mathematics **233**(8), 2044–2061 (2010)
37. Chen, Di.-R., Han, H., Riemenschneider, S.D.: Construction of multivariate biorthogonal wavelets with arbitrary vanishing moments, Adv. Comput. Math. **13** (2000), no. 2, 131–165
38. Chui, C.K., Lian, J.: Construction of compactly supported symmetric and antisymmetric orthonormal wavelets with scale = 3. Appl. Comput. Harmon. Anal. **2**(1), 21–51 (1995)
39. Chui, C.K., He, W.: Compactly supported tight frames associated with refinable functions. Appl. and Comp. Harm. Anal. **8**, 293–319 (2000)
40. Chui, C.K., He, W.: Construction of multivariate tight frames via Kronecker products. Appl. Comput. Harmon. Anal. **11**, 305–312 (2001)
41. Chui, C.K., He, W., Stöckler, J.: Compactly supported tight and sibling frames with maximum vanishing moments. Appl. Comput. Harmon. Anal. **13**, 224–262 (2002)
42. Chui, C.K., Czaja, W., Maggioni, M., Weiss, G.: Characterization of General Tight Wavelet Frames with Matrix Dilations and Tightness Preserving Oversampling. Journal of Fourier. Analysis and Applications **8**(2), 173–200 (2002)
43. Cohen, A.: Ondelettes, analyses multirësolutions et filtres miroir enquadrature, Ann inst. H. Poincarë, Anal. non linëaire, **7** (1990), 439-459
44. Cohen, A., Daubechies, I., Feauveau, J.C.: Biorthogonal Bases of Compactly Supported Wavelets, pp. 485–560. XLV, Communications on Pure and Applied Mathematics (1992)
45. Cohen, A., Daubechies, I.: A stability criterion for biorthogonal wavelet bases and their related subband coding schemes. Duke Math. J. **68**, 313–335 (1992)
46. Cohen, A., Daubechies, I.: Nonseparable bi-dimentional wavelet bases, Revista Mat. Iberoamericana (I), 51–137 (1993)
47. Cohen, A., Dyn, N.: Nonstationary subdivision schemes and multiresolution analysis. SIAM J. Math. Anal. **27**, 1745–1769 (1996)
48. Cohen, A., Schlenker, J.-M.: Compactly supported bidimensional wavelet bases with hexagonal symmetry. Constructive Approximation **9**(2), 209–236 (1993)

49. Collela, D., Heil, C.: Dilation equations and the smoothness of compactly supported wavelets, in Wavelets: Mathematics and applications., J. Benedetto, M. Frazier, eds., CRC Press, Bosa Raton, FL 1993, 161–200
50. Daubechies, I.: Orthonormal basis of compactly supported wavelets. Comm. Pure Appl. Math. **46**, 909–996 (1988)
51. Daubechies, I., Han, B., Ron, A., Shen, Z.: Framelets: MRA-based constructions of wavelet frames. Appl. Comput. Harmon. Anal. **14**(1), 1–46 (2003)
52. Daubechies, I., Han, B.: Pairs of dual wavelet frames from any two refinable functions. Constr. Approx. **20**(3), 325–352 (2004)
53. Daubechies, I., Lagarias, J.: Two-scale difference equations. I. Global regularity of solutions, SIAM. J. Math. Anal., **22**, 1388–1410 (1991)
54. Daubechies, I., Lagarias, J.: Corrigendum/addendum to: Sets of matrices all infinite products of which converge. Linear Alg. Appl. **327**, 69–83 (2001)
55. Derfel, G.A., Dyn, N., Levin, D.: Generalized refinement equations and subdivision processes. Journal of Approx. Theory **80**, 272–297 (1995)
56. Deslauriers, G., Dubuc, S.: Symmetric iterative interpolation processes. Constr. Approx. **5**, 49–68 (1989)
57. Duffin, R.J., Schaeffer, A.S.: A class of nonharmonic Fourier series. Trans. Amer. Math. Soc. **72**, 341–366 (1952)
58. Dyn, N., Gregory, J.A., Levin, D.: Analysis of linear binary subdivision schemes for curve design. Constr. Approx. **7**, 127–147 (1991)
59. Dyn, N., Levin, D.: Interpolatory subdivision schemes for the generation of curves and surfaces, Multivariate approximation and interpolation, 1990, (Duisburg 1989), 91–106
60. Dyn, N., Levin, D.: Subdivision schemes in geometric modelling. Acta Numer. **11**, 73–144 (2002)
61. Dyn, N., Skopina, M.: Decompositions of trigonometric polynomials with applications to multivariate subdivision schemes. Advances in Computational Mathematics **38**(2), 321–349 (2013)
62. Ehler, M., Han, B.: Wavelet bi-frames with few generators from multivariate reinable functions. Appl. Comput. Harmon. Anal. **25**(3), 407–414 (2008)
63. Ehler, M.: On multivariate compactly supported bi-frames. J. Fourier Anal. Appl. **13**(5), 511–532 (2007)
64. Ehler, M.: Compactly supported multivariate pairs of dual wavelet frames obtained by convolution, Int. J. Wavelets, Multiresolut. Inf. Process., **6** (2008), no. 2, 183–208
65. Ehler, M., Koch, K.: The construction of multiwavelet bi-frames and applications to variational image denoising, Int. J. Wavelets, Multiresolut. Inf. Process., **8** (2010), no. 3, 431–455
66. Entezari, A., Moller, T., Vaisey, J.: Subsampling matrices for wavelet decompositions on body centered cubic lattices. IEEE Signal Processing Letters **11**(9), 733–735 (2004)
67. Geronimo, J.S., Woerdeman, H.J.: Positive extensions. Fejér-Riesz factorization and autoregressive filters in two variables, Annals of Mathematics **160**(3), 839–906 (2004)
68. Goh, S.S., Lim, Z.Y., Shen, Z.: Symmetric and antisymmetric tight wavelet frames. Appl. Comput. Harmon. Anal. **20**, 411–421 (2006)
69. Gripenberg, G.: Computing the joint spectral radius. Lin. Alg. Appl. **234**, 43–60 (1996)
70. Gröchenig, K., Madych, W.R.: Maltiresolution analysis, Haar bases and self-similar tillings of R^n. IEEE Trans. Inform. Theory. **38**, 556–568 (1992)
71. Gröchenig, K., Haas, A.: Self-similar lattice tilings. J. Fourier Anal. Appl. **2**, 131–170 (1994)
72. Guglielmi, N., Zennaro, M.: On the zero-stability of variable stepsize multistep methods: the spectral radius approach. Numer. Math. **88**, 445–458 (2001)
73. Guo, W., Lai, M.-J.: Box spline wavelet frames for image edge analysis. SIAM Journal on Imaging Sciences **6**(3), 1553–1578 (2013)
74. Fujinoki, K., Vasilyev, O.V.: Triangular Wavelets: An Isotropic Image Representation with Hexagonal Symmetry. EURASIP Journal on Image and Video Processing (2009). doi:10.1155/2009/248581
75. Haar, A.: Sur Theorie de orthogonalen Funktionensysteme. Math. Ann. **69**, 331–371 (1910)

76. Han, B.: On dual wavelet tight frames. Appl. Comput. Harmon. Anal. **4**, 380–413 (1997)
77. Han, B., Jia, R.Q.: Multivariate refinement equations and convergence of subdivision schemes. SIAM J. Math. Anal. **29**(5), 1177–1199 (1998)
78. Han, B.: Symmetric orthonormal scaling functions and wavelets with dilation factor 4. Advances in Computational Mathematics **8**(3), 221–247 (1998)
79. Han, B.: Analysis and construction of optimal multivariate biorthogonal wavelets with compact support. SIAM J. Math. Anal. **31**(2), 274–304 (1999)
80. Han, B., Jia, R.Q.: Optimal interpolatory subdivision schemes in multidimensional spaces. SIAM Journal on Numerical Analysis **36**(1), 105–124 (1999)
81. Han, B.: Construction of multivariate biorthogonal wavelets by CBC algorithm, Wavelet analysis and multiresolution methods (Urbana-Champaign, IL, 1999), 105–143, Lecture Notes in Pure and Appl. Math., 212, Dekker, New York, 2000
82. Han, B.: Symmetry property and construction of wavelets with a general dilation matrix. Linear Alg. Appl. **353**, 207–225 (2002)
83. Han, B., Jia, R.Q.: Quincunx fundamental refinable functions and Quincunx biorthogonal wavelets. J. Math. Comp. **71**(237), 165–196 (2002)
84. Han, B.: Compactly supported tight wavelet frames and orthonormal wavelets of exponential decay with a general dilation matrix. J. Comput. Appl. Math. **155**, 43–67 (2003)
85. Han, B.: Symmetric multivariate orthogonal refinable functions. Appl. Comput. Harmon. Anal. **17**, 277–292 (2004)
86. Han, B., Mo, Q.: Symmetric MRA tight wavelet frames with three generators and high vanishing moments. Appl. Comput. Harmon. Anal. **18**, 67–93 (2005)
87. Han, B.: Matrix extension with symmetry and applications to symmetric orthonormal complex M-wavelets. J. Fourier Anal. Appl. **15**, 684–705 (2009)
88. Han, B., Shen, Z.: Characterization of Sobolev spaces of arbitrary smoothness using nonstationary tight wavelet frames. Israel J. Math. **172**, 371–398 (2009)
89. Han, B.: Symmetric orthonormal complex wavelets with masks of arbitrarily high linear-phase moments and sum rules. Adv. Comput. Math. **32**, 209–237 (2010)
90. Han, B.: Pairs of frequency-based nonhomogeneous dual wavelet frames in the distribution space. Appl. Comput. Harmon. Anal. **29**(3), 330–353 (2010)
91. Han, B., Zhuang, X.S.: Matrix extension with symmetry and its application to symmetric orthonormal multiwavelets. SIAM J. Math. Anal. **42**, 2297–2317 (2010)
92. Han, B.: Symmetric orthogonal filters and wavelets with linear-phase moments. J. Comput. Appl. Math. **236**, 482–503 (2011)
93. Han, B.: Nonhomogeneous wavelet systems in high dimensions. Springer Proceedings in Mathematics **13**, 121–161 (2012)
94. Han, B.: Properties of Discrete Framelet Transforms. Math. Model. Nat. Phenom. **8**(1), 18–47 (2013)
95. Han, B.: Algorithm for constructing symmetric dual framelet filter banks. Math. Comp. **84**, 767–801 (2015)
96. Jetter, K., Zhou, D.X.: Order of linear approximation from shift invariant spaces. Constr. Approx. **11**(4), 423–438 (1995)
97. Jia, R.Q.: Refinable shift-invariant spaces: From splines to wavelets, [CA] Chui, C.K. (ed.) et al., Approximation theory VIII. Vol. 2. Wavelets and multilevel approximation. Papers from the 8th Texas international conference, College Station, TX, USA, January 8–12, 1995. Singapore: World Scientific. Ser. Approx. Decompos. **6**, 179–208 (1995)
98. Jia, R.Q.: Approximation properties of multivariate wavelets. Math. Comp. **67**, 647–655 (1998)
99. Jia, R.Q.: Convergence rates of cascade algorithms. Proc. Amer. Math. Soc. **131**, 1739–1749 (2003)
100. Jia, R.Q.: Approximation by quasi-projection operators in Besov spaces. J. Approx. Theory **162**(1), 186–200 (2010)
101. Jia, R.Q., Micchelli, C.A.: Using the refinement equations for the construction of pre-wavelets II: Powers of two, Curves and Surfaces (P.J. Laurent, A. Le M haut é and L.L. Schumaker, eds.), Academic Press, New York. 1991, 209–246

102. Jia, R.Q., Micchelli, C.A.: Using the refinement equations for the construction of pre-wavelets V: extesibility of trigonometric polynomials. Computing **48**, 61–72 (1992)
103. Jia, R.Q., Shen, Z.: Multiresolution and wavelets. Proceedings of the Edinburgh Mathematical Society **37**, 271–300 (1994)
104. Jia, R.Q., Wang, J.: Stability and linear independence associated with wavelet decomposition, Proc.Amer.Math.Soc., **117**, 1115–1124 (1993)
105. Jiang, Q.T.: Orthogonal and Biorthogonal FIR Hexagonal Filter Banks With Sixfold Symmetry. IEEE Transactions on Signal Processing **56**(12), 5861–5873 (2008)
106. Jiang, Q.T.: FIR filter banks for hexagonal data processing. IEEE Trans. Image Proc. **17**, 1512–1521 (2008)
107. Jiang, Q.T.: Biorthogonal wavelets with 6-fold axial symmetry for hexagonal data and triangle surface multiresolution processing, Int. J. Wavelets, Multiresolution Info. Proc., **9** (2011), no. 5, 773–812
108. Jiang, Q.T.: Bi-frames with 4-fold axial symmetry for quadrilateral surface multiresolution processing. J. Comput. Appl. Math. **234**(12), 3303–3325 (2010)
109. Jiang, Q.T., Pounds, D.K.: Highly symmetric bi-frames for triangle surface multiresolution processing. Appl. Comput. Harmonic Anal. **31**, 370–391 (2011)
110. Ji, H., Riemenschneider, S.D.: Shen Z. Multivariate compactly supported fundamental refinable functions, dual and biorthogonal wavelets, Studies of Applied Mathematics **102**, 173–204 (1999)
111. Karakaz'yan, S., Skopina, M., Tchobanou, M.: Symmetric multivariate wavelets, Int. J. Wavelets, Mult. Inform. Proc., **7** (2009), no. 3, 1–28
112. Kashin, B.S., Kulikova, T.Yu.: A Note on the description of frames of general form. Mathematical Notes **72**, 281–284 (2002)
113. Koch, K.: Multivariate symmetric interpolating scaling vectors with duals. J. of Fourier Anal. and Apps. **15**, 1–30 (2009)
114. Kotelnikov, V.A.: On the transmission capacity of the 'ether' and of cables in electrical communications (in Russian), Proceedings of the first All-Union Conference, Upravlenie svyazi, RKK, 1933
115. Krivoshein, A.V.: On construction of multivariate symmetric MRA-based wavelets. Appl. Comput. Harmon. Anal. **36**(2), 215–238 (2014)
116. Krivoshein, A.: Multivariate symmetric refinable functions and function vectors, Int. J. Wavelets Multiresolut. Inf. Process, **14** (2016). doi:10.1142/S021969131650034X
117. Krivoshein, A.: Symmetric Interpolatory dual wavelet frames, St. Petersburg Math. J., Algebra i Analiz 28 (2016), no. 3, 36-66 (in Russian); English translation in St. Petersburg Math. J. (to appear)
118. Krivoshein, A.V., Ogneva, M.A.: Symmetric orthogonal wavelets with dilation factor M=3. J. of Math. Sciences **194**, 667–677 (2013)
119. Krivoshein, A., Skopina, M.: Approximation by frame-like wavelet systems. Appl. Comput. Harmon. Anal. **31**, 410–428 (2011)
120. Krivoshein, A.V., Skopina, M.A.: Construction of multivariate frames using the polyphase method, Matem. Zametki **100** (2016), no. 3, 473–476 (in Russian); English translation in Mathematical Notes, **100** (2016), no. 3,
121. Lagarias, J.C., Wang, Y.: The finiteness conjecture for the generalized spectral radius. Linear Alg. Appl. **21**, 17–42 (1995)
122. Lai, M.-J., Petukhov, A.: Method of Virtual Components for Constructing Redundant Filter and Wavelet Frames. Appl. and Comp. Harm. Anal. **22**(3), 304–318 (2007)
123. Lai, M.-J., Stöcler, J.: Construction of multivariate compactly supported tight wavelet frames **21**, 324–348 (2006)
124. Lawton, W.: Necessary and sufficient conditions for constructing orthonormal wavelet bases. Math. Phys. **32**, 57–81 (1991)
125. Lawton, W.: Tight frames of compactly supported affine wavelets. J. Math. Phys. **31**, 1898–1901 (1990)

126. Lawton, W., Lee, S.N., Shen, Z.: Stability and orthonormality of multivariate refinable functions. SIAM J. Math. Anal. **28**(4), 999–1114 (1997)
127. Lawton, W.M.: Necessary and sufficient conditions for constructing orthonormal wavelets. J. Math. Phys. **32**, 57–61 (1991)
128. Lemarier, P.G., Meyer, Y.: Ondelettes et bases Hilbertiennes. Rev. Math. Iber. **2**, 1–18 (1987)
129. Madych, W.R.: Some elementary properties of multiresolution analysis – a tutiroial in theory and applications, C.K. Chui ed. Academic Press. 1992, 259–294
130. Maesumi, M.: An efficient lower bound for the generalized spectral radius. Linear Alg. Appl. **240**, 1–7 (1996)
131. Maximenko, I.E.: Biorthogonality of multivariate refinable functions, in book "Voprosy sovremennoi teorii approksimatsii", SPbGU Publishing, 2004, 132–145 (in Russian)
132. Maximenko, I.E., Skopina, M.A.: Multivariate Periodic Wavelets. St. Petersburg Math. J. **15**(2), 165–190 (2004)
133. Mallat, S.: Multiresolution approximation and wavelets. Trans. AMS. **315**, 69–88 (1989)
134. Mallat, S.: A theory of multiresolution signal decomposition: the wavelets representation. IEEE Trans. Pattern Anal. Machine Intell **11**, 674–693 (1989)
135. Mallat, S.: Multiresolution approximation and wavelets. Trans. Amer. Math. Soc. **315**, 69–88 (1989)
136. Meyer, Y.: Ondelettes and fonctions splines. Seminaire EDP, Paris (December (1986)
137. Meyer, Y.: Principle d'incertitude, bases hilbertiennes et algebres d'operateurs, Seminaire Bourbaki. 1985–1986. V. 38. no. 662
138. Micchelli, C.A., Prautzsch, H.: Uniform refinement of curves. Linear Alg. Appl. **114**(115), 841–870 (1989)
139. Moision, B.E., Orlitsky, A., Siegel, P.N.: On codes that avoid specified differences. IEEE Trans. Inform. Theory **47**(1), 433–442 (2001)
140. Möller, H.M., Sauer, T.: Multivariate refinable functions of high approximation order via quotient ideals of Laurent polynomials. Adv. Comput. Math. **20**(1–3), 205–228 (2004)
141. Packer, J.A., Rieffel, M.A.: Wavelet filter functions, Matrix completion problem, and projective modules over $C(\mathbb{T}^n)$. J. Fourier. Anal. Appl. **9**(3), 101–116 (2003)
142. Petukhov, A.: Explicit construction of framelets. Appl. Comp. Harm. Anal. **11**, 313–327 (2001)
143. Petukhov, A.: Symmetric framelets. Constr. Approx. **19**, 309–328 (2003)
144. Petukhov, A.: Construction of symmetric orthogonal bases of wavelets and tight wavelet frames with integer dilation factor. Appl. Comput. Harmon. Anal. **17**, 198–210 (2004)
145. Petukhov, A.: Explicit construction of framelets. Appl. Comput. Harmon. Anal. **11**, 313–327 (2006)
146. Protasov, V.Yu.: The joint spectral radius and invariant sets of linear operators. Fundam. Prikl. Mat. **2**, 205–231 (1996)
147. Protasov, V.Yu.: Fractal curves and their applications to wavelets, Proceeding of the International Workshop on self-similar systems, July 30 - August 7, 1998, pp. 120–125. Russia, Dubna (1999)
148. Protasov, V.Yu.: Refinement equations with nonnegative coefficients. J. Fourier Anal. Appl. **6**(6), 55–77 (2000)
149. Protasov, V.Yu.: A complete solution characterizing smooth refinable functions. SIAM Journal of Math. Analysis **31**(6), 1332–1350 (2000)
150. Protasov, V.Yu.: The stability of subdivision operator at its fixed point. SIAM J. Math. Analysis **33**(2), 448–460 (2001)
151. Protasov, V.Yu.: Refinement equations and corresponding linear operators, Int. J. Wavelets, Mult. Inform. Proc., **4** (2004), no. 3, 461–474
152. Riemenschneider, S.D., Shen, Z.W.: Multidimensional interpolatory subdivision schemes. SIAM J. Numer. Anal. **34**(6), 2357–2381 (1997)
153. Riemenschneider, S.D., Shen, Z.W.: Construction of compactly supported biorthogonal wavelets in $L_2(\mathbb{R}^s)$ I, Physics and modern topics in mechanical and electrical engineering. Scientific and Engineering Society Press, 1999, 201–206

154. Riemenschneider, S.D., Shen, Z.W.: Construction of compactly supported biorthogonal wavelets in $L_2(\mathbb{R}^s)$ II, Wavelet applications signal and image Processing VII. Proceedings of SPIE. **3813**, 264–272 (1999)
155. Ron, A., Shen, Z.: Gramian analysis of affine bases and affine frames, in book: Approximation Theory VIII, V. 2: Wavelets (C.K. Chui and L. Schumaker, eds) World Scientific Publishing Co. Inc (Singapure), 1995, 375–382
156. Ron, A., Shen, Z.: Frame and stable bases for shift-invariant subspaces of $L_2(R^d)$. Canad. J. Math. **47**(5), 1051–1094 (1995)
157. Ron, A., Shen, Z.: Affine systems in $L_2(R^d)$: dual systems. J. Fourier. Anal. Appl. **3**, 617–637 (1997)
158. Ron, A., Shen, Z.: The Sobolev regularity of refinable functions. J. Approx. Theory **106**, 185–225 (2000)
159. San Antolin, A., Zalik, R.A.: Some Smooth Compactly Supported Tight Wavelet Frames with Vanishing Moments, J. Fourier. Anal. Appl., to appear
160. San Antolin, A., Zalik, R.A.: Some smooth compactly supported tight framelets associated to the quincux matrix, J. Mathematical Analysis and Applications, to appear
161. Shannon, C.E.: Communication in the presence of noise. Proc. of the IRE. **37**, 10–21 (1949)
162. Shen, Z.: Extension of matrices with Laurent polynomial entries, Proceedings of the 15th IMACS World Congress on Scientific Computation Modeling and Applied Mathematics, Ashim Syclow eds. 1997, 57–61
163. Skopina, M.: Local convergence of Fourier series with respect to periodized wavelets. J. Approx. Theory **94**, 191–202 (1998)
164. Skopina, M.: Wavelet approximation of periodic functions. J. Approx. Theory **104**, 302–329 (2000)
165. Skopina, M.: Localization principle for wavelet expansions, Proceedings of the International Workshop (July 30 - August 7, 1998, Dubna, Russia), (JINR, E5-99-38, Dubna, 1999), 125–130
166. Skopina, M.: On Construction of Multivariate Wavelets with Vanishing Moments. Appl. Comput. Harmon. Anal. **20**(3), 375–390 (2006)
167. Skopina, M.: Tight wavelet frames. Dokl. Ross. Akad. Nauk **419**(1), 26–29 (2008)
168. Skopina, M.: On construction of multivariate wavelet frames. Appl. Comput. Harmon. Anal. **27**(1), 55–72 (2009)
169. Stanhill, D., Zeevi, Y.Y.: Two-Dimensional Orthogonal Filter Banks and Wavelets with Linear Phase. IEEE Trans. on Sign. Proc. **46**(1), 183–190 (1998)
170. Sweldens, W.: The Lifting Scheme: A Custom-Design Construction of Biorthogonal Wavelets. Appl. Comput. Harmon. Anal. **2**, 186–200 (1996)
171. Tay, D.B.H.: Analytical design of 3-D wavelet filter banks using the multivariate Bernstein polynomial. IEE Proceedings - Vision Image and Signal Processing **147**(2), 122–130 (2000)
172. Tchobanou, M.K.: Design and implementation of multidimensional multirate systems, Midwest Symposium on Circuits and Systems, **2** (2004), II, 541–544
173. Tchobanou, M.K.: Polynomial methods for multi-dimensional filter banks' design, European Signal Processing Conference, (2015) art. no. 7075514
174. Unser, M., Chenouard, N., De Ville, D.V.: Steerable Pyramids and Tight Wavelet Frames in $L_2(\mathbb{R}^d)$. IEEE TRANSACTIONS ON IMAGE PROCESSING **20**(10), 2705–2721 (2011)
175. Weinmann, A.: Subdivision schemes with general dilation in the geometric and nonlinear setting. Journal of Approximation Theory **164**(1), 105–137 (2012)
176. Wellend, G.V., Lundberg, M.: Construction of compact p-wavelets. Constr. Approx. **9**, 347–370 (1993)
177. Yang, J., Li, S.: Smoothness of multivariate refinable functions with infinitely supported masks. J. Approx. Theory **162**(6), 1279–1293 (2010)
178. Zhuang, X.S.: Matrix extension with symmetry and construction of biorthogonal multiwavelets with any integer dilation. Appl. Comput. Harmon. Anal. **33**, 159–181 (2012)
179. Zhou, D.X.: Stability of refinable functions, multiresolution analysis and Haar bases. SIAM J. Math. Anal. **27**(3), 891–904 (1996)

180. Zhou, D.X.: The p-norm joint spectral radius and its applications in wavelet analysis. International conference in wavelet analysis and its applications, AMS/IP Studies in Advanced Mathematics **25**, 305–326 (2002)
181. Zhou, D.X.: The p-norm joint spectral radius for even integers. Methds Appl. Anal. **5**, 39–54 (1998)

Index

A. Krivoshein et al., *Multivariate Wavelet Frames*,
Industrial and Applied Mathematics, DOI 10.1007/978-981-10-3205-9

Zeitfracht Medien GmbH
Ferdinand-Jühlke-Straße 7
99095 Erfurt, Deutschland
produktsicherheit@kolibri360.de